Water Deficits

plant responses from cell to community

ENVIRONMENTAL PLANT BIOLOGY series

Editor: W.J. Davies
Institute of Environmental and Biological Sciences, Division of Biological Sciences, University of Lancaster, Lancaster LA1 4YQ, UK

Abscisic Acid: physiology and biochemistry

Carbon Partitioning: within and between organisms

Pests and Pathogens: plant responses to foliar attack

Water Deficits: plant responses from cell to community

Forthcoming titles include:

Photoinhibition of Photosynthesis: from molecular mechanisms to the field

Water Deficits
plant responses from cell to community

J.A.C. Smith
Department of Plant Sciences, University of Oxford, Oxford OX1 3RB, and
Fellow of Magdalen College, Oxford, UK

H. Griffiths
Department of Agricultural and Environmental Science, University of
Newcastle, Newcastle upon Tyne NE1 7RU, UK

*β*IOS
SCIENTIFIC
PUBLISHERS

© BIOS Scientific Publishers Limited, 1993

First published in the United Kingdom 1993 by
BIOS Scientific Publishers Limited,
St Thomas House, Becket Street, Oxford OX1 1SJ, UK.

A CIP catalogue record for this book is available from the British Library.

ISBN 1 872748 06 6

Typeset by Alden Multimedia, Northampton, UK.
Printed by Information Press Ltd, Oxford, UK.

Preface

Current awareness of the importance of plant responses to water deficits is strongly dependent on the popular media. All too frequently we find ourselves confronted by networked images of drought and famine, providing us with a glimpse of the global scale of this problem. But within the scientific community, researchers are still trying to understand the complex ways in which water deficits can limit plant growth. This is currently one of the most active areas of environmental physiology, encompassing diverse component disciplines from molecular and cellular physiology, through whole-plant physiology and ecology, to the domains of agronomy, hydrology and climatology.

In this book, we have brought together specialists in all these topics to present a comprehensive, up-to-date review of plant responses to water deficits. An important aim was to combine for the first time a discussion of cell water relations with the larger scales of whole plants and plant communities. Collecting such diverse contributions in a single volume helps to show how different approaches can be integrated, even if our understanding of the underlying processes is still imperfect. Despite this broad scope, we have not been able to include any detailed consideration of the extremes of this spectrum – the molecular biology of changes in gene expression caused by water deficits at the subcellular level, and the application of remote-sensing technology to studies of water budgets at the global level. Some topics, such as differences in plant water use associated with the biochemistry of carbon fixation, are also not treated explicitly because of detailed coverage elsewhere. Instead, we hope this volume will provide an in-depth review of recent progress in plant water relations and the implications for plant growth of environmental water deficits.

This book arose from a 3-day discussion meeting organized by the Environmental Physiology Group of the Society for Experimental Biology at the University of Lancaster in April 1992. As with previous volumes in the Environmental Plant Biology series, the aim has been to provide an authoritative review of a subject area rather than a report of a meeting. We hope that the varied backgrounds of the authors will prove a stimulating interdisciplinary approach, particularly as a basis for future studies. The often daunting complexity of terminology in these diverse areas of water relations has been resolved, as far as possible, to provide symbols and nomenclature that are consistent throughout the text. In certain places a particular notation has been retained to allow uniformity with an author's style in earlier publications.

We are very grateful for the financial support of the Society for Experimental Biology, the British Ecological Society and Safeway plc, which enabled us to bring together the distinguished contributors represented in this volume. In addition, we would like to thank the Series Editor, Bill Davies, for his support, and Jonathan Ray and Rachel Robinson at BIOS Scientific Publishers for their hard work towards the production of this book.

Andrew Smith (*Oxford*)
Howard Griffiths (*Newcastle upon Tyne*)

Contents

Contributors

Baker, N.R. Department of Biology, University of Essex, Colchester CO4 3SQ, UK

Benkert, R. Lehrstuhl für Biotechnologie der Universität Würzburg, Biozentrum, Am Hubland, D-8700 Würzburg, Germany

Chaves, M.M. Departamento de Engenharia Florestal, Universidade Técnica de Lisboa, Tapada da Ajuda, P-1399 Lisboa Codex, Portugal

Condon, A.G. CSIRO, Division of Plant Industry, PO Box 1600, Canberra, ACT 2601, Australia

Davies, W.J. Division of Biological Sciences, University of Lancaster, Lancaster LA1 4YQ, UK

Dolman, A.J. The Winand Staring Centre, Marijkeweg 11-22, PO Box 125, NL-6700 AC, Wageningen, The Netherlands

Ehleringer, J.R. Department of Biology, University of Utah, Salt Lake City, UT 84112, USA

Grace, J. Institute of Ecology and Resource Management, University of Edinburgh, Edinburgh EH9 3JU, UK

Griffiths, H. Department of Agricultural and Environmental Science, University of Newcastle, Newcastle upon Tyne NE1 7RU, UK

Jarvis, P.G. Institute of Ecology and Resource Management, University of Edinburgh, Edinburgh EH9 3JU, UK

Jones, H.G. Horticulture Research International, Wellesbourne, Warwick CV35 9EF, UK

Nobel, P.S. Department of Biology and Laboratory of Biomedical and Environmental Sciences, University of California, Los Angeles, CA 90024, USA

North, G.B. Department of Biology and Laboratory of Biomedical and Environmental Sciences, University of California, Los Angeles, CA 90024, USA

Ortega, J.K.E. Department of Mechanical Engineering, University of Colorado at Denver, Denver, CO 80217-3364, USA

Passioura, J.B. CSIRO, Division of Plant Industry, Institute of Plant Production and Processing, PO Box 1600, Canberra, ACT 2601, Australia

Pereira, J.S. Departamento de Engenharia Florestal, Universidade Técnica de Lisboa, Tapada da Ajuda, P-1399 Lisboa Codex, Portugal

Pritchard J. Ysgol Gwyddorau Bioleg, Coleg y Brifysgol, Bangor, Gwynedd LL57 2UW, UK

Raven, J.A. Department of Biological Sciences, University of Dundee, Dundee DD1 4HN, UK

Richards, R.A. CSIRO, Division of Plant Industry, PO Box 1600, Canberra, ACT 2601, Australia

Rygol, J. Lehrstuhl für Biotechnologie der Universität Würzburg, Biozentrum, Am Hubland, D-8700 Würzburg, Germany

Saab, I.N. Department of Agronomy, University of Missouri, Columbia, MO 65211, USA

Schneider, H. Lehrstuhl für Biotechnologie der Universität Würzburg, Biozentrum, Am Hubland, D-8700 Würzburg, Germany

Schulte, P.J. Department of Biological Sciences, University of Nevada, Las Vegas, NV 89154, USA

Schulze, E.-D. Lehrstuhl für Pflanzenökologie, Universität Bayreuth, Universitätstraße 30, Postfach 101251, D-8580 Bayreuth, Germany

Sharp, R.E. Department of Agronomy, University of Missouri, Columbia, MO 65211, USA

Smith, J.A.C. Department of Plant Sciences, University of Oxford, South Parks Road, Oxford OX1 3RB, UK

Spollen, W.G. Department of Agronomy, University of Missouri, Columbia, MO 65211, USA

Sprent, J.I. Department of Biological Sciences, University of Dundee, Dundee DD1 4HN, UK

Steudle, E. Lehrstuhl für Pflanzenökologie, Universität Bayreuth, Universitätstraße 30, Postfach 101251, D-8580 Bayreuth, Germany

Tardieu, E. INRA, Laboratoire d'Agronomie, F-78850 Thivernal Grignon, France

Tomos, D. Ysgol Gwyddorau Bioleg, Coleg y Brifysgol, Bangor, Gwynedd LL57 2UW, UK

Wu, Y. Department of Agronomy, University of Missouri, Columbia, MO 65211, USA

Zimmermann, G. Lehrstuhl für Biotechnologie der Universität Würzburg, Biozentrum, Am Hubland, D-8700 Würzburg, Germany

Zimmermann, U. Lehrstuhl für Biotechnologie der Universität Würzburg, Biozentrum, Am Hubland, D-8700 Würzburg, Germany

Zhu, J.J. Lehrstuhl für Biotechnologie der Universität Würzburg, Biozentrum, Am Hubland, D-8700 Würzburg, Germany

Abbreviations

A	area subscript: A_r root surface area
A	net photosynthesis, assimilation rate (mol m^{-2} s^{-1})
ABA	abscisic acid
AE	acoustic emission
AL	actinic light
c	concentration (mol mol^{-3}) subscripts: c_a atmospheric (CO_2); c_i intercellular space (CO_2); c_s solute superscripts: c^i internal; c^o external
C	capacitance (MPa m^{-3}) subscripts: C_s storage; C_x xylem
c_p	molar heat capacity of dry air at constant pressure (J mol^{-1} K^{-1})
CAM	crassulacean acid metabolism
cDNA	copy DNA
d	diameter (m)
D	water vapour saturation deficit (Pa) subscripts: D_a ambient air; D_{surf} air at the surface
D_{wv}	diffusion coefficient of water vapour in air
DW	dry weight
e	water vapour pressure (Pa) subscripts: e_a air; e_l saturation at leaf temperature superscript: e^* saturation vapour pressure of water vapour in air
E	transpiration, or evaporation rate (mol m^{-2} s^{-1}, or kg m^{-2} s^{-1}, or m^3 s^{-1}) subscripts: E_{eq} equilibrium transpiration rate; E_{imp} imposed transpiration rate; E_I evaporation of intercepted water; E_p potential evaporation rate
F	fluorescence (relative units) subscripts: F_m maximum, F_o minimum, F_v variable (all following dark adaptation); F_o' minimum, F_m' maximum, F_v' variable (during steady-state photosynthesis); F_s steady state
FLU	fluridone
FR	far red light
g	conductance (mol m^{-2} s^{-1}, or m s^{-1}) subscripts: g_a aerodynamic conductance of canopy; g_A stand boundary layer; g_{at} boundary layer conductance per tree crown; g_i isothermal (climatological); g_s stomatal; g_{surf} surface superscript: g' CO_2
GA	gibberellic acid
GCM	general circulation model
H	sensible heat flux density (W m^{-2})

HI	harvest index
IAA	indole-3-acetic acid
J	flux density (mol m^{-2} s^{-1}, or m s^{-1})
	subscripts: J_V volumetric flux density; J_w water
	superscript: J^{trans} transpiration
k	rate constant (s^{-1}), or experimental constant
	subscript: k_r root; k_s solute; k_w water exchange
K	energy-transfer coefficient (m^2 s^{-1})
K_h	hydraulic conductance per unit length (m^4 s^{-1} MPa^{-1})
l	length (m)
L	leaf area index (dimensionless)
L_{gap}	water vapour conductance of root–soil air gap (m s^{-1} MPa^{-1})
L_p	hydraulic conductivity (m s^{-1} MPa^{-1})
	subscripts: L_{Pr} root; L_{Px} axial (xylem)
	superscripts: L_P^{apo} apoplast; L_P^{cc} cell-to-cell
L_R	radial hydraulic conductivity (m s^{-1} MPa^{-1})
L_{soil}	hydraulic conductivity coefficient of the soil (m^2 s^{-1} MPa^{-1})
m	coefficient of extensibility
NMR	nuclear magnetic resonance
p	partial pressure (Pa)
	subscripts: p_a ambient; p_i intercellular space; p_Γ internal
P	pressure (Pa)
	subscripts: P_a atmospheric pressure (absolute); P_c cell; P_i initial; P_r root; P_x xylem; P_Y yield threshold
P_s	permeability coefficient of solute s (m s^{-1})
	subscript: P_{sr} root
PAR	photosynthetically active radiation (400–700 nm)
PEG	polyethyleneglycol
PEPc	phosphoenolpyruvate carboxylase
PPFD	photosynthetic photon flux density (400–700 nm)
Q_A, Q_B	primary and secondary quinone acceptors of PS2
Q_n	net radiation absorbed, or radiant flux density (W m^{-2})
Q_V	volumetric flow of water (m^3 s^{-1})
qE	non-photochemical quenching component
qP	photochemical quenching coefficient
r	correlation coefficient
r	radius (m)
r	relative growth rate (dimensionless)
r	resistance, gas phase
	subscripts: r_a boundary layer; r_e excess; r_m mesophyll; r_s stomatal
	superscript: r' CO$_2$
R	resistance, liquid phase (MPa s m^{-1})
	subscripts: R_p plant; R_{sp} soil–plant
R	gas constant (8.314 J mol^{-1} K^{-1} or m^3 Pa mol^{-1} K^{-1})
RH	relative humidity
Rubisco	ribulose bisphosphate carboxylase-oxygenase
RWC	relative water content
s	slope of relation between saturation water vapour pressure and temperature (Pa) $= \mathrm{d}e^*(T)/\mathrm{d}T$
S	soil heat flux (W m^{-2})

SiB	simple biosphere (model)
SLA	specific leaf area
SP	saturating pulse
SPS	sucrose phosphate synthase
SVAT	soil vegetation atmosphere transfer
T	temperature (°C or K)
	subscripts: T_d dry bulb (air); T_{surf} leaf surface temperature; T_w wet bulb
T	relative transpiration rate (dimensionless)
u	windspeed (m s^{-1})
	subscript: u_* friction velocity
UDP	uridine diphosphate
UV-A	ultraviolet radiation (320–400 nm)
UV-B	ultraviolet radiation (280–320 nm)
v	relative rate of change in volume
	subscripts: v_c cell wall; v_w water
V	volume
	subscript: V_x xylem
$\bar{V}_w$	partial molal volume of water (18.048 $\times$ 10^{-6} m^3 mol^{-1} at 20°C)
VA	vesicular–arbuscular
VPD	vapour pressure deficit
W	mass (kg)
	subscripts: W_C mass of water on canopy; W_S mass of water on canopy after drip has ceased (canopy storage capacity)
W_T	transpiration efficiency
WUE	water-use efficiency
XET	xyloglucan endotransglycosylase
Y	yield, grain
Y	yield threshold turgor pressure (MPa)
z	distance, height, or length (m)
β	Bowen ratio (dimensionless) $= \mathbf{H}/\lambda E$
β	elastic coefficient (Pa m^{-3})
γ	fractional contribution of pathway
	superscripts: γ^{apo} apoplast; γ^{cc} cell-to-cell
γ	psychrometric constant (Pa K^{-1}) $= c_p P_a/\lambda$
Δ	carbon isotope discrimination (‰)
δD	deuterium isotope composition (‰)
δe	water vapour pressure deficit (Pa)
ε	coefficient for change in sensible and latent heat contents of air with respect to temperature (K) $= s/\gamma$
ε	volumetric elastic modulus (MPa)
	subscripts: ε_{av} average; ε_m maximum; ε_0 zero turgor
η	dynamic viscosity (Pa s)
κ	coefficient of compressibility (Pa^{-1})
θ	potential temperature of dry air (K or °C)
λ	molar latent heat of vaporization of water (J mol^{-1})
π	osmotic pressure (MPa) $= -$osmotic potential
	subscripts: π_c cell; π_o outside medium; π_x xylem
ρ_a	molar density of dry air (mol m^{-3})

σ reflection coefficient (dimensionless)
subscripts: σ_c cell; σ_{sr} solute, root; σ_x xylem
superscripts: σ^{apo} apoplast; σ^{cc} cell-to-cell

τ time constant (s) [also shoot:root ratio in Chapter 17]
superscript: τ_w water exchange

ϕ quantum yield
subscripts: ϕ_{CO_2} CO$_2$ assimilation; ϕ_{PS1} photosystem 1;
ϕ_{PS2} photosystem 2

ϕ yield coefficient, or extensibility (s^{-1} Pa^{-1})

Ψ water potential (MPa)
subscripts: Ψ_d pre-dawn; Ψ_l leaf; Ψ_r root; Ψ_w water

Ω decoupling coefficient (dimensionless)

Integrating plant water deficits from cell to community

J.A.C. Smith and H. Griffiths

Plant growth and productivity is limited in many regions of the world by water deficits. In the face of mounting population pressure, these limitations will be of increasing severity for world agriculture in the decades ahead. Rates of desertification in tropical and subtropical areas are already alarmingly high, and are likely to be compounded by the widely anticipated effects of global climate change. Plant scientists who study such effects are concerned with problems of water availability at many levels, right through to the agronomists, farmers and retailers who provide plant products for consumption. Trying to understand how plants respond to water deficits and, in certain instances, are able to tolerate them should lead us eventually to ways of optimizing plant productivity in marginal environments.

Any study of plant water deficits obviously requires good methods for measuring them. As with many areas of science, this field has been profoundly influenced in recent years by important technical advances. Standard equipment, such as the pressure chamber, psychrometers and infra-red gas analysers, has now been joined by newer technologies such as the pressure probe, acoustic emission sensors, portable photosynthesis and fluorescence meters with solid-state circuitry, and mass spectrometers for analysing stable isotopes, in addition to sophisticated modelling techniques for extrapolation to larger scales. These methods allow us to study underlying processes at a range of levels from the cellular through the whole plant to entire plant communities and ecosystems. Equipped with this array of methods, we are obtaining the first glimpses of an integrated view of the effects of water deficits on plant growth.

Nowhere has the impact of a new technique been more far-reaching than with the advent of the pressure probe. Since the introduction of the miniaturized version of the probe in 1978, the water relations of individual cells in higher plants have been accessible to direct measurement. Some of the findings from the probe that were initially surprising – such as the very short half-times for water exchange of individual cells – have long since become established as typical properties of higher-plant cells. But the applications of this technique have been proliferating and progress in this field to date has not been reviewed in a multi-author volume. Thus the first part of this book forms a detailed assessment of the pressure probe, focusing on the principles behind the technique (Chapter 2), its applications to study cell growth and expansion as influenced by water deficits (Chapters 3 and 4), as well as methods for determining transpirational water loss from individual cells (Chapter 5).

As befits the inventiveness of the originators of the modern pressure probe, further important developments of the technique have continued to be made, allowing modified forms of the instrument to be used for studying the osmotic behaviour of whole organs, such as root systems (Chapter 2). Also, conversion of the cell pressure probe to an entirely water-filled system has allowed its application to the long-standing problem of measuring negative pressures in the xylem directly (Chapter 6). Needless to say, the results emerging from these studies are challenging some widely accepted notions in plant water relations. These include questions such as the cellular pathway of radial water movement across the root and the location of the osmotically important barriers to radial water movement (Chapters 2 and 4). More provocatively, there are suggestions that the pressures measurable in the xylem are insufficiently low (negative) to be consistent with the cohesion theory of sap ascent (Chapter 6; see also Passioura, 1991, for a critique). Under at least certain conditions, there can be excellent agreement between turgor pressures measured with the pressure probe and those derived from psychrometry (Nonami *et al.*, 1987) or the pressure chamber (Murphy and Smith, 1993), so the resolution of apparently conflicting data must await further studies.

Because of the tensions under which it operates, the 'vulnerable pipeline' of the xylem is potentially very sensitive to water deficits and the danger of cavitation (Chapter 7). While it is accepted that air embolisms may form quite commonly, it is much less clear how embolized xylem elements are refilled, especially in tall trees where root pressures are inadequate. Transport through the xylem is likely to be involved in long-distance signalling within the plant via chemical messengers, for which a very strong candidate is abscisic acid (ABA; Chapters 8 and 9). In this model, the xylem would play a vital role in transducing the water stress signal by root–shoot communication: ABA produced in the root tips as the soil dries would be transported to the shoot, bringing about stomatal closure. Although this conceptual framework is now generally accepted, many facets of it require further scrutiny, such as whether substances other than ABA are physiologically active, and whether ABA is involved in the response of stomata to atmospheric humidity.

The interface between the plant and its environment most dramatically affected by changing water availability is the soil–root boundary. Transfer resistances here involve both soil and plant components, and the former can be very important in desert environments in restricting reverse water flow out of the root into a drying soil (Chapter 10). Water flow through the plant can now be modelled faithfully by electrical analogues using powerful circuit-simulation programs. These can be valuable tools in assessing the relative importance of flow resistances, tissue storage capacitances and cell osmotic pressure in whole-plant water relations (Chapter 11). In fact, these approaches allow the mechanistic understanding derived from studies of physiology and anatomy to be tested quantitatively against plant performance in the field.

To move up from the level of cellular processes to that of the whole plant poses questions about the description and classification of drought-tolerance mechanisms. These have often been discussed in terms of the observed water-use efficiency (WUE) of carbon assimilation. However, there are now arguments for believing that stomata may operate to maximize assimilation per unit water *available* while keeping damage to the conducting system within acceptable limits (Chapter 12). Under these conditions, available inorganic nutrient sources (especially N and P) may be exploited in a way that optimizes assimilatory costs of nitrogen and the effective rooting volume established through mycorrhizal infection (Chapter 13). In shoot tissues, when water deficits limit the supply of CO_2 for photosynthesis, a number of reversible mechanisms down-regulate the efficiency of excitation-energy transfer to photosystem 2, as can be monitored by modulated fluorescence techniques (Chapter 14; Horton and Ruban, 1992).

These approaches can be applied to the study of specific plant communities in natural environments that experience seasonal water deficits. In Mediterranean-type climates, the well-known midday depression of photosynthesis and associated change in carboxylation efficiency (Tenhunen *et al.*, 1984) has been shown to be accompanied by a change in photochemical efficiency (Demmig-Adams *et al.*, 1989). Thus, it is now possible to enquire after the feedback controls that link both light reactions and carbon-assimilation reactions to the severity of ambient water deficits (Chapter 15). In the agronomic context, selection of cultivars for Mediterranean or semi-arid environments may need to consider the timing of leaf area development in response to water availability (Chapter 16). Thus, improvements in crop yield may yet be brought about by a combination of genetic selection and agronomic practice as a result of research at the physiological level. It is intriguing, however, that the concepts of 'conservative' and 'profligate' water-use strategies find parallels in the natural and agronomic environments.

Carbon isotope compositions of plant organic material determined by mass spectrometry are now finding widespread use as long-term integrals of c_i, the intercellular CO_2 concentration. This can be viewed as a measure of the metabolic set point for gas exchange, and serves to relate carbon assimilation to water utilization over seasonal time-scales (Chapter 17). In addition, measurement of a combination of stable isotopes (^{2}H and ^{18}O in water) can be used to trace source water (Chapter 17), and may also have the potential to reveal the quantitative contribution to atmospheric fluxes of transpirational water derived from vegetation (Ehleringer *et al.*, 1993; Farquhar *et al.*, 1993). Other approaches such as mass-balance techniques, micrometeorological methods, and measurement of liquid flow and vapour fluxes can be used as a basis for scaling up to the level of canopies and stands of vegetation (Chapter 18). Calculated canopy conductances can show close correlations with changes in ambient water deficits and climatic conditions, but there is also a major problem of extrapolating to a mosaic of non-uniform vegetation types (Chapter 19). By refining models to account for stratification in vegetation and heterogeneity across land surfaces, it should nevertheless prove possible to incorporate data for particular regions into general circulation models (GCMs) of the atmosphere.

The need to understand plant responses to water deficits has never been more acute. Changes in land use and regional climate brought about by anthropogenic incursions mean that crop productivity in such marginal areas will be of increasing importance in world agriculture. This book deliberately focuses on the processes underlying plant growth in water-limited environments. But it is already clear that further progress towards practical benefits will depend on close collaboration between research scientists and many others concerned with the problems of water availability.

References

Demmig-Adams, B., Adams, W.W.,III, Winter, K., Meyer, A., Schreiber, U., Pereira, J.S., Krüger, A., Czygnan, F.-C. and Lange, O.L. (1989) Photochemical efficiency of photosystem II, photon yield of O_2 evolution, photosynthetic capacity and carotenoid composition during the midday depression of net CO_2 uptake in *Arbutus unedo* growing in Portugal. *Planta*, **177,** 377–387.

Ehleringer, J.R., Farquhar, G.D. and Hall, A.E. (1993) *Perspectives of Plant Carbon and Water Relations From Stable Isotopes.* Academic Press, San Diego, in press.

Farquhar, G.D., Lloyd, J., Taylor, J.A., Flanagan, L.B., Syvertsen, J.P., Hubick, K.T., Wong, S.C. and Ehleringer, J.R. (1993) Vegetation effects on the isotope composition of oxygen in atmospheric CO_2. *Nature*, **363,** 439–443.

Horton, P. and Ruban, A.V. (1992) Regulation of photosystem II. *Photosyn. Res.*, **34,** 375–385.

Murphy, R. and Smith, J.A.C. (1993) A critical comparison of the pressure-probe and pressure-chamber techniques for estimating leaf-cell turgor pressure in *Kalanchoë daigremontiana. Plant Cell Environ.*, in press.

Nonami, H., Boyer, J.S. and Steudle E. (1987) Pressure probe and isopiestic pyschrometer measure similar turgor. *Plant Physiol.*, **83,** 592–595.

Passioura, J.B. (1991) An impasse in plant water relations? *Bot. Acta*, **104,** 405–411.

Tenhunen, J.D., Lange, O.L., Gebel, J., Beyschlag, W. and Weber, J.A. (1984) Changes in photosynthetic capacity, carboxylation efficiency, and CO_2 compensation point associated with midday stomatal closure and midday depression of net CO_2 exchange of leaves of *Quercus suber. Planta*, **162,** 193–203.

Pressure probe techniques: basic principles and application to studies of water and solute relations at the cell, tissue and organ level

E. Steudle

2.1 Introduction

In the past 10 years, the pressure probe technique originally introduced to measure turgor and water relations of higher plant cells (Hüsken *et al.*, 1978; Steudle and Zimmermann, 1971; Zimmermann *et al.*, 1969) has been applied in diverse ways, both by extending the technique and by further developing the theoretical background. One extension was its use for measuring solute flows in addition to water flow in isolated cells and tissues (Rüdinger *et al.*, 1992; Steudle 1989a, 1990; Steudle and Tyerman, 1983). Another field of application was in studies of the biophysics of extension growth, in order to characterize the importance of water transport and mechanical properties of cell walls besides the solute relations (Boyer, 1985; Cosgrove, 1986; Cosgrove and Steudle, 1981; Meshcheryakov *et al.*, 1992; Steudle, 1985; Steudle and Boyer, 1985; see Chapters 2, 3 and 4). Attempts have been also made to measure negative pressures in the xylem (Balling and Zimmermann, 1990; Heydt and Steudle, 1988, 1991; see Chapter 6). The detailed knowledge of water relations parameters of individual cells has led to extended models of tissue water relations (Steudle, 1989a, 1992, 1993), which include couplings between water and solute flows and effects of compartmentation.

The application of the pressure probe technique to excised roots ('root pressure probe') combined both the problem of an appropriate mathematical description of tissue (organ) water relations and of interactions between water and solutes (nutrients). In the root pressure probe technique, measurements at the organ level have been combined with those at the cell level to obtain detailed information about transport pathways of water across the root cylinder. These data allow modelling of the efficiency of roots in absorbing water and solutes. The advantage of the method is that, starting at the level of individual cells, root zones and segments as well as entire root systems can be measured. Radial hydraulic resistances can be separated from those to axial transport in the root xylem along developing roots. The technique can be used in hydroponics as well as in soil, thus allowing study of important environmental effects on plant nutrition, such as drought, low temperature, salinity, toxic metals and low pH, anoxia or mycorrhiza.

Rather than giving a complete overview of the pressure probe, this chapter concentrates on basic principles of measurement and on some important applications. A particular focus will be on the recent development of the root pressure probe technique. For space reasons, measurements on extension growth will be omitted, but this subject is considered further in Chapters 4 and 5. Further information about the technique and other applications can be found in the older reviews by Dainty (1976) and Zimmermann and Steudle (1978), and in the more recent ones of Tomos (1988), Zimmermann (1989) and Steudle (1989a,b,c, 1990, 1992, 1993).

2.2 Cell pressure probes

Turgor pressure (P) is a physiological variable of fundamental importance. Because it is a component of water potential (Ψ), turgor is a measure of water status. It is also a measure of water content since it is directly related to cell volume by the elastic properties of cell walls. For mature, turgid cells of higher plants, changes of water potential are largely reflected in changes of turgor, i.e. $\Delta P/\Delta \Psi \approx 1$ will hold because of the high rigidity of walls. Furthermore, turgor is the driving force during extension growth of plants where it causes tensions in the primary wall and a viscous (irreversible) flow of wall material.

For a long time, the direct measurement of turgor in plant cells was not possible. Estimates of turgor have been obtained indirectly by determining the water potential of tissues by psychrometry, pressure chambers, etc. and measuring the osmotic pressure (Π) after killing the tissue (e.g. by freezing) to obtain turgor by difference $(\Psi = P - \Pi)$. These indirect methods have considerable disadvantages because of the mixing between cell sap and apoplasmic water, and also because they require water-potential equilibrium within the tissue. However, the most important disadvantage would be that plant samples would be composed of different tissues with different types of cells exhibiting quite different turgidity, osmotic pressure, elasticity and size, so that some kind of averaged value of turgor would be obtained — the physiological meaning of which would be unclear. It has been shown that, even in fairly homogeneous tissue, turgor may vary considerably from cell to cell (Nonami *et al.*, 1987).

2.2.1 *Precursors of cell pressure probes: Gutknecht's and Green's cell manometers*

To my knowledge, the first attempts to measure turgor directly within cells were made by Gutknecht (1968) and Green (1968), who inserted microcapillaries into the large coenoblasts of *Valonia* and internodes of *Nitella*, respectively. Gutknecht counterbalanced turgor (and applied changes thereof) with the aid of a column of mercury, and Green measured turgor from gas compression. The capillary introduced into a *Nitella* internode was closed at one end and contained water and an air bubble that changed in size in response to pressure according to the gas laws. With large cells, Green's technique worked in principle, but had several drawbacks:

(1) The system had a very high compressibility, which would have tended to artificially lower the turgor present in the intact cell upon puncturing.
(2) The gas (air) would have tended to dissolve in water, this tendency being enhanced at higher pressure (Henry's law).
(3) The method would be difficult to apply to higher plant cells because their cell volume is six to seven orders of magnitude smaller than that of a *Nitella* internode.

Injection of defined amounts of gas into cells to monitor the size of a bubble under the microscope would be even more difficult for the same reasons. Also, a cell modified in this way would exhibit a much lower (apparent) elastic modulus and rate of water exchange than a cell that is not modified.

2.2.2 *Cell pressure probes for giant cells*

In 1969, the first version of a cell pressure probe was developed. This was a manometer adapted for use with giant-celled algae such as *Valonia* spp. or characean internodes (Zimmermann *et al.*, 1969; Figure 2.1a). The instrument consisted of a microcapillary with a tip of diameter 100–200 μm which was inserted into the cell. The capillary was connected to a pressure chamber, and both the capillary and chamber were filled with a silicone oil of low viscosity and surface tension that was not miscible with water. The volume (pressure) in the system could be regulated by means of a rod which distorted a polymer membrane to adjust the oil right at the tip of the capillary. Thus, the measuring principle was to balance cell turgor by an opposing oil pressure, which was then measured by an electronic pressure transducer. The transducer consisted of a silicon membrane containing a Wheatstone bridge which, when the membrane was distorted, resulted in a voltage proportional to the pressure applied.

This first version of the pressure probe was used just to monitor turgor in plant cells (giant algae), which was possible over fairly long periods of time (several hours). It was also possible to follow changes of turgor caused by the addition of osmotica to the medium (Steudle and Zimmermann, 1974; Zimmermann and Steudle, 1970). The first probe had considerable disadvantages in that the internal volume was quite large and fairly compressible, so it was not applicable to the cells of higher plants. In the next step of development, the internal volume of the system was therefore reduced, and a means was provided of quantitatively manipulating the internal volume of the system with a movable rod (Figure 2.1b; Steudle and Zimmermann, 1971). With this second version of the probe, it was possible to induce volume changes in plant cells (ΔV) and to measure the corresponding changes in turgor (ΔP) which allowed evaluation of the elastic coefficient, or volumetric elastic modulus, of cells (ϵ), i.e.

$$\epsilon = V \frac{dP}{dV} \approx V \frac{\Delta P}{\Delta V}.$$

2.1

The system was applicable for the measurement of the rate of water exchange (half-times, $T_{1/2}^w$) and hydraulic conductivity (L_p) of cells exhibiting cell volumes of the order of microlitres ($\mu l = mm^3$). However, with the exception of giant cells of higher plants (Steudle *et al.*, 1975), the second type of cell pressure probe could still not be applied to higher plant cells because of its high compressibility. Besides rubber seals, transducers and other parts, the compressibility of the probe could be largely identified with that of the silicone oil. Since the coefficient of compressibility of the oil (κ_{oil}) is given by:

$$\kappa_{oil} = -\frac{1}{V_{app}} \frac{dV_{app}}{dP},$$

2.2

the overall change in volume per change in turgor of the system (cell plus oil) was:

$$\frac{dV_{total}}{dP} = \frac{V}{\epsilon} - \kappa_{oil} \cdot V_{app}$$

2.3

where V_{app} is the internal volume of the probe. Thus, despite the low compressibility factor of the oil ($\approx -10^{-3}$ MPa^{-1}), the second term on the right-hand side of Equation 2.3 — and not the extensibility of the cell (V/ϵ) — dominated the volume changes measured from the movement of the rod because the internal volume of the apparatus (V_{app}) was much larger than that of the

Figure 2.1. *Cell pressure probes for measuring the water relations of plant cells. (a) First, the simple version constructed for the measurement of turgor in giant-celled algae. (b) Second version also used for large cells, but equipped with a means of quantitatively manipulating the cell volume (measurement of elastic modulus of cell) and inducing hydrostatic pressure relaxations. (c) Third version adapted for use with higher plant cells. The internal volume is reduced further, and the oil/cell sap meniscus in the tip of the probe is used as the point of reference (see inset). The meniscus is adjusted by, and changes of cell volume are produced with, a motor driving the metal rod in the probe. For further explanation, see the text.*

(a)

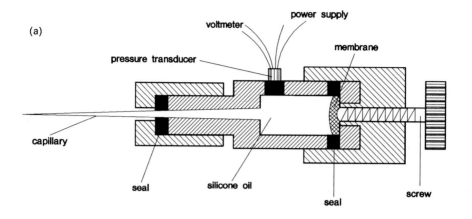

power supply
voltmeter
pressure transducer
membrane
capillary
seal
silicone oil
seal
screw

(b)

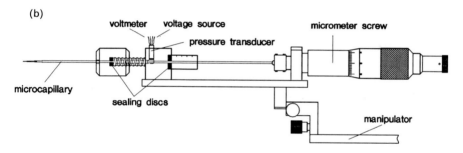

voltmeter   voltage source
pressure transducer
micrometer screw
microcapillary
sealing discs
manipulator

(c)

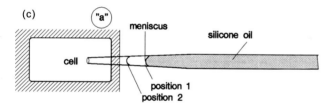

"a"
meniscus
silicone oil
cell
position 1
position 2

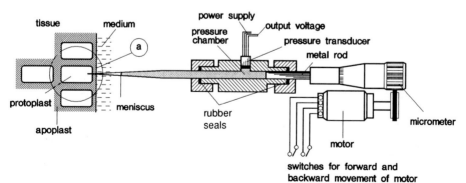

tissue   medium
power supply
pressure chamber   output voltage
pressure transducer
metal rod
a
protoplast
meniscus
rubber seals
micrometer
apoplast
motor
switches for forward and
backward movement of motor

cells (V). In other words, the measuring system was not sufficiently rigid compared with that of a cell. For constructing cell pressure probes it is important that:

$$\frac{V}{\epsilon} \gg \kappa_{\text{oil}} \cdot V_{app} \quad \text{or} \quad \frac{V_{app}}{V} \ll (\kappa_{\text{oil}} \cdot \epsilon)^{-1}. \qquad 2.4$$

For a typical higher plant cell, it may hold that $\epsilon = 5\,\text{MPa}$ and $V = 100\,\text{pl}(= 10^{-13}\,\text{m}^3)$, which would mean that, from Equation 2.4, $V_{app} \ll 20\,\text{nl}$ $(2 \times 10^{-11}\,\text{m}^3)$ should also hold. However, V_{app} was about $5\,\mu\text{l}$ in the second version of pressure probes, i.e. it was two to three orders of magnitude larger than this calculated value.

2.2.3 Pressure probe for the cells of higher plants

The criterion given in Equation 2.4 led directly to the construction of the cell pressure probe for higher plant cells (third version: Figure 2.1c; Hüsken et al., 1978). The volume of the pressure chamber of the probe was reduced further and, more importantly, the effective volume of the probe (effective V_{app}) was reduced by several orders of magnitude. Now the meniscus formed in the tip of the probe during the measurements was used as the point of reference, i.e. V_{app} in Equation 2.4, became identical with that of the small volume of water or cell sap in the tip of the probe. Using this procedure, higher plant cells could be measured (turgor; half-time of water exchange; elastic modulus; hydraulic conductivity). However, measurement required adjustment of the meniscus under the microscope and the ability to follow and measure the small volume changes produced by shifting the meniscus. This needed to be done by a motor (Figure 2.1c) rather than manually (1) to achieve a sufficiently fine adjustment of the position of the meniscus, but also (2) to avoid vibrations that would cause leakages around the tip inserted into a cell. An electronic feedback system was used in early prototypes of this version of the probe to adjust the meniscus automatically (Hüsken et al., 1978). However, this way of regulating the position of the meniscus was fairly time-consuming because it had to be established for each new tip, so it was later omitted. Apparently, this was also the fate of some other mechanisms introduced over the years to achieve automatic adjustment of the meniscus (for references, see Zimmermann, 1989).

Despite reduction of the effective volume, the entire volume of the equipment still played an important role during the measurements because the sealing of the cell membrane around the capillary appeared to depend on whether turgor immediately dropped to zero when puncturing cells, which in turn depended on the internal volume of the system, as shown above. The way in which the cell membrane seals around the tip is still not clear. However, sealing appears to be very effective, since detailed studies in which the fraction of the 'wounded' area of a cell was varied widely did not show any effect on the measured value of the hydraulic conductivity of the cell membranes (Zimmermann and Hüsken, 1979).

Over the past 14 years, the modified (third) version of the cell pressure probe (without electronic feedback) has been applied to quite different types of cells in many laboratories, with about 20 species to date (Steudle, 1989a). The method has become the conventional one for the measurement of cell water relations. There are differences in the 'stability' of the cells to be measured: in some cases, measurements on an individual cell can be extended over several hours, whereas in others the stability is only sufficient for up to 0.5 h (e.g. Steudle and Wienecke, 1985). This appears to be due to the effects of osmoregulatory processes, in particular, in excised tissue, or to leakages around the tip caused, for example, by vibrations.

Diameters of cells measured with the probe have ranged between $10 \, \mu m$ and about $100 \, \mu m$. The tip of the probe has to be adapted to the cell size and usually ranges between 2 and $7 \, \mu m$. In principle, there might be a lower limit of cell size to which the technique can be applied, since the diameter of the tip is normally considerably smaller than that of the cell. Thus, when reducing cell size, the hydraulic resistance of the cell $(1/(L_p \cdot A)$, where L_p is the hydraulic conductivity of cell membrane and A is the cell surface area) would increase. However, the hydraulic resistance of the tip increases much more rapidly (with the fourth power of the tip diameter according to Poiseuille's law). Since both resistances are in series, this may become problematic if the resistance of the tip is no longer negligible compared with that of the cell. However, calculations and measurements show that this effect is not important, even for small cells. The *calculated limit* would not be reached even with rather long and narrow tips. With narrow tips, the *practical limit* would be due to clogging of the tip; in particular, this would happen with cells rich in cytoplasm and organelles. However, such difficulties can easily be detected during the measurement because they result in a strongly reduced 'mobility' of the oil/cell sap meniscus in the tip.

2.2.4 *Measurement of negative pressure*

Attempts have also been made to use the cell pressure probe in the range of negative pressures (i.e. pressures lower than vacuum) to determine tensions in the xylem (Balling and Zimmermann, 1990; see also Chapter 6). Since there were cavitation problems caused by the oil/water boundary under tension, a probe had to be used filled only with water. However, despite this precaution, pressures could be only measured down to $-0.25 \, MPa$ (-2.5 bar) of absolute pressure. The most striking result was that the xylem pressures measured in intact plants of *Nicotiana rustica* did not react when transpiration changed, which was explained by an immediate adjustment of the hydraulic conductance of the root to the demand for water from the shoot. This finding is at variance with that of Heydt and Steudle (1991), who found a strong and immediate reaction of the pressure in the xylem of roots when adapting the root pressure probe (see below) for use in the range of negative pressures. Since the elastic modulus of the xylem should be very high (at least much larger than that of unlignified cells), the pressure in the xylem measured by the cell pressure probe of Balling and Zimmermann could have been quite different from that in the intact system, since the requirements of Equation 2.4 were not fulfilled. The significance of the negative pressures measured using this technique is discussed more extensively in Chapter 6.

2.2.5 Short half-times of water exchange and unstirred layers

Usually, the solutes present in a cell permeate the cell membranes 3 to 8 orders of magnitude more slowly than water. That is, the reflection coefficient (σ) of the solutes may be taken as unity to a good approximation. The water (volume) flow (J_V) across the cell membrane is thus given by:

$$J_V = -\frac{1}{A}\frac{dV}{dt} = L_p[P - RT(c^i - c^o)] = L_p \cdot \Delta\Psi, \qquad 2.5$$

where $RT \cdot c^{i,o}$ represents the internal (and external) osmotic pressures, respectively (R being the gas constant, T the absolute temperature and c the molar solute concentration), and A is the cell surface area. It can be seen from Equation 2.5 that the driving force for water flow in this case is the difference in water potential between cell interior and surroundings ($\Delta\Psi$). It can be shown that a cell equilibrated to a certain water potential would respond to a change in Ψ by an exponential change of both water potential and turgor with time, i.e. the cell will perform a 'relaxation of cell turgor'. The half-time ($T^w_{1/2}$) or time-constant (τ^w) of the process will be determined by the hydraulic conductivity (L_p) and the volumetric elastic modulus (ϵ) besides the cell geometry, i.e.

$$T^w_{1/2} = \tau^w \cdot \ln(2) = \frac{\ln(2)}{k_w} = \frac{V \cdot \ln(2)}{L_p \cdot A \cdot (\epsilon + \Pi^i)}, \qquad 2.6$$

where $\Pi^i = RT \cdot c^i$ (estimated form the stationary turgor) and k_w is the rate constant for water exchange.

The application of the cell pressure probe with higher plant cells has shown that, due to both high L_p and ϵ, water exchange of individual cells with their immediate surroundings is usually a rapid process, with half-times of the order of a few seconds to a few tens of seconds. There are also cells exhibiting half-times of less than 1 second (Cosgrove and Steudle, 1981). In the latter cases, the changes of pressure performed with the aid of the pressure probe would have to be very short to avoid difficulties caused by an outflow of water while changing pressure, i.e. the time-constant of the change must be much shorter than that of water flow across the membrane. However, the rate of movement of the metal rod cannot be increased arbitrarily because this may damage the cell or cause difficulties in regulating the meniscus properly. Therefore, the outflow of water has to be taken into account when calculating ϵ values of cells exhibiting short $T^w_{1/2}$ (Steudle et al., 1980).

Problems of unstirred layers arise during the measurement of water flows because a flow of water will 'sweep away' solution right at the membrane surface. On the opposite side of the membrane, solutes will be concentrated. This type of concentration polarization ('sweep-away effect') tends to reduce water flow and the apparent L_p of membranes. It has been extensively discussed by Dainty (1963, 1976). In the presence of rapidly permeating solutes, another effect would be important. In this case, prior to membrane permeation, solutes would have to diffuse to the membrane surface. If the rate of diffusion towards the barrier is comparable to that for crossing it, the actual gradient across the barrier would be affected ('gradient-dissipation effect'; Barry and Diamond, 1984). This would also result in a reduced measured permeability of water (and solutes) if water flow is driven by osmotic gradients. In other words, solute diffusion could limit osmotic water transport if the diffusion paths involved are long.

It has been shown that during pressure relaxation experiments the effect of sweep away would usually be small for plant tissue cells when using the cell pressure probe (Steudle et al., 1980). This is so because the recording technique is very sensitive and only small amounts of water are moved across the cell membrane. However, control of water flow by solute diffusion could be very important during osmotic experiments with tissues, where it could limit the osmotic process. Therefore, osmotic experiments can be only performed with isolated cells or monolayers of tissue. Also, it has to be shown that diffusion of solutes does not become limiting (Tyerman and Steudle, 1982). In some of the older literature (e.g. Dainty, 1976) it has been claimed that, in osmotic experiments with tissues, the rate of solute diffusion may be faster than that of the bulk flow of water and that the hydraulic conductivity of cells may be evaluated from such experiments. This is generally not true. Data obtained from such experiments have to be interpreted very cautiously. The same refers to measurements in which labelled (isotopic) water or certain nuclear magnetic resonance (NMR) techniques were used to determine the permeability coefficient of water (for references, see Steudle, 1989a). Due to the rapid movement of water across membranes, gradient-dissipation effects would play an important, if not dominating, role in these experiments (Dainty, 1976; Steudle, 1989a).

2.2.6 *Polarity of water flow: pressure and concentration dependence of* L_p

A polarity of water flow across cell membranes, i.e. a difference in the absolute value of L_p depending on the direction of flow, has been observed using both the technique of transcellular osmosis (Dainty and Ginzburg, 1964; Tazawa and Kamiya, 1966) and the cell pressure probe (Steudle and Zimmermann, 1974). The phenomenon seems to be restricted to cells of the Characeae, although a polarity during the swelling or shrinking of protoplasts has been reported (for literature, see Kiyosawa and Tazawa, 1973). However, the latter finding might have been, in part, due to unstirred-layer effects. The phenomenon of a polarity seems to be related to the concentration dependence of L_p, also found with the Characeae (see below). Using transcellular osmosis and the cell pressure probe, it has been shown that polarity would increase with increasing external solute concentration. The experiments of Kiyosawa and Tazawa (1973) indicate that during transcellular osmosis the cell membrane(s) at the endosmotic side would be more hydrated than at the opposite exosmotic side, causing the effect by a dehydration of the membrane. However, the authors also showed that the L_p at the endosmotic side would depend more strongly on a change of the driving concentration difference ($\Delta\Pi$) than at the exosmotic side. Nevertheless, this explanation would not hold in hydrostatic experiments using the cell pressure probe where concentration changes (e.g. due to concentration polarization) would be too small to account for differences between exosmotic and endosmotic L_p values of 10–15% (Steudle and Zimmermann, 1974). However, it has to be noted that, when using smaller flows, there was no polarity observed (Steudle and Tyerman, 1983). Thus, the polarity seems to be also influenced by the absolute value of water flow (J_V) induced during the experiment, and the problem of polarity still remains to be solved.

To date, a concentration dependence of L_p has only been observed with characean species. It has been shown that this effect can be separated from the effect of turgor pressure on L_p (Steudle and Tyerman, 1983), which is also largely confined to algal cells (for references, see Zimmermann and Steudle, 1978). Interestingly, L_p increases in the range of low pressures towards plasmolysis, i.e. in a range of pressures where the elastic modulus decreases. A decrease of ϵ would increase the extensibility of the cell in this range. Therefore, it is conceivable that the phenomenon is related to the extension of the membrane, which is pinned to the wall and strictly follows wall extension (Coster et al., 1976). Another possibility is that mechanical compression of the membrane is involved (Zimmermann, 1978). A pressure dependence of L_p has also been found in the leaf cells of the higher plant, *Elodea densa* (Steudle et al., 1982).

2.2.7 *Variations of the cell pressure probe technique*

Pressure vs. volume relaxations: pressure clamp. In this variation of the technique, a 'relaxation of volume' rather than the usual 'pressure relaxation' is measured (Wendler and Zimmermann, 1982). The typical procedure would be to first establish a water-flow equilibrium for a given cell while the meniscus is at a fixed position. Then a step change of pressure is created — either as a step up or step down — which causes an exosmotic or endosmotic flow of water, respectively. However, turgor pressure is then clamped in the new position, which requires moving the meniscus in the tip of the capillary by turning the metal rod to compensate for flow across the membrane (Figure 2.1). What is measured is the volume change along the capillary, which proceeds with an exponential time-course. The half-time of the process will be given by:

$$T_{1/2}^{w} = \frac{\ln(2) \cdot V}{A \cdot L_p \cdot \Pi^i}.$$ 2.7

Equation 2.7 states that, compared with the pressure relaxation (see Equation 2.6), $T_{1/2}^{w}$ is larger by a factor $(\epsilon + \Pi^i)/\Pi^i$, which could be quite a high value since $\epsilon \gg \Pi^i$ would usually hold. The process will be independent of ϵ because the cell will not shrink or swell when clamped at a fixed turgor pressure. Thus, the pressure clamp allows measurement of L_p without knowing ϵ. Other advantages are that (1) half-times are longer, which avoids problems of precisely regulating the meniscus with cells having very short $T_{1/2}^{w}$ (see above), and that (2) the cell volume can also be estimated from volume relaxations. The latter point is important because the determination of V (and A) usually causes the largest errors in the cell pressure probe technique. However, there are also disadvantages with the pressure clamp technique. For example, following the meniscus in narrow tips with time can be difficult and requires having a large amount of cell sap sucked into the tip prior to the experiment. Furthermore, the large amounts of water moved across the membrane may cause problems with unstirred layers, especially when there are high solute concentrations in the apoplast. For practical reasons, pressure clamp has rarely been applied to higher plant cells (Zhang and Tyerman, 1991).

It should be mentioned that, in principle, cell volume may also be calculated from pressure relaxations, i.e. from the measured original (P_0), initial (P_A), and final (P_E) pressures (see Figure 2.2a). However, this procedure causes fairly large errors because of uncertainties in determining the small difference between P_E and P_0, again because $\epsilon \gg \Pi^i$. Since, at the present state of development, neither pressure clamp nor the procedure just mentioned are really satisfactory techniques for determining the volume of tissue cells of higher plants, there is an urgent need for such techniques. Normally, the volume and surface area of individual cells of higher plants embedded in a tissue are estimated from cross-sections and will, therefore, refer to average values for the tissue under consideration. But as already mentioned, the geometric factors (V, A) are the parameters that cause the largest error when calculating L_p and ϵ from cell pressure probe data.

Measurement of solute transport: permeability and reflection coefficients.
A considerable extension of the cell pressure probe technique was achieved by incorporating the measurement of solute transport besides the water, and of couplings between water and solute relations (Rüdinger *et al.*, 1992; Steudle and Tyerman, 1983; Tyerman and Steudle, 1982; see reviews in Steudle, 1989a, 1990, 1992, 1993). In principle, active transport of solutes can be measured, because changes of the osmotic concentration in a cell or in the cell surroundings cause corresponding changes of turgor. Such

measurements were attempted in early experiments with the marine alga *Valonia*, which performs turgor regulation due to pressure-dependent potassium transport (for references see Zimmermann and Steudle, 1978).

Passive solute transport across cell membranes has been intensively studied using both isolated algal cells and isolated tissue of higher plants, employing versions two and three of the cell pressure probe (Rüdinger *et al.*, 1992; Steudle and Tyerman, 1983; Tyerman and Steudle, 1982). In these experiments, rapidly permeating solutes, such as low molecular weight alcohols, ketones, amides and the like, were offered to the cells at certain osmotic concentrations and the responses in turgor were measured. From the $P(t)$ curves obtained, the permeability and reflection coefficients of the solutes could be evaluated besides the hydraulic conductivity. A modified Equation 2.5 was used for the volume (water) flow that included the reflection coefficient of the solute (σ_s). In the presence of non-permeating solutes and one permeating solute (denoted by subscript 's'), Equation 2.5 was re-written as:

$$J_V = -\frac{1}{A}\frac{dV}{dt} = L_p[P - RT \cdot (c^i - c^o) - \sigma_s \cdot RT(c_s^i - c_s^o)], \qquad 2.8$$

where $RT \cdot c_s^{o,i}$ represents osmotic pressures caused by the permeating solute 's'. According to the flow equations of irreversible thermodynamics, the solute flow (J_s in $\mathrm{mol\,m^{-2}\,s^{-1}}$) of the permeating solute 's' is given by:

$$J_s = -\frac{1}{A}\frac{dn_s^i}{dt} = P_s(c_s^i - c_s^o) + (1 - \sigma_s) \cdot \bar{c}_s \cdot J_V + J_s^*. \qquad 2.9$$

n_s^i is the amount (mole) of solutes in the cell; P_s, the permeability coefficient of solute s; and $\bar{c}_s$, the mean concentration of solute s in the membrane ($\approx (c_s^i + c_s^o)/2$). The first term on the right-hand side of Equation 2.9 denotes the diffusional flow of solute 's', and the second term the effect of 'solvent drag', i.e. the direct interaction between water and solutes as they cross the membrane. The third term (J_s^*) denotes the active component of solute flow, which is usually not present in the exchange experiments with the rapidly permeating solutes mentioned above.

In a typical experiment (Figure 2.2), a permeating solute is added to an

---

Figure 2.2. *Measurement of water and solute transport in plant cells using the cell pressure probe. In (a), measurements are shown on an isolated giant internode of* Chara corallina *and in (b) on cells of the isolated leaf epidermis of* Tradescantia virginiana. *Hydrostatic experiments and experiments with non-permeating solutes (mannitol, sucrose) gave monophasic responses in cell turgor pressure from which the hydraulic conductivity could be evaluated (Equation 2.6). Permeating solutes (formamide, methanol, ethanol and n-propanol) produced 'biphasic' responses from which permeability (P_s) and reflection (σ_s) coefficients of the solutes were also evaluated. The second phase represented the permeation of solute into the cell. Note that in one of two cells of* Tradescantia epidermis *(cell (b)), the cell responded with an increase of turgor when subjected to n-propanol as the osmotic solute. This means that for this solute the cell showed an 'anomalous osmosis' (negative reflection coefficient, σ_s). Effects were completely reversible. APW, artificial pond water.*

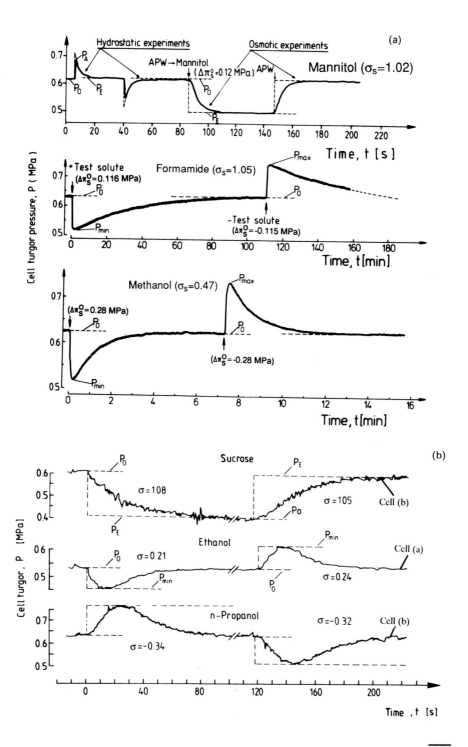

isolated cell or tissue, which causes a fairly rapid decline of turgor pressure due to an exosmotic water flow. This is because the water potential of the medium is lowered or, more precisely, $\sigma_s \cdot \Pi_s^o$ is increased. However, there will only be a transient equilibration of water flow during the experiment ($J_V = 0$), and turgor will increase again due to the passive flow of solute into the cell tending to equilibrate the concentration of permeating solute in both compartments. As solutes are taken up, water will follow the changes in water potential or, more precisely, $(P - \sigma_s \cdot \Delta\Pi_s)$. It can be seen from Figure 2.2 that, eventually, $P_0 = P_E$ will be established. When changing back to the original solution, a symmetrical response occurred in the opposite direction with a maximum in turgor, as expected.

It has been shown that for isolated cells (like internodes of *Chara*) the responses shown in Figure 2.2 were exactly as predicted from theory (Equations 2.8 and 2.9). This means that a simple two-compartment model was sufficient to fit the data (Rüdinger *et al.*, 1992). By integrating Equations 2.8 and 2.9, an analytical expression was obtained which exactly described the measured $P(t)$ curves (Steudle and Tyerman, 1983). The contribution of the solvent drag term in Equation 2.9 was usually negligible, even for solutes for which σ_s was low. Recently, osmotic responses have also been analysed in the presence of more than one osmotic solute, i.e. with both a permeating and non-permeating solute. For this process, the integrated flow equations have been given as well (Rüdinger *et al.*, 1992). Since the maximum responses in turgor $(P_0 - P_{min})$ and $(P_0 - P_{max})$ are proportional to the concentration changes of both the permeating and non-permeating solutes present, respectively, in a solution following a change of medium (see Equation 2.11, below), the procedure has been used to measure concentrations in mixtures. This means that plant cells can be used as osmotic biosensors to determine the concentrations of two different osmotic solutes in a solution at the same time (Rüdinger *et al.*, 1992). Furthermore, using reverse-osmosis membranes of certain selectivity, artificial osmotic cells have been constructed to simulate the osmotic processes in living cells and to provide selective artificial sensors for certain solutes, such as non-electrolytes of low molecular weight (Steudle and Stumpf, 1989; Steudle *et al.*, 1988). Artificial osmotic cells have also been used as test systems for studying negative pressures and freezing phenomena (Steudle and Heydt, 1988; Zhu *et al.*, 1989).

As already mentioned, the processes occurring in the presence of permeating solutes are characterized by a rapid (first) water phase, mainly determined by water relations parameters such as the hydraulic conductivity (L_p) and the elastic modulus of the cell (ϵ) (Equation 2.6), and by a second, slower phase determined by solute movement. It can be shown that the rate constant of the second phase (k_s) is given by:

$$k_s = P_s \frac{A}{V}. \qquad\qquad 2.10$$

On the other hand, the reflection coefficient (σ_s), defined as $(dP/d\Pi_s)_{J_V=0}$, will be given by:

$$\sigma_s = \frac{P_0 - P_{min}}{\Delta\Pi_s^o} \frac{\epsilon + \Pi^i}{\epsilon} \exp(k_s \cdot t_{min}), \qquad 2.11$$

where t_{min} is the time to reach the minimum turgor pressure, and P_{min} is the minimum turgor (Steudle and Tyerman, 1983). Thus, measuring k_s, t_{min} and $(P_0 - P_{min})$ for a given change of osmotic pressure of the medium $(\Delta\Pi_s^o)$ yields both P_s and σ_s (besides L_p) from the observed biphasic kinetics.

The osmotic experiments illustrated in Figure 2.2 have shown that the theory of irreversible thermodynamics used to describe the osmotic processes measured with cell pressure probes exactly fits the experimental results, including the coupling between water and solute transport. The two-compartment model of the cell used also fits very well, although the 'osmotic barrier' involved in plant cells in the exchange of water and solutes is rather complicated and consists of the cell wall, plasma membrane, cytoplasm and tonoplast arranged in series. However, the measurement of solute transport is only possible for completely isolated cells because of unstirred-layer problems (Tyerman and Steudle, 1982). Fortunately, the latter is not true for the measurement of water relations parameters of cells embedded in tissues.

2.3 Root pressure probes

In 1983, Steudle and Jeschke successfully applied the cell pressure probe technique to the measurement of root pressure and of water and solute relations parameters of excised roots. The 'root pressure probe technique' showed some differences compared with the cell pressure probe, both in the practical performance of experiments and in the theoretical treatment of the results. Root systems as well as single root segments (root tips) could be fixed to the equipment and measured (Figure 2.3). The technique with roots relies on the fact that excised roots will exude xylem sap due to the active accumulation of solutes in the xylem. If exudation is stopped by attaching a 'root manometer', a root pressure will develop in the system and can be measured continuously. In fact, during the measurement with the root pressure probe, roots can be treated as osmometers comparable with cells (Weatherley, 1982). This principle can be used in the measurement of root water relations including solutes, again analogous (though not identical) to the theory outlined above for cells. For details of the root pressure probe technique, the reader is referred to original papers (e.g. Azaizeh and Steudle, 1991; Frensch and Steudle, 1989; Heydt and Steudle, 1991; Peterson and Steudle, 1993; Steudle and Frensch, 1989; Steudle and Jeschke, 1983; Steudle et al., 1987; Zhu and Steudle, 1991) and reviews (Steudle 1989a,b,c, 1990, 1992, 1993).

In the following, the root pressure probe technique will be outlined. Its possibilities and limitations will be discussed. Some important results will be summarized. The discussion will also include combined measurements

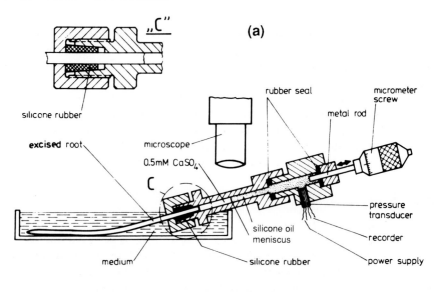

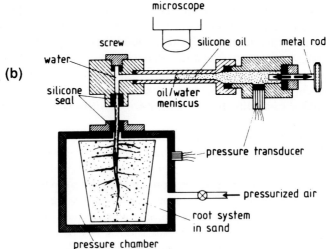

Figure 2.3. *Root pressure probes for measuring water and solute relations of roots (schematic). (a) Version for use with root segments (root tips). (b) Version for use with root systems. The root segment or root system is tightly connected to the root pressure probe by silicone seals so that the root pressure develops in the measuring system consisting of the root, the seal, a measuring capillary and a small pressure chamber. Half of the system is filled with silicone oil and the other with 0.5 mM CaCl$_2$ solution, so that a meniscus forms in the measuring capillary. This meniscus is used as a point of reference during the measurements. When a steady root pressure (P$_{ro}$) is established, water flows can be induced either by changing the pressure in the system using a metal rod (hydrostatic experiments), or by changing the osmotic pressure of the medium or soil solution (osmotic experiments). In (b), the root system is enclosed in a pressure chamber, which allows application of pneumatic pressures to the root and measurement of the resulting flow of water into the xylem by means of separate application of hydrostatic pressure gradients. From osmotic experiments (see text), reflection and permeability coefficients of solutes can be measured in an analogous way to the experiments with isolated cells (Figure 2.2). Typical experiments with roots are shown in Figure 2.4.*

using the cell and root pressure probes, i.e. transport studies at the cell and tissue (organ) level, as well as the use of the technique to model root water and solute transport and to quantify the contribution of the different pathways to overall transport across roots. Results will be reported from recent studies in which the effects of external factors (salinity, anoxia) as well as modifications of roots were studied. A 'composite membrane model of the root' has been derived from the results, which treats the root as a system of (mainly) parallel 'membrane-like' elements (pathways) of different selectivity according to basic theories of the thermodynamics of irreversible processes.

2.3.1 *Root pressure probe technique*

As shown in Figure 2.3a and b, a root (segment or root system) is connected with the root pressure probe by silicone seals which compress the root cortex over a length of 10–20 mm to form a pressure-tight seal, but without compressing the xylem of young roots. The latter would cause high hydraulic resistances in the sealing zone, which would interfere with measurement of the root hydraulic resistance or could even limit the water transport measured. Thus, the hydraulic resistance of the seal has to be carefully checked in the root pressure probe technique. This is done by cutting off the root right at the seal after the experiment and measuring the hydraulic resistance of the remaining open stump sitting in the silicone seal, i.e. the hydraulic resistance in the sealing zone. The latter resistance should be at least one order of magnitude smaller (the half-time of water exchange shorter) than with the intact root.

Also, it must be verified that the apoplasmic bypass at the seal for both water and solutes is negligible, i.e. the hydraulic or permeation resistance of this parallel path should be very high. This is checked (1) by cutting experiments, or (2) by injecting dyes from the top of the cut surface. Furthermore, during fixation of the root at the probe, the tightness of the seal is followed quantitatively by measuring the stationary root pressure (P_{ro}) when tightening the screw with the silicone seal (Figure 2.3). When the system is tight, turning the screw will not increase P_{ro} further.

When a root is properly sealed to the probe, the root pressure built up in the system is stationary for considerable time intervals (1–10 days, depending on the size of the root system), although diurnal rhythms of root pressure have been observed with root systems (Kramer, 1983). Half of the interior of the root pressure probe is filled with silicone oil and the other half with water (or 0.5 mM $CaSO_4$; Figure 2.3) so that a meniscus is formed in a measuring capillary (diameter 200–600 μm, depending on the size of the root or root system). The meniscus serves as a point of reference (as in the cell pressure probe technique) and is monitored with the aid of a stereomicroscope. When a steady root pressure (P_{ro}) is attained, the measurements are started. First, the elastic coefficient of the measuring system (β) will be determined, which is a parameter analogous to the volumetric elastic modulus (ϵ) in cell measurements. However, unlike ϵ, β is largely a property of the measuring system (probe) rather than the root xylem, which is quite rigid. β is defined by (P_r, the root pressure; and V_S, the volume of the measuring system):

$$\beta = \frac{\Delta P_r}{\Delta V_S}. \qquad 2.12$$

β is simply measured by rapidly changing the position of the meniscus in the measuring capillary (ΔV_S) and recording the response in pressure (ΔP_r). As with isolated cells, water flow across the root can be induced either by manipulating P_r with the aid of the metal rod, thereby creating additional gradients of hydrostatic pressure across the root, or by exchanging the medium or soil solution with an osmoticum to create an osmotic gradient. Thus, as with isolated cells but unlike intact tissues, two types of experiments can be performed with roots to evaluate water and solute relations parameters. Pressure relaxations as well as pressure-clamp experiments have been performed.

A third type of experiment is also indicated in Figure 2.3b. Step changes in hydrostatic pressure can be applied across the root by applying step changes of pneumatic pressure to the soil solution or to a medium. This will cause a root pressure relaxation in the xylem equivalent to the experiment in which root pressure was manipulated with the aid of the probe from inside. The results (not shown) demonstrate that both types of 'hydrostatic' root pressure relaxations result in the same time constants and hydraulic conductivities (Steudle, 1993).

Typically, as with cells, pressure relaxations (osmotic and hydrostatic) are performed with roots which are then analysed to work out the hydraulic conductivity of roots (L_{pr}) and solute parameters (P_{sr} and σ_{sr}). For the evaluation of parameters, the root is treated as a two-compartment system (as with cells). However, there are differences in the theoretical treatment, since the xylem (the root interior) is open to the interior of the probe at the cut surface and, furthermore, the volume of mature xylem is usually not known. It has been shown elsewhere that the former consideration would cause apparent differences between exosmotic and endosmotic water flow (Steudle and Jeschke, 1983; Steudle et al., 1987). The rate constant of water flow across the root (k_{wr}) will be related to the hydraulic conductivity of the root (L_{pr}) by:

$$k_{wr} = \frac{\ln (2)}{T^w_{r1/2}} = A_r \cdot \beta \cdot L_{pr}, \qquad 2.13$$

which is analogous to Equation 2.6 ($T^w_{r1/2}$, half-time of water exchange across root; A_r, root surface area). It is evident that the volume of the xylem (V_x) does not need to be known to calculate L_{pr}. However, when the solute permeability coefficient of roots (P_{sr}) is evaluated from biphasic root pressure relaxation curves (rate constant of solute phase, k_{sr}), V_x needs to be determined from cross-sections of the roots. k_{sr} will be given by (cf.

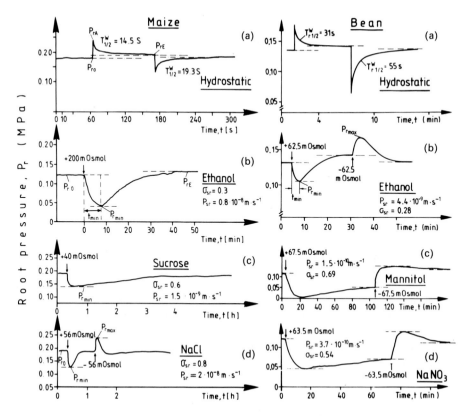

Figure 2.4. *Typical root pressure relaxations of roots of maize (Zea mays L.) and bean (Phaseolus coccineus) measured using the root pressure probe. Either hydrostatic (upper traces) or osmotic pressure gradients have been applied. In osmotic experiments, biphasic responses of root pressure were obtained, from which reflection (σ_{sr}) and permeability (P_{sr}) coefficients of the roots could also be determined in addition to half-times of water exchange ($T^w_{r1/2}$) and hydraulic conductivities (L_{pr}; see also Figure 2.2).*

Equation 2.10):

$$k_{sr} = P_{sr} \cdot \frac{A_r}{V_x}.$$
2.14

Typical examples of measurements of root water and solute parameters are illustrated in Figure 2.4 for young roots of maize and bean. Figure 2.4 shows that reflection coefficients of roots (σ_{sr}) may also be determined in a similar way to that described for cells (see Equation 2.11).

2.3.2 Two-compartment, membrane-equivalent model of the root

In the analysis of transport processes across roots, it has usually been assumed that the root behaves like an osmometer comparable to a cell (Steudle and Brinckmann, 1989; Weatherley, 1982). An interior (xylem) is separated from the outside (soil solution, medium) by a membrane-like barrier, which is usually identified with the endodermis. This simple membrane-equivalent model of the root may be questioned, because in the root cylinder other compartments (e.g. cortex, stele) also have to be considered for transport of water and solutes across the root (Dainty, 1985; Newman, 1976). In fact, each cell layer of the root cylinder could be regarded as a different compartment, which would result in a multicompartment model of the root. The situation becomes even more complicated if axial hydraulic resistances in the xylem are considered as well (see below).

As long as mature xylem is considered, effects of longitudinal transport appear to be small. This has been clearly demonstrated using the root pressure probe, which allows measurement of axial as well as radial resistances (not shown here; Frensch and Steudle, 1989; Melchior and Steudle, 1993). These findings have been confirmed by work from other laboratories (Cruz et al., 1992; North and Nobel, 1991). However, the root pressure probe work also showed that in the tip region of roots the situation may be different. In young maize roots grown in hydroponic solutions, xylem vessels developed in the apical 20 mm. In this region, the longitudinal hydraulic resistance was very high due to the existence of living cells and of cross-walls, so this part of the root was practically hydraulically isolated from the rest of the root (Frensch and Steudle, 1989; Peterson and Steudle, 1993). However, Häussling et al. (1987) showed by micropotometry on intact spruce seedlings that the major uptake of water was in the very tip of the root. It is possible that, in spruce roots, L_{pr} rapidly decreased towards the apex, which would mean that the relative contribution of the tip region to overall uptake was still rather high. Since potometry would only measure the uptake ($J_{Vr} \cdot A_r$) rather than hydraulic conductivity and driving forces, this possibility cannot be excluded from the measurements.

Since the radial hydraulic resistance is usually much larger than the longitudinal resistance, it is justified to apply the two-compartment model. If the condition does not hold, the more detailed theory of Landsberg and Fowkes (1978) should be used. This theory assumes constant longitudinal and radial hydraulic resistances along the developing root, but may also be adapted to the more realistic situation of variable parameters along the root (Frensch and Steudle, 1989; Melchior and Steudle, 1993). The influence of the longitudinal component may be important in plants with tracheids in the xylem, such as the roots of conifers.

2.3.3 Composite membrane model of the root

As outlined above, current models favour the idea of a root being a rather perfect (semipermeable) osmometer (i.e. $P_{sr} \approx 0$; $\sigma_{sr} \approx 1$). However, there is evidence, both experimental and theoretical, that this is an oversimplification. Measurements using the root pressure probe show that roots deviate substantially from the ideal behaviour, exhibiting reflection coefficients

$\sigma_{sr} < 1$ at fairly low (but not negligible) values of permeability coefficients (P_{sr}). Thus, roots are not leaky, despite a rather low σ_{sr}. A low σ_{sr} is also indicated from experiments using other techniques (e.g. Mees and Weatherley, 1957; Miller, 1985), although there are, for technical reasons, few data for this parameter obtained by techniques other than the root pressure probe.

Root pressure probe experiments have indicated that, depending on the nature of the driving force used to induce water flows, there may be differences — or apparent differences — in L_{pr}. In osmotic experiments, L_{pr} was much smaller than in hydrostatic experiments, though not for all species. To some extent, this is true not only in root pressure probe work (for references, see Cruz et al., 1992; Steudle et al., 1987). Another finding is that, when varying water flow across the root, L_{pr} increased with increasing water flow, i.e. the root hydraulic resistance is variable. To date, the latter phenomenon has not been observed in root pressure probe experiments which, at the present state of development, do not allow large flows to be induced. The variability of root L_{pr} has been explained in terms of a dilution effect (Fiscus, 1975), i.e. to be apparent only, and to be caused, by changes in the osmotic driving force rather than by changes in L_{pr}. Another explanation of variable L_{pr} assumes that there are valve-like mechanisms at the plasmodesmata of root cells (Passioura, 1988, 1992): plasmodesmata might open at certain pressure gradients across the root, which would then cause the variable L_{pr}. Both explanations appear to be unsatisfactory. An alternative explanation, which would also explain both the apparent dependence of L_{pr} on the driving force (J_{Vr}) and the low σ_{sr} at low P_{sr}, is given below.

The results summarized above are readily explained in terms of a 'composite membrane model of the root', which considers the root to be composed of 'membrane-like elements' arranged both in series (such as the rhizodermis, exodermis, cortex, endodermis, etc.) and in parallel (such as for the different pathways of water and solute flows in the root). The equations of irreversible thermodynamics may be applied to such a system (Kedem and Katchalsky, 1963a,b). This application, as well as basic structural facts, forms the basis of the model.

In the composite membrane model, the parallel arrangement of pathways appears to be more important than the serial. Parallel pathways could be envisaged in roots on a microscopic scale, i.e. there are the apoplasmic, symplasmic (via plasmodesmata) and transcellular (vacuolar) paths. The latter two may be summarized as the cell-to-cell path because, to date, they cannot be separated experimentally (Steudle, 1989a, 1992, 1993). On a macroscopic scale, there would be zones of different permeability along the developing root, which has been shown to exhibit changes in transport coefficients, e.g. during the development of the endodermis (L_{pr}, P_{sr}, σ_{sr}; Frensch and Steudle, 1989; Melchior and Steudle, 1993).

A parallel apoplasmic path could be via root primordia (secondary root initials), which have been shown to provide a diffusional path for apoplasmic tracers (Peterson *et al.*, 1981), but the apoplasmic path is not necessarily confined to these structures (Hanson *et al.*, 1985; Steudle *et al.*, 1993; Yeo *et al.*, 1987). The low root σ_{sr} is readily explained by considering the overall σ_{sr} of a root in terms of the Kedem and Katchalsky (1963a) theory to be composed of an apoplasmic and a parallel cell-to-cell component. This would yield:

$$\sigma_{sr} = \gamma^{cc} \cdot \sigma_s^{cc} \cdot \frac{L_p^{cc}}{L_{pr}} + \gamma^{apo} \cdot \sigma_s^{apo} \cdot \frac{L_p^{apo}}{L_{pr}}. \qquad 2.15$$

σ_s^{apo} and σ_s^{cc} represent the reflection coefficient of the apoplast and cell-to-cell path, respectively; γ^{apo} and γ^{cc}, the fractional contribution of pathways to overall cross-sectional area ($\gamma^{apo} + \gamma^{cc} = 1$); and L_p^{apo} and L_p^{cc}, the hydraulic conductivity of pathways. It holds that $L_{pr} = \gamma^{cc} \cdot L_p^{cc} + \gamma^{apo} L_p^{apo}$.

It should hold that for the membrane-bound path, $\sigma_{sr} \approx 1$, whereas the reflection coefficient of the apoplast should be close to zero (no selectivity between solutes and water). However, even if $\gamma^{apo} \ll \gamma^{cc}$, the contribution of the apoplasmic component to the overall σ_{sr} could be large (and σ_{sr} substantially smaller than unity), if $L_p^{apo} \gg L_p^{cc}$. Thus, σ_{sr} represents a weighted mean of σ_s^{apo} and σ_s^{cc}. At the same time, the permeability coefficient of roots (P_{sr}) could be low due to the fact that $\gamma^{cc} \gg \gamma^{apo}$ and the entire barrier is rather thick. The absolute values of P_{sr} measured using the root pressure probe for typical solutes such as salts are rather low, and are of the same order as those given in the literature for root cell membranes (for references, see reviews by Steudle, 1989a,b,c, 1990, 1992, 1993).

Recently, the composite membrane model of the root and the principle of the root pressure probe have been questioned by Zimmermann *et al.* (1992). These authors measured water relations parameters (P, $T_{1/2}^w$, ϵ, L_p) of cortical cells of young roots of *Aster trifolium*, but could not determine the equivalent parameters at the whole-root level. One of the concerns of Zimmermann *et al.* (1992) was that, when using the root pressure probe, the elastic modulus of mature xylem (ϵ_x) may not be much larger than that of the equipment. The authors measured an ϵ_x of the root xylem that was comparable to cells of the marine alga *Halicystis parvula* ($\epsilon = 0.05$–$0.25\,MPa$; Zimmermann and Hüsken, 1980). This would mean that the xylem would change its internal volume by 100% with pressure changes of only 0.5–2 bar (Equation 2.1), which is unlikely. Zimmermann *et al.* (1992) neglected the fact that the extensibility of the xylem (even if it were to contribute significantly to the overall extensibility of the measuring system) would be incorporated in the elastic coefficient, β (Equation 2.12). Another concern of the authors referred to unstirred layers, which may lower the absolute value of the measured $(\Delta P_r / \Delta \Pi_s^o)_{J_V=0}$, and hence result in an underestimation of σ_{sr}. However, these effects have been carefully examined in the root pressure probe technique and cannot completely account for the finding of a $\sigma_{sr} < 1$ (Steudle and Frensch, 1989). Zimmermann *et al.* (1992), furthermore, reject the idea of treating a structure such as the root in terms of the basic theory of irreversible thermodynamics. However, this theory has been applied to composite structures (Kedem and Katchalsky, 1963a,b) and has been used with roots for some time (Dainty, 1985; Dalton *et al.*, 1975; Fiscus, 1975; Schambil and Woermann, 1989). Several studies confirm that there are low values of σ_{sr} as well as apparent differences between osmotic and hydrostatic water flow in roots (for references, see Cruz *et al.*, 1992; Steudle *et al.*, 1987).

Some of the concerns of Zimmermann *et al.* (1992) arise from their own cell pressure probe work on roots. These results showed that the young roots of *Aster*, equilibrated either in the control medium or in 300 mM NaCl (≈ 1.5 MPa $= 15$ bar osmotic pressure), exhibited a water potential across the root cortex that was constant and close to zero under both conditions. This was so despite the fact that the root cells as well as root tissue equilibrated rapidly with their surroundings. The authors explained this apparent anomaly by assuming that solutes were excluded from the root apoplast. This, however, does not help to resolve the contradiction, since, according to the data presented, the equilibration of water potential across the root would also take place rapidly along the cell-to-cell path. The finding of a complete exclusion of salt (NaCl) from the root cortical apoplast is not consistent with the general finding that salts freely diffuse into the cortical apoplast of young roots, as also indicated by the plasmolysis experiment of Figure 2c of Zimmermann *et al.* (1992). It remains unclear why, with respect to water relations, the roots of *Aster* behaved so unusually for Zimmermann *et al.* (1992).

2.3.4 *Variable hydraulic resistance of roots: osmotic vs. hydrostatic driving forces*

The composite membrane model of the root can explain differences between osmotic and hydrostatic L_{pr}. In the case of osmotic experiments, the driving force along the cell-to-cell path would be fully converted into a gradient of osmotic pressure, since $\sigma_s^{cc} \approx 1$. However, along the apoplast $\sigma_s^{apo} \cdot \Delta \Pi_s$ would be much smaller, since $\sigma_s^{apo} \approx 0$. In hydrostatic experiments, the driving force would be fully exerted along both parallel pathways. Thus, the differences in L_{pr} values could, in part, be apparent and due to differences in the effective driving forces along the pathways (Steudle *et al.*, 1987). In fact, thermodynamic theory predicts that, in the presence of two parallel pathways exhibiting differences of σ_s, opposing water flows will be set up across the barrier in the pathways. This is because the osmotic gradient would draw water into the xylem along the cell-to-cell path to build up a hydrostatic pressure (root pressure) larger than that in the medium. In turn, this pressure will cause a backflow of water (solution) along the non-selective apoplasmic path. At steady pressure ($J_{Vr} = 0$), the two opposite flows would cancel. In other words, at low or vanishing J_{Vr}, there will be a circulation flow of water in the root.

Opposing water flows also mean that there will be a high apparent hydraulic resistance in the root at low J_{Vr} when the flows are opposite. On the other hand, as the tension in the xylem increases in the intact plant, flows will soon be in the same direction and the high hydraulic conductance of the apoplast will be fully used for water uptake. Hence, in terms of the composite membrane model of the root, there should be a variable hydraulic resistance as observed. The explanation is different to those of Fiscus and Passioura given above. Testing the model would require varying J_{Vr} in the root pressure probe over larger ranges which, in turn, would also require creating tensions in the probe driving the water flow. This is, in principle, possible, but requires certain precautions and a set-up that will minimize cavitations (Heydt and Steudle, 1991). In this context, it is significant that the osmotic L_{pr} value as well as σ_{sr} and P_{sr} measured under conditions of tension (negative pressure) in the xylem were the same as those obtained in the range of positive pressures. However, under tension, the hydrostatic L_{pr} could not be determined because of cavitations occurring immediately in the system when moving the metal rod.

2.3.5 Comparison between cell L_p and root L_{pr}: combination of cell and root pressure probe

In several studies, the cell and root pressure probe techniques have been combined to study the relative contribution of pathways (cell-to-cell vs. apoplasmic) to the overall radial water uptake (Steudle and Brinckmann, 1989; Steudle and Jeschke, 1983; Steudle et al., 1987; Zhu and Steudle, 1991). L_{pr} was calculated from the cell L_p assuming that there was no apoplasmic component, and these values were compared with the measured L_{pr}. For some species (barley and bean), there was a substantial cell-to-cell component for water for both hydrostatic and osmotic gradients, which resulted in the same L_{pr} for both driving forces. The apoplasmic component of water flow appeared to be small, possibly indicating rather tight Casparian bands in these species. For other species (maize, onion and beech), large differences were found in hydrostatic and osmotic experiments, as discussed above. For maize roots, the comparison between the root and cell levels showed that the (hydrostatic) root L_{pr} and L_p of cortical cells were similar, or L_{pr} was larger. This can only be explained in terms of considerable amounts of water bypassing the protoplasts.

A similar result has been obtained by Radin and Matthews (1989) for cotton roots. The authors used the cell pressure probe to measure water transport at the cell level, but a technique different from the root pressure probe for measuring hydrostatic root L_{pr}. Radin and Matthews (1989) concluded that about 50% of the water was moving across the root in the apoplast. In 1983, Jones et al. measured the cell L_p of wheat roots and compared this value with the root L_{pr} determined by an osmotic stop-flow technique and claimed that there was a predominant flow from cell to cell. However, later measurements revised this result and showed that for both wheat and maize the water was predominantly apoplasmic (Jones et al., 1988). These results show that a dependence of L_{pr} on the driving force is also indicated by experiments using techniques besides the pressure probe. Unfortunately, with other techniques it is not straightforward to measure solute relations parameters in addition to those of water transport.

A comparison of data on root hydraulics obtained by different methods indicates that there is good agreement with the root pressure probe data. This applies to experiments with pressurized exudation (Fiscus, 1975) as well as to measurements in which stop-flow techniques were used (Miller, 1985; Pitman *et al.*, 1981). These data are also indicated by the more recent results from Nobel's and Drew's groups (Cruz *et al.*, 1992; North and Nobel, 1991), who induced water flows by applying low pressure to the cut surface of excised roots (cf. Mees and Weatherley, 1957). With the latter techniques, effects of growing conditions (drought) have also been studied on both the radial (L_{pr}) and axial (L_x) hydraulic conductivities.

2.3.6 *Effects of external factors on root* L_{pr} *and cell* L_p: *high salinity and anoxia*

The root pressure probe has now been used to study the effects of important ecological and physiological parameters on root function. Effects of drought, low temperature, anoxia, high salinity, toxic metals and low pH, and mycorrhiza are of special interest. So far, detailed studies have been performed on the effects of high salinity and low oxygen on young maize roots (Azaizeh and Steudle, 1991; Azaizeh *et al.*, 1992; Birner and Steudle, 1993) to determine how they affect the relative contribution of pathways to overall transport in the root.

High soil salinity represents an environmental stress that inhibits plant growth and development. The detailed mechanisms of the adverse effects at the root level are still not fully understood. They may include osmotic effects, as well as effects caused by the specific toxicity of certain ions (Rengel, 1992). Reductions of growth could be caused by an inhibition of water transport (root hydraulic conductivity, L_{pr}), which may result from a reduced hydraulic conductivity of root cell membranes. It is known that negative effects of high salinity may be reversed by calcium (Lynch *et al.*, 1987), which is also thought to be a membrane effect, with Na^+ competing for Ca^{2+} at high salinity. Azaizeh and Steudle (1991) and Azaizeh *et al.* (1992) have tested this using both the cell and root pressure probe. Maize seedlings grown in a nutrient solution plus 100 mM NaCl had an L_{pr} that was reduced 30–60% compared with the control. Increased levels of calcium had an ameliorative effect on root L_{pr}. It was shown that the positive effect was much larger at the cell than at the whole-root level, which was taken to mean that most of the water flow across the root was around cells under hydrostatic conditions. These findings are in line with those of Munns and Passioura (1984) and O'Leary (1969) for bean and lupin, but are at variance with those of Munns and Passioura (1984) and Shalhevet *et al.* (1976) for barley, tomato and sunflower. Also, for tobacco, Tyerman *et al.* (1989) found no change of cell L_p at high salinity measured with the cell pressure probe.

As with high salinity, anoxia also had an adverse effect on root L_{pr}, as shown in root pressure probe experiments (Birner and Steudle, 1993). The decrease in L_{pr} could have been due to a reduced cell L_p, but this has not yet been measured. In addition to changes of root L_{pr} reported in the literature (Everard and Drew, 1987; Zhang and Tyerman, 1991), the root pressure probe also allowed measurement of reductions of root pressure caused by anoxia, as well as changes of P_{sr} and of the root reflection coefficient (σ_{sr}) in addition to effects on the water transport. The stationary root pressure and P_{sr} were reduced by anoxia at constant σ_{sr}, which suggested that the active pumping of ions rather than leakage from the root was affected by low oxygen (Birner and Steudle, 1993).

2.3.7 *Modified roots: steaming, scraping, dissecting and puncturing experiments*

The use of modified roots is another possibility for using the root pressure probe technique to gain more insight into the mechanisms of water and solute uptake into roots. Modifications of the endodermis would be most interesting, besides that of the exodermis or of suberization of outer root layers. These could involve the application of chemicals to entire roots or to distinct root zones, the use of heat, or of mechanical wounding, dissecting or puncturing. To date, detailed results are available for roots modified by steam-ringing and by different kinds of wounding (Peterson and Steudle, 1993; Peterson *et al.*, 1993; Steudle *et al.*, 1993).

In steam-ringing experiments, distinct, short zones of roots have been killed by steaming or by applying hot water to eliminate the hydraulic resistance of living tissue, including the endodermis. In young maize roots, the contribution of the lateral hydraulic resistance of xylem vessels (L_{px}) to the overall radial hydraulic resistance was evaluated. Killing a zone of a few millimetres of a maize root of about 100 mm in length caused the root pressure to drop to a value close to zero. The hydraulic conductivity of the steamed zone, which was calculated from hydrostatic relaxation experiments before and after steaming, corresponded to that of the vessel walls and could be compared with that of the intact root cylinder. Although L_{px} was much larger than L_{pr}, the contribution of the lateral hydraulic resistance of vessel walls was not completely negligible.

Puncturing experiments offer the possibility of creating additional 'apoplasmic bypasses' that, by area, exhibit a fraction of only 10^{-3} to 10^{-2}% of the entire surface area of the endodermis of the young maize roots used in the experiments (Steudle *et al.*, 1993). Nevertheless, it was expected that this would affect L_{pr} as well as P_{sr} and σ_{sr}. More importantly, the effects on the absolute magnitude of transport coefficients, and on σ_{sr} in particular, should allow extrapolation to the possible size of apoplasmic bypasses in the intact root. The puncturing experiments indicated that even a small bypass of the size given above was sufficient to reduce turgor to a low value by a leakage of solutes out of the root. The reflection coefficient was also reduced, but was still substantially larger than zero in the punctured root. Changes of (hydrostatic) L_{pr} were not detectable, indicating that the small hole had no effect on the hydraulic conductivity, which was already rather high in the intact system. However, P_{sr} was substantially increased. The results of the puncturing experiments indicate that the small bypasses present in the intact system, such as those across root primordia (Peterson *et al.*, 1981), across the root tip or some other leakiness of the Casparian bands, could be sufficient to cause a value of $\sigma_{sr} < 1$ as found. Thus, the puncturing experiments support the composite membrane model of the root. They indicate that the endodermis is a very efficient barrier for ions, which is in line with the traditional view of the function of this structure. However, the experiments also indicate that the endodermis should present no significant barrier for water, at least in a state where the suberin lamellae have not yet developed (state II of root endodermis). This latter view is somewhat different from the conventional picture of the endodermis. Scraping and dissecting the root cortex also supported the findings obtained in the puncturing experiments, with root pressure being unaffected as long as the endodermis remained intact.

Experiments with modified roots appear to be a powerful tool to obtain more detailed information on the limiting resistances for transport across the root cylinder. In particular, they should reveal information about the transport properties of the endodermis. They should also provide a tool for further checking and development of the composite membrane model of the root.

Acknowledgements

Between 1987 and 1992, the work reviewed in this article was supported by grants from the Deutsche Forschungsgemeinschaft, Sonderforschungsbereich 137, and Ste 319/2-1 as well as by EUROSILVA (project no. 39473C). The contributions of several guest scientists (Drs Hassan Azaizeh, Technion, Israel Institute of Technology, Haifa, Israel; Benito Gunse, Universidad Autonoma de Barcelona, Spain; Anatoli B. Meshcheryakov, Russian Academy of Sciences, Moscow; Carol A. Peterson, University of Waterloo, Canada; and Guo-Li Zhu, Agricultural University of Peking, China) are gratefully acknowledged, as well as those of different co-workers (T. Birner, Dr J. Frensch, H. Heydt, R. Lütgenau, W. Melchior, M. Murrmann and M. Rüdinger). The author is also grateful for the expert technical assistance of Burkhard Stumpf.

References

Azaizeh, H. and Steudle, E. (1991) Effects of salinity on water transport of excised maize (*Zea mays* L.) roots. *Plant Physiol.*, **97**, 1136–1145.
Azaizeh, H., Gunse, B. and Steudle, E. (1992) Effects of NaCl and CaCl$_2$ on water transport across root cells of maize (*Zea mays* L.) seedlings. *Plant Physiol.*, **99**, 886–894.

Balling, A. and Zimmermann, U. (1990) Comparative measurements of the xylem pressure of *Nicotiana* plants by means of the pressure bomb and pressure probe. *Planta*, **182**, 325–338.

Barry, P.H. and Diamond, J.M. (1984) Effects of unstirred layers on membrane phenomena. *Physiol. Rev.*, **64**, 763–872.

Birner, T.P. and Steudle, E. (1993) Effects of anaerobic conditions on water and solute relations and active transport in roots of maize (*Zea mays* L.). *Planta*, in press.

Boyer, J.S. (1985) Water transport. *Ann. Rev. Plant Physiol.*, **36**, 473–516.

Cosgrove, D.J. (1986) Biophysical control of plant cell growth. *Ann. Rev. Plant Physiol.*, **37**, 377–405.

Cosgrove, D.J. and Steudle, E. (1981) Water relations of growing pea epicotyl segments. *Planta*, **153**, 343–350.

Coster, H.G.L., Steudle, E. and Zimmermann, U. (1976) Turgor pressure sensing in plant cell membranes. *Plant Physiol.*, **58**, 636–643.

Cruz, R.T., Jordan, W.R. and Drew, M.C. (1992) Structural changes and associated reduction of hydraulic conductance in roots of *Sorghum bicolor* L. following exposure to water deficit. *Plant Physiol.*, **99**, 203–212.

Dainty, J. (1963) Water relations of plant cells. *Adv. Bot. Res.*, **1**, 279–326.

Dainty, J. (1976) Water relations of plant cells. In: *Encyclopedia of Plant Physiology*, Vol. 2, Part A, (eds U. Lüttge and M.G. Pitman). Springer-Verlag, Berlin, pp. 12–35.

Dainty, J. (1985) Water transport through the root. *Acta Hort.*, **171**, 21–31.

Dainty, J. and Ginzburg, B.Z. (1964) The measurement of hydraulic conductivity (osmotic permeability to water) of internodal characean cells by means of transcellular osmosis. *Biochim. Biophys. Acta*, **79**, 102–111.

Dalton, F.N., Raats, P.A.C. and Gardner, W.R. (1975) Simultaneous uptake of water and solutes by plants. *Agron. J.*, **67**, 334–339.

Everard, J.D. and Drew, M.C. (1987) Mechanisms of inhibition of water movement in anaerobically treated roots of *Zea mays* L. *J. Exp. Bot.*, **38**, 1154–1165.

Fiscus, E.L. (1975) The interaction between osmotic- and pressure-induced water flow in plant roots. *Plant Physiol.*, **55**, 917–922.

Frensch, J. and Steudle, E. (1989) Axial and radial hydraulic resistance to roots of maize (*Zea mays* L.). *Plant Physiol.*, **91**, 719–726.

Green, P.B. (1968) Growth physics of *Nitella*: a method for continuous *in vivo* analysis of extensibility based on a micro-manometer technique for turgor pressure. *Plant Physiol.*, **43**, 1169–1184.

Gutknecht, J. (1968) Salt transport in *Valonia*: inhibition of potassium uptake by small hydrostatic pressures. *Science*, **160**, 68–70.

Hanson, P.J., Sucoff, E.I. and Markhart, A.H. (1985) Quantifying apoplastic flux through red pine root system using trisodium, 3-hydroxy-5,8,10-pyrenetrisulfonate. *Plant Physiol.*, **77**, 21–24.

Häussling, M., Jorns, C.A., Lehmbecker, G., Hecht-Buchholz, C.H. and Marschner, H. (1988) Ion and water uptake in relation to root development in Norway spruce (*Picea abies* (L.) Karst). *J. Plant Physiol.*, **133**, 486–491.

Heydt, H. and Steudle, E. (1991) Measurement of negative root pressure in the xylem of excised roots. Effects on water and solute relations. *Planta*, **184**, 389–396.

Hüsken, D., Steudle, E. and Zimmermann, U. (1978) Pressure probe technique for measuring water relations of cells in higher plants. *Plant Physiol.*, **61**, 158–163.

Jones, H., Tomos, A.D., Leigh, R.A., Wyn Jones, R.G. (1983) Water-relation parameters of epidermal and cortical cells in the primary root of *Triticum aestivum* L. *Planta*, **158**, 230–236.

Jones, H., Leigh, R.A., Wyn Jones, R.G. and Tomos, A.D. (1988) The integration of whole-root and cellular hydraulic conductivities in cereal roots. *Planta*, **174**, 1–7.

Kedem, O. and Katchalsky, A. (1963a) Permeability of composite membranes. Part 2: Parallel elements. *Trans. Far. Soc.*, **59**, 1931–1940.

Kedem, O. and Katchalsky, A. (1963b) Permeability of composite membranes. Part 3: Series array of elements. *Trans. Far. Soc.*, **59**, 1941–1953.

Kiyosawa, K. and Tazawa, M. (1973) Rectification characteristics of *Nitella* membranes in respect to water permeability. *Protoplasma*, **78**, 203–214.

Kramer, P.J. (1983) *Water Relations of Plants.* Academic Press, Orlando.

Landsberg, J.J. and Fowkes, N.D. (1978) Water movement through plant roots. *Ann. Bot.* **42**, 493–508.

Lynch, J., Cramer, G.R. and Läuchli, A. (1987) Salinity reduces membrane-associated calcium in corn root protoplasts. *Plant Physiol.*, **83**, 390–394.

Mees, G.C. and Weatherley, P.E. (1957) The mechanism of water absorption by roots. I. Preliminary studies on the effects of hydrostatic pressure gradients. *Proc. R. Soc. Lond. B*, **147**, 367–380.

Melchior, W. and Steudle, E. (1993) Water transport in onion (*Allium cepa* L.) roots. Changes of axial and radial hydraulic conductivity during root development. *Plant Physiol.* **101**, 1305–1315.

Meshcheryakov, A., Steudle, E. and Komor, E. (1992) Gradients of turgor, osmotic pressure, and water potential in the cortex of the hypocotyl of growing *Ricinus* seedlings. Effects of the supply of water from the xylem and of solutes from the phloem. *Plant Physiol.* **98**, 840–852.

Miller, D.M. (1985) Studies of root function in *Zea mays*. III. Xylem sap composition at maximum root pressure provides evidence of active transport in the xylem and a measurement of the reflection coefficient of the root. *Plant Physiol.*, **77**, 162–167.

Munns, R. and Passioura, J.B. (1984) Hydraulic resistance of plants. III. Effects of NaCl in barley and lupin. *Aust. J. Plant Physiol.*, **11**, 351–359.

Newman, E.I. (1976) Interaction between osmotic- and pressure-induced water flow in plant roots. *Plant Physiol.*, **57**, 738–739.

Nonami, H., Boyer, J.S. and Steudle, E. (1987) Pressure probe and isopiestic psychrometer measure similar turgor. *Plant Physiol.*, **83**, 592–595.

North, G.B. and Nobel, P.S. (1991) Changes in hydraulic conductivity and anatomy caused by drying and rewetting roots of *Agave deserti* (Agavaceae). *Am. J. Bot.*, **78**, 906–915.

O'Leary, J.W. (1969) The effect of salinity on permeability of roots to water. *Isr. J. Bot.*, **18**, 1–9.

Passioura, J.B. (1988) Water transport in and to the root. *Ann. Rev. Plant Physiol. Plant Mol. Biol.*, **39**, 245–265.

Passioura, J.B. (1992) An impasse in plant water relations? *Bot. Acta*, **104**, 405–411.

Peterson, C.A. (1988) Exodermal Casparian bands: their significance for ion uptake in roots. *Physiol. Plant.*, **72**, 204–208.

Peterson, C.A., Emanuel, G.B. and Humphreys, G.B. (1981) Pathway of movement of apoplastic fluorescent dye tracers through the endodermis at the site of secondary root formation in corn (*Zea mays*) and broad bean (*Vicia faba*). *Can. J. Bot.*, **59**, 618–625.

Peterson, C.A., Murrmann, M. and Steudle, E. (1993) Location of the major barriers to water and ion movement in young roots of *Zea mays* L. *Planta*, **190**, 127–136.

Peterson, C.A. and Steudle, E. (1993) Lateral hydraulic conductivity of early metaxylem vessels in *Zea mays* L. roots. *Planta*, **189**, 288–297.

Pitman, M.G., Wellfare, D. and Carter, C. (1981) Reduction of hydraulic conductivity during inhibition of exudation from excised maize and barley roots. *Plant Physiol.*, **67**, 802–808.

Radin, J.W. and Matthews, M.A. (1989) Water transport properties of cells in the root cortex of nitrogen- and phosphorus-deficient cotton seedlings, *Plant Physiol.*, **89**, 264–268.

Rengel, Z. (1992) The role of calcium in salt toxicity. *Plant Cell Environ.*, **15**, 625–632.

Rüdinger, M., Hierling, P. and Steudle, E. (1992) Osmotic biosensors. How to use a characean internode for measuring the alcohol content of beer. *Bot. Acta*, **105**, 3–12.

Schambil, F. and Woermann, D. (1989) Radial transport of water across cortical sleeves of excised roots of *Zea mays* L. *Planta*, **178**, 488–494.

Shalhevet, J., Maass, E.V., Hoffmann, G.J. and Ogata, G. (1976) Salinity and the hydraulic conductance of roots. *Physiol. Plant.*, **38**, 224–232.

Steudle, E. (1985) Water transport as a limiting factor in extension growth. In: *Control of Leaf Growth* (eds N.R. Baker, W.J. Davies and C.K. Ong). Cambridge University Press, Cambridge, pp. 35–55.

Steudle, E. (1989a) Water flow in plants and its coupling to other processes: an overview. *Methods of Enzymol.*, **174**, 183–225.

Steudle, E. (1989b) Water transport in roots. In: *Structural and Functional Aspects of Transport in Roots* (eds B.C. Loughman, O. Gasparikova and J. Kolek). Kluwer Academic Publishers, Amsterdam, pp. 139–145.

Steudle, E. (1989c) Water transport in roots. In: *Plant Water Relations and Growth Under Stress* (eds M. Tazawa, M. Katsumi, M. Masuda and Y. Okamoto). Yamada Science Foundation, Osaka, and Myu, K.K., Tokyo, pp. 253–260.

Steudle, E. (1990) Methods for studying water relations of plant cells and tissues. In: *Measurement Techniques in Plant Sciences* (eds Y. Hashimoto, P.J. Kramer, H. Nonami and B.R. Strain). Academic Press, New York, pp. 113–150.

Steudle, E. (1992) The biophysics of plant water: compartmentation, coupling with metabolic processes, and water flow in plant roots. In: *Water and Life: Comparative Analysis of Water Relationships at the Organismic, Cellular, and Molecular Levels* (eds G.N. Somero, C.B. Osmond and C.L. Bolis). Springer-Verlag, Berlin, pp. 173–204.

Steudle, E. (1993) The regulation of plant water at the cell, tissue and organ level: role of active processes and of compartmentation. In: *Flux Control in Biological Systems: From the Enzyme to the Ecosystem Level* (ed. E.-D. Schulze). Academic Press, New York, in press.

Steudle, E. and Boyer, J.S. (1985) Hydraulic resistance to radial water flow in growing hypocotyl of soybean measured by a new pressure-perfusion technique. *Planta*, **164**, 189–200.

Steudle, E. and Brinckmann, E. (1989) The osmometer model of the root: water and solute relations of *Phaseolus coccineus*. *Bot. Acta*, **102**, 85–95.

Steudle, E. and Frensch, J. (1989) Osmotic responses of maize roots. Water and solute relations. *Planta*, **177**, 281–295.

Steudle, E. and Heydt, H. (1988) An artificial osmotic cell: a model system for simulating osmotic processes and for studying phenomena of negative pressure in plants. *Plant Cell Environ.*, **11**, 629–637.

Steudle, E. and Jeschke, W.D. (1983) Water transport in barley roots. *Planta*, **158**, 237–248.

Steudle, E. and Stumpf, B. (1989) Verfahren und Einrichtung zur Bestimmung des Gehaltes von in einem Lösungsmittel gelösten Stoffen mittels eines Osmometers. German Patent DE 38 25 208, and corresponding patents in other countries. (Date of German patent: 2 November 1989.)

Steudle, E. and Tyerman, S.D. (1983) Determination of permeability coefficients, reflection coefficients and hydraulic conductivity of *Chara corallina* using the pressure probe: effects of solute concentrations. *J. Membrane Biol.*, **75**, 85–96.

Steudle, E. and Wienecke, J. (1985) Changes in water relations and elastic properties of apple fruit cells during growth and development. *J. Am. Soc. Hort. Sci.*, **110**, 824–829.

Steudle, E. and Zimmermann, U. (1971) Hydraulische Leitfähigkeit von *Valonia utricularis*. *Z. Naturforsch.*, **26b**, 1302–1311.

Steudle, E. and Zimmermann, U. (1974) Determination of hydraulic conductivity and of reflection coefficients in *Nitella flexilis* by means of direct cell-turgor pressure measurements. *Biochim. Biophys. Acta*, **322**, 399–412.

Steudle, E., Lüttge, U. and Zimmermann, U. (1975) Water relations of the epidermal bladder cells of the halophytic species *Mesembryanthemum crystallinum*: direct measurement of hydrostatic pressure and hydraulic conductivity. *Planta*, **126**, 229–246.

Steudle, E., Smith, J.A.C. and Lüttge, U. (1980) Water relation parameters of individual mesophyll cells of the crassulacean acid metabolism plant *Kalanchoe daigremontiana*. *Plant Physiol.*, **66**, 1155–1163.

Steudle, E., Zimmermann, U. and Zillikens, J. (1982) Effect of cell turgor on hydraulic conductivity and elastic modulus of *Elodea* leaf cells. *Planta*, **154**, 371–380.

Steudle, E., Oren, R. and Schulze, E.-D. (1987) Water transport in maize roots. *Plant Physiol.*, **84**, 1220–1232.

Steudle, E., Zillikens, J. and Böling, J. (1988) Verfahren zur Bestimmung des Gehaltes von mindestens zwei in einem Lösungsmittel gelösten Stoffen. German Patent DE 35 25 668, and corresponding patents in other countries. (Date of German patent: 17 November 1988.)

Steudle, E., Murmann, M. and Peterson, C.A. (1993) Transport of water and solutes across maize roots modified by puncturing the endodermis: further evidence for the composite transport model of the root. *Plant Physiol.*, in press.

Tazawa, M. and Kamiya, N. (1966) Water permeability of a characean internodal cell with special reference to its polarity. *Aust. J. Biol. Sci.*, **19**, 399–419.

Tomos, A.D. (1988) Cellular water relations of plants. In: *Water Science Reviews*, Vol. 3, (ed. F. Franks). Cambridge University Press, Cambridge, pp. 186–277.

Tyerman, S.D. and Steudle, E. (1982) Comparison between osmotic and hydrostatic water flow in a higher plant cell: determination of hydraulic conductivities and reflection coefficients in isolated epidermis of *Tradescantia virginiana*. *Aust. J. Plant Physiol.*, **9**, 461–479.

Tyerman, S.D., Oats, P., Gibbs, J., Dracup, M. and Greenway, H. (1989) Turgor-volume regulation and cellular water relations of *Nicotiana tabacum* roots grown in high salinities. *Aust. J. Plant Physiol.*, **16**, 517–531.

Weatherley, P.E. (1982) Water uptake and flow in roots. In: *Encyclopedia of Plant Physiology*, Vol. 12B (eds O.L. Lange, P.S. Nobel, C.B. Osmond, and H. Ziegler). Springer-Verlag, Berlin, pp. 79–109.

Wendler, S. and Zimmermann, U. (1982) A new method for the determination of hydraulic conductivity and cell volume of plant cells by pressure clamp. *Plant Physiol.*, **69**, 998–1003.

Yeo, A.R., Yeo, M.E. and Flowers, T.J. (1987) The contribution of an apoplastic pathway to sodium uptake by rice roots in saline conditions. *J. Exp. Bot.*, **38**, 1141–1153.

Zhang, W.H. and Tyerman, S.D. (1991) Effect of low O_2 concentration and azide on hydraulic conductivity and osmotic volume of the cortical cells of wheat roots. *Aust. J. Plant Physiol.*, **18**, 603–613.

Zhu, G.L. and Steudle, E. (1991) Water transport across maize roots: simultaneous measurement of flows at the cell and root level by double pressure probe technique. *Plant Physiol.*, **95**, 305–315.

Zhu, J.J., Steudle, E. and Beck, E. (1989) Negative pressure produced in an artificial osmotic cell by extracellular freezing. *Plant Physiol.*, **91**, 1454–1459.

Zimmermann, U. (1978) Physics of turgor- and osmoregulation. *Ann. Rev. Plant Physiol.*, **29**, 121–148.

Zimmermann, U. (1989) Water relations of plant cells: pressure probe technique. *Methods of Enzymol.*, **174**, 338–366.

Zimmermann, U. and Hüsken, D. (1979) Theoretical and experimental exclusion of errors in the determination of the elasticity and water transport parameters of plant cells by the pressure probe technique. *Plant Physiol.*, **64**, 18–24.

Zimmermann, U. and Hüsken, D. (1980) Turgor pressure and cell volume relaxation in *Halicystis parvula*. *J. Membrane Biol.*, **56**, 55–64.

Zimmermann, U. and Steudle, E. (1970) Bestimmung von Reflexionskoeffizienten an der Membran der Alge *Valonia utricularis*. *Z. Naturforsch.*, **25b**, 500–504.

Zimmermann, U. and Steudle, E. (1974) The pressure-dependence of the hydraulic conductivity, the membrane resistance, and membrane potential during turgor pressure regulation in *Valonia utricularis*. *J. Membrane Biol.*, **16**, 331–352.

Zimmermann, U. and Steudle, E. (1978) Physical aspects of water relations of plant cells. *Adv. Bot. Res.*, **6**, 45–117.

Zimmermann, U., Räde, H. and Steudle, E. (1969) Kontinuierliche Druckmessung in Pflanzenzellen. *Naturwissenschaften*, **56**, 634.

Zimmermann, U., Rygol, J., Balling, A., Klöck, G., Metzler, A. and Haase, A. (1992) Radial turgor and osmotic pressure gradients in intact and excised roots of *Aster trifolium*. *Plant Physiol.*, **99**, 186–196.

3

Regulation of cell expansion in roots and shoots at low water potentials

W.G. Spollen, R.E. Sharp, I.N. Saab and Y. Wu

3.1 Introduction

Root growth is often less inhibited than shoot growth when the supply of soil water is limited (Sharp *et al.*, 1988; Weaver, 1926; Westgate and Boyer, 1985). This has been observed at various developmental stages and is illustrated in Figure 3.1 for seedlings of four species grown over a range of water potentials (Ψ_w) in vermiculite. Primary root elongation in each species continued at a much lower Ψ_w than that which completely inhibited shoot growth.

The differential response of root and shoot growth to low Ψ_w is thought to be a means by which plants avoid excessive dehydration (Sharp and Davies, 1989). As roots continue to grow they tap into supplies of water deeper in the soil, while the inhibition of shoot growth, together with stomatal closure, restricts transpiration. In seedlings, the response is of obvious advantage for successful establishment under dry conditions. Despite the importance of these morphological changes, however, the underlying regulatory mechanisms are only beginning to be understood.

This chapter focuses on two aspects of the control of cell expansion in roots and shoots at low Ψ_w: (1) changes in cell wall yielding properties, and (2) effects of the hormone abscisic acid (ABA). Emphasis is placed on the spatial distribution of elongation within the organs, as this approach has recently allowed several advances to be made. It is important to note that cell and organ expansion rates may be governed by several factors at any one time and we do not cover all of them here. In particular, readers are referred to the comprehensive review by Boyer (1985) on water transport.

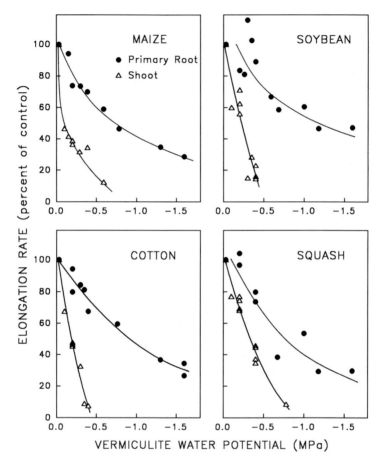

Figure 3.1. *Elongation rates of the primary root (●) and shoot (△) of four species growing in vermiculite of various Ψ_w. Seedlings were transplanted to the different Ψ_w 36 h after planting, and were grown at 29°C in darkness and near saturation humidity to minimize transpiration. For roots, data were evaluated at root lengths of 5 cm, when elongation rates were approximately steady. For shoots, data represent maximum elongation rates obtained after transplanting. Data are means from 10–40 seedlings, and are plotted as a percentage of the mean rates at high Ψ_w. (R.E. Sharp and G.S. Voetberg, unpublished data.)*

3.2 Cell wall yielding properties

3.2.1 *Turgor and the Lockhart equation*

Turgor provides the driving force for cell and hence organ expansion, stretching the cell walls irreversibly at a rate determined by their yielding properties. Cell wall yielding properties may be quantified by use of an equation originally derived by Lockhart (1965) to describe the growth of single cells, but which has also been used to describe the growth of whole organs. The relative growth rate (r) is related to turgor (P) via two yielding properties of the cell walls, the yield coefficient (or extensibility) (ϕ), and the yield threshold (Y), as follows: $r = \phi(P - Y)$. The difference between P and Y is the amount of turgor that drives growth. (For a more extensive discussion of Lockhart's equation and some novel developments of it see Ortega, Chapter 5, this volume.)

It was once considered that the inhibition of shoot growth at low Ψ_w resulted solely from reduced turgor in the expanding cells. Measurements in the growing zone of maize leaves (Michelena and Boyer, 1982) and soybean hypocotyls (Nonami and Boyer, 1989), however, showed that growth was inhibited at low Ψ_w despite the complete maintenance of turgor as a result of osmotic adjustment. This suggested that the cell walls had become less yielding. Subsequently, the yielding properties of the cell walls in the soybean hypocotyl were quantified and were shown to be a contributing factor in growth inhibition at low Ψ_w (Nonami and Boyer, 1990a,b).

Several studies suggested that the opposite was true in roots. In several species, root elongation recovered more rapidly than root tip turgor following immersion in solutions of low Ψ_w, suggesting that the cell walls became more yielding to allow growth maintenance at low Ψ_w (Hsiao and Jing, 1987; Itoh et al., 1987; Kuzmanoff and Evans, 1981).

In most applications of the Lockhart model to the growth of plant organs, consideration has not been given to spatial variation of the parameters within the elongation zone of the organ in question. Therefore, the above-mentioned assessments of changes in cell wall yielding properties at low Ψ_w largely reflect average responses for tissues of varying elongation rates and stages of development. Importantly, several recent studies showed that cell expansion was differentially sensitive to low Ψ_w at different locations within the elongation zones of both shoots (Paolillo, 1989; Saab et al., 1992; Schultz and Matthews, 1988) and roots (Pritchard et al., 1991; Sharp et al., 1988; Spollen and Sharp, 1991). In the maize primary root, for example, longitudinal expansion rates at low Ψ_w are maintained preferentially toward the root apex, resulting in a shorter elongation zone than at high Ψ_w (Figure 3.2a). To gain an understanding of the underlying regulatory processes in such cases, it is obviously essential to examine adjustments in wall yielding properties with a high degree of spatial resolution.

The cell pressure microprobe has made it possible to measure turgor directly in individual cells of higher plants (see Steudle, Chapter 2, this volume), and hence the spatial distribution of turgor within the elongation zone of an organ can be accurately determined. Using this approach, Spollen and Sharp (1991) showed that in maize primary roots growing at low Ψ_w ($-1.6\,\mathrm{MPa}$) in vermiculite, turgor throughout the elongation zone was about 50% less than at high Ψ_w (Figure 3.2b). Thus, osmotic adjustment, while substantial (Sharp et al., 1990), was insufficient to maintain turgor at well-watered values.

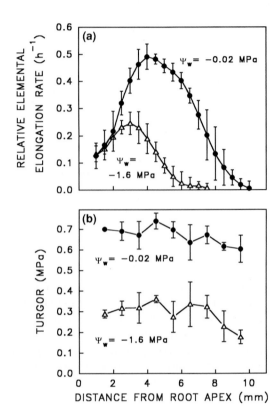

Figure 3.2. *Spatial distribution of (a) relative elemental elongation rate (local longitudinal growth rate) and (b) turgor in the apical 10 mm of maize primary roots growing at high or low Ψ_w in vermiculite. Turgor measurements were made with a pressure probe in individual cells of both the cortex and stele; data are means ± SD. (Modified from Spollen and Sharp, 1991, with permission from the American Society of Plant Physiologists.)*

Turgor was relatively uniform throughout the elongation zone at both high and low Ψ_w (Figure 3.2b), indicating that turgor alone did not control the observed growth patterns (Figure 3.2a). Thus, the cell wall yielding properties must have varied with distance from the root apex. In addition, the spatial resolution of growth and turgor allowed important inferences to be made about adaptation to low Ψ_w. Elongation was completely maintained in the apical few millimetres of the roots at low Ψ_w despite the decrease in turgor. The approximate values obtained for the relative elongation rate and turgor at 2 mm from the apex at high and low Ψ_w are plotted as triangles in Figure 3.3. A line drawn through these data has a slope (ϕ) of zero and no x-intercept (Y). For growth to occur, however, ϕ must be positive and Y must be greater than or equal to zero. The likely explanation for these results is that ϕ or Y, or both, have changed in the apical region at low Ψ_w to allow greater longitudinal expansion per unit of turgor.

In some earlier studies of roots the length of the elongation zone was assumed to be constant at different Ψ_w (Greacen and Oh, 1972; Pritchard *et al.*, 1990), and a plot of overall root elongation rate versus root tip turgor was used to investigate the wall yielding properties. This gave an

Figure 3.3. *Using data from Figure 3.2, line (1) shows the approximate relationship between relative elemental elongation rate and turgor at 2 mm from the primary root apex when maize seedlings were grown at high Ψ_w (solid symbols) or low Ψ_w (open symbols) in vermiculite. The slope of zero (i.e. an apparent lack of dependence on turgor) indicates that the cell wall yielding properties were probably altered at low Ψ_w to allow greater longitudinal expansion per unit of turgor. This conclusion is missed when whole-root elongation rate is plotted against the average turgor of the root growth zone, shown by line (2). See the text for further discussion.*

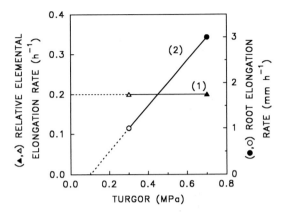

approximately linear relationship with a positive slope and *x*-intercept. It was concluded, therefore, that ϕ and Y were unaffected at low Ψ_w. From our own work, a similar relationship can be generated for the maize primary root; these data are plotted as circles in Figure 3.3. However, this approach does not provide useful information on the values of ϕ and Y, because, as shown already, these parameters have probably changed at low Ψ_w, at least in the apical region. This indicates clearly the inadequacy of a growth–turgor plot of the whole organ to assess accurately the effects of low Ψ_w on cell wall yielding properties.

3.2.2 *A molecular model of growth*

Lockhart's model has been useful in showing that the cell wall is dynamic, but it does not reveal the mechanism by which the wall yields or how the yielding properties may change in response to altered environmental conditions. A recent molecular model (Passioura and Fry, 1992) makes quantitative predictions that agree with experimental observations of responses of plant expansive growth to changes in turgor. This model is based on an earlier one of Fry (1989) and studies by Green *et al.* (1971), which suggested that the cell wall simultaneously 'tightened' and 'loosened' during cell expansion. Hemicelluloses are thought to bind to and tether adjacent cellulose microfibrils, thereby bearing much of the stress in the cell wall due to turgor. The number of taut tethers is a result of two processes that have opposing effects on growth: the activity of enzymes that cleave tethers and allow expansion (wall loosening), and the rate at which expansion causes slack tethers to become taut (wall tightening). The cell wall yielding properties of the Lockhart model can then be considered a function of the number of taut tethers. Dynamic control of the number of taut tethers could account for the observation that cell or organ growth rates sometimes seem independent of turgor. The Passioura and Fry model suggests that at low Ψ_w the number of taut tethers increases in shoots and decreases in roots, since in shoots growth is inhibited while turgor is maintained, whereas in roots growth continues even when turgor is substantially decreased. The number of taut tethers in shoots and roots is probably differentially adjusted by altered metabolism at low Ψ_w.

3.2.3 *Xyloglucan endotransglycosylase and growth maintenance*

Much recent work on the structure and dynamics of the expanding cell wall has focused on the role of xyloglucan, a hemicellulose abundant in cell walls of higher plants with the exception of the Gramineae. Graminaceous plants also contain xyloglucan in the cell walls but mixed-linkage β-glucan, another hemicellulose, is more abundant. Both have properties that would allow them to fill the role of a tether in the model described above, i.e. they bind strongly to cellulose and are usually long enough to allow them to straddle adjacent microfibrils (Fry, 1989). An enzyme activity, xyloglucan endotransglycosylase (XET), has recently been characterized that has the property of both cutting and rejoining xyloglucan polymers (Fry *et al.*, 1992). The cutting activity alone qualifies it as a candidate for involvement in wall yielding, and the added ability to rejoin one cut end with another allows formation of new tethers, thus conserving xyloglucan. XET activity has now been found in several species across the plant kingdom and, surprisingly, some of the highest activity is in extracts from shoots of some of the Gramineae (Fry *et al.*, 1992).

The spatial distribution of XET activity in the apical few millimetres of the maize primary root was recently measured to see how it might be associated with the maintenance of elongation at low Ψ_w. Figure 3.4a shows that the spatial distribution of XET activity was similar to that of elongation at high Ψ_w. Activity was significantly enhanced in the apical few millimetres at low Ψ_w where, as discussed above, wall yielding is thought to increase. Furthermore, because cell wall mass per unit root length was less in roots grown at low Ψ_w (roots were thinner than at high Ψ_w; Sharp *et al.*, 1988), XET activity was greatly increased at low compared to high Ψ_w throughout the entire length of the elongation zone when expressed on a cell wall dry weight basis (Figure 3.4b; it should be noted, however, that the fraction of XET activity in the cell walls is unknown). Activity was also increased at low Ψ_w if expressed on either a fresh weight or soluble protein basis. These data suggest that an increase in XET activity may play an important role in increasing cell wall yielding and allowing continued root elongation at low Ψ_w despite incomplete turgor maintenance. In contrast, Pritchard and Tomos (Chapter 4, this volume) showed that XET activity was not altered (on a fresh weight basis) in root tips of maize growing at a Ψ_w of -1.0 MPa in mannitol solution. The difference between their findings and ours is probably due to the difference in level of stress. Root tip turgor was completely maintained in their study although, interestingly, the elongation zone was still shorter than at high Ψ_w.

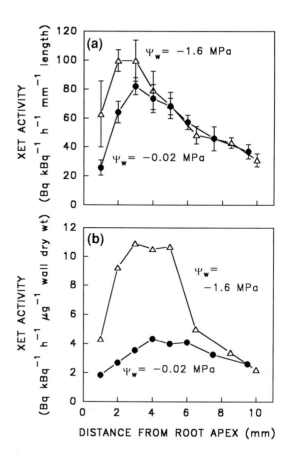

Figure 3.4. *Spatial distribution of xyloglucan endotransglycosylase (XET) activity in the apical 10 mm of maize primary roots growing at high or low Ψ_w in vermiculite. (a) Activities expressed per mm of root length ($\pm SD$, n = 4). (b) Mean activities expressed per μg cell wall dry weight. (Y. Wu, W.G. Spollen, R.E. Sharp and S.C. Fry, unpublished data.)*

3.2.4 *Peroxidase and the cessation of growth*

Cells are continually displaced through the elongation zone of an organ by the production and expansion of younger cells. Therefore, the velocity of displacement away from the meristematic region increases with distance, and the transition from maximal rates of elongation to growth cessation occurs rapidly, often within only a few hours (Figure 3.5).

Deceleration of cell expansion may be related to increased activity of peroxidase, an enzyme that is believed to cross-link cell wall components, making the wall less yielding (Fry, 1986). Support for this idea was recently provided by a study of tall fescue leaves growing at high Ψ_w (MacAdam *et al.*, 1992). Spatial distributions of both elongation rate and apoplastic peroxidase activity were determined in the elongation zone at the leaf base. When viewed temporally, a dramatic increase in peroxidase activity was closely correlated with the onset of growth deceleration (Figure 3.5b). In grasses, peroxidase is thought to join two arabinoxylan-linked feruloyl groups to form a diferuloyl ester, thus cross-linking the arabinoxylans. Localization studies using UV-

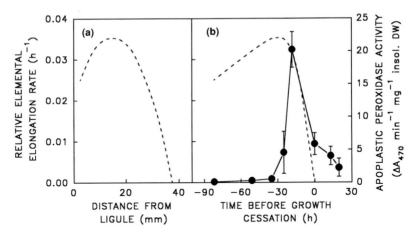

Figure 3.5. *(a) Spatial distribution of relative elemental elongation rate in the basal 40 mm of leaf blades of tall fescue growing at high Ψ_w. (b) Elongation rates (dashed line) were replotted as a function of time before growth cessation to represent the values associated with a particular tissue element as it was displaced through the elongation zone. The solid line shows the corresponding values of peroxidase activity measured in apoplastic fluid extracted from leaf blade segments by vacuum infiltration and centrifugation. (Modified from MacAdam et al., 1992, with permission from the American Society for Plant Physiologists.)*

fluorescence showed that ferulic acid was associated with apoplastic peroxidase activity in tall fescue leaves (Ahn and MacAdam, 1992), indicating that production of substrate and enzyme activity were co-ordinated with deceleration of cell expansion.

Whether or not regulation of peroxidase activity plays a role in either the shortening of the root elongation zone (Figure 3.2a) or the inhibition of shoot growth at low Ψ_w has not yet been addressed.

3.2.5 *Changes in cell wall components at low Ψ_w*

Only a few studies of changes in cell wall synthesis and composition in tissues grown at low Ψ_w have been conducted, and the main results are summarized briefly below. It is not clear whether any of these changes are causally related to the effects of low Ψ_w on cell expansion, however.

Cell wall synthesis is necessary for continued growth under all conditions, and the inhibition of net synthesis of cellulose and of other wall polysaccharides has been reported at low Ψ_w in shoots (Ordin, 1960; Sakurai *et al.*, 1987; Sweet *et al.*, 1990) and with salinity stress in roots (Zhong and Läuchli, 1988). Salt-extractable protein content increased in cell walls of the elongation zone in soybean hypocotyls growing at low Ψ_w, and a 28 kDa wall protein was specifically increased (Bozarth *et al.*, 1987). Differential expression of the gene for this and a related 31 kDa protein has been reported in soybean seedlings growing under well-watered or water-limited conditions (Surowy and Boyer, 1991).

Both genes were expressed in hypocotyls at high Ψ_w, but only the 28 kDa message increased at low Ψ_w. Conversely, only the 31 kDa mRNA accumulated in roots at low Ψ_w. The cDNAs for these messages were homologous to those of leaf vegetative storage proteins. The authors suggest that the proteins may also have a role in growth, as a dual function for a storage protein is not without precedent.

In addition, expression of a H^+-ATPase mRNA increased in roots but not in shoots when soybean seedlings were grown at low Ψ_w (Surowy and Boyer, 1991). A H^+-ATPase could play a role in acidification of the apoplast to maintain growth (the acid growth hypothesis), regulation of osmotic potential or adaptation to nutrient deficiencies that may be associated with water deficits.

Decreased and increased expression of several genes was found in the root and shoot elongation zones of soybean seedlings at low Ψ_w (Creelman and Mullet, 1991). Based on sequence analysis, one of these (*sbPRP1*) is thought to code for a proline-rich cell wall protein that may serve to decrease wall yielding. Interestingly, the mRNA for sbPRP1 increased in the elongation zone of hypocotyls and decreased in root tips at low Ψ_w.

3.3 The role of ABA

ABA accumulates to high concentrations in various organs of plants growing at low Ψ_w. Recent studies show that ABA plays important roles in both root growth maintenance and shoot growth inhibition in maize seedlings growing at low Ψ_w (Saab *et al.*, 1990, 1992). Three aspects of the experimental approach used in these studies are highlighted below and should also apply in assessing the involvement of other hormones in growth regulation at low Ψ_w.

(1) Assessment of the role of ABA was performed on plants growing at low Ψ_w, and not by extrapolation from the effects of applied ABA on well-watered plants. This approach avoided possible problems in interpretation due to variation in sensitivity and/or response to ABA with tissue water status. Endogenous ABA concentrations in the elongation zones of the primary root and mesocotyl at low Ψ_w were decreased by treatment with fluridone, an inhibitor of carotenoid (and ABA) synthesis (see Figure 3.7b,d). The fluridone treatment severely inhibited root elongation but promoted shoot elongation at low Ψ_w, while having little effect on growth of either organ at high Ψ_w (Figure 3.6). Similar results were obtained using the *vp5* mutant of maize, which carries a lesion at the same step as that inhibited by fluridone. Thus, the effects of fluridone at low Ψ_w were not caused by side-effects of the inhibitor on growth processes.

Additional experiments confirmed that the high ABA concentrations present at low Ψ_w are required to maintain root elongation. Addition of ABA to fluridone-treated roots growing at low Ψ_w was shown to restore the root elongation rate to that of untreated roots growing at the same Ψ_w (R.E. Sharp, G.S. Voetberg, Y. Wu and I.N. Saab, unpublished data). Restoration

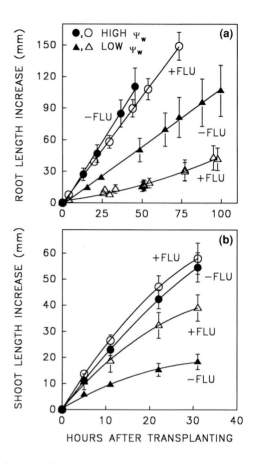

Figure 3.6. *Effect of fluridone (FLU) on elongation of the primary root (a) and shoot (b) of maize seedlings after transplanting to high Ψ_w (circles, −0.03 MPa) or low Ψ_w (triangles) vermiculite. The low Ψ_w were approximately −1.6 MPa and −0.3 MPa, respectively, for root and shoot experiments. Fluridone (10 µM) was mixed with the vermiculite into which the seeds were germinated and into which the seedlings were transplanted, and was used to inhibit the accumulation of ABA at low Ψ_w (see Figure 3.7). (Modified from Saab et al., 1990, with permission from the American Society for Plant Physiologists.)*

of root elongation rate was achieved when the internal ABA level was restored to the usual stressed level. When the internal ABA concentration was increased substantially above the usual level, however, root elongation was inhibited, indicating that the roots grew at or near optimal ABA concentration at low Ψ_w. Moreover, root growth at high Ψ_w was inhibited when the root tip ABA level was increased to that of the stressed roots, suggesting that the root growth response to ABA does depend on tissue water status.

(2) The effects of ABA on root and shoot elongation at low Ψ_w were determined with spatial resolution (Figure 3.7) since, as discussed above, responses of cell expansion to low Ψ_w vary with location in both root and shoot elongation zones. In the root, ABA concentrations at low Ψ_w were increased throughout the elongation zone, with the highest concentrations found in apical regions where elongation was maintained (Figure 3.7a,b). Fluridone treatment reduced the ABA concentration in all regions and caused further

PRIMARY ROOT MESOCOTYL

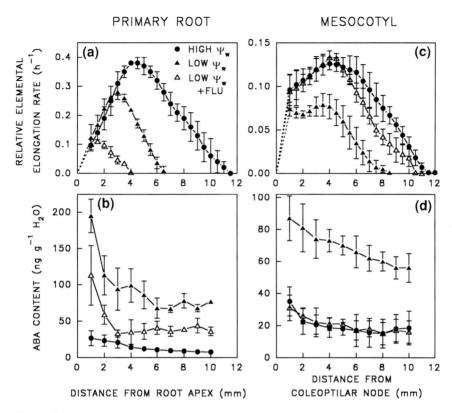

Figure 3.7. *Spatial distribution of relative elemental elongation rate (a,c) and ABA content (b,d) in the primary root (a,b) and mesocotyl (c,d) of maize seedlings growing in vermiculite at high Ψ_w (●), low Ψ_w (▲), or low Ψ_w plus fluridone (FLU) (△). The low Ψ_w were approximately −1.6 MPa and −0.3 MPa, respectively, for root and mesocotyl experiments. (Modified from Saab et al., 1992, with permission from the American Society for Plant Physiologists.)*

shortening of the growth zone towards the apex. In the mesocotyl, low Ψ_w also shortened the length of the elongation zone (Figure 3.7c) while increasing ABA concentration in all regions (Figure 3.7d). Fluridone treatment decreased the ABA concentration and largely restored elongation rates throughout the growth zone to well-watered values. These results suggest that a developmental gradient in responsiveness to ABA exists within each organ. Confirmation of this, however, will require knowledge of the active-site ABA concentrations.

(3) It was considered that the differential effects of ABA on root and shoot growth at low Ψ_w could be partly indirect, i.e. ABA might act on only one organ directly, altering its competitive ability to acquire water and nutrients, and thus indirectly affecting growth of the other organ. It was shown, however, that the effects of altered ABA concentrations on growth of both organs were largely independent (Saab *et al.*, 1992). First, the root experiments were performed at a Ψ_w of −1.6 MPa which completely

inhibited shoot growth regardless of ABA status. Secondly, in the shoot, the increased gain in water and dry weight due to treatment with fluridone at low Ψ_w was not accompanied by a decrease in water and dry weight gain in the roots.

3.3.1 *How does ABA regulate growth at low Ψ_w?*

Most studies of the mechanisms of ABA-induced changes in growth have been conducted at high Ψ_w, and thus may not adequately explain the responses observed at low Ψ_w. Nevertheless, these studies still suggest possible mechanisms of action. ABA-induced inhibition of cell wall extensibility was observed in bean leaves (Van Volkenburgh and Davies, 1983). The site where ABA inhibits shoot growth appears to be the cell walls of the epidermis in segments of maize coleoptile (Kutschera *et al.*, 1987) and squash hypocotyl (Wakabayashi *et al.*, 1989). In both cases, growth inhibition by ABA was not observed in segments from which the epidermis had been peeled. Two suggestions from the literature on how ABA might affect epidermal cell wall yielding properties follow.

First, the frequency of transverse microtubule orientation in the tangential wall of the cucumber hypocotyl epidermis was closely associated with growth rate modulated by ABA, GA$_4$ or both (Ishida and Katsumi, 1992). Microtubule orientation may help determine the orientation of microfibrils, thereby affecting tension in the cell wall and the rate and polarity of cell expansion. Secondly, ABA inhibited the incorporation of [^{14}C]glucose into cellulose and hemicellulose of whole squash hypocotyls prior to the inhibition of growth (Wakabayashi *et al.*, 1989). Incorporation of label into UDP-sugar was not inhibited by ABA, however, suggesting that the inhibition of wall synthesis by ABA occurred later in the synthetic pathway (Wakabayashi *et al.*, 1991). It is not known if this occurs in the epidermal wall itself, however.

What role ABA may play in the maintenance of root growth at low Ψ_w is unclear. Preliminary studies indicate that the enhancement of XET activity in the root elongation zone at low Ψ_w (Figure 3.4) is inhibited when ABA accumulation is reduced by fluridone (Y. Wu, W.G. Spollen, R.E. Sharp, S.C. Fry and R. Hetherington, unpublished data). Thus, ABA may have some role in increasing wall yielding in roots at low Ψ_w. ABA might also regulate osmotic adjustment. Evidence for the short-term regulation of osmotic potential by ABA is abundant in the case of stomatal guard cells, where extracellular ABA promotes an efflux of potassium that lowers cell turgor and closes the stomata. Long-term regulation of osmotic potential by ABA in growing tissue might also involve an effect on potassium transport, but must integrate all the changes that occur in response to this over time (Hetherington and Quatrano, 1991). One long-term effect might be the accumulation of proline. In maize primary roots growing at a Ψ_w of -1.6 MPa, proline accumulated in the apical few millimetres to concentrations sufficient to account for the level of turgor (Voetberg and Sharp, 1991). Treatment with fluridone decreased the proline content throughout the root tip (E.S. Ober and R.E. Sharp, unpublished data), indicating that ABA may play a regulatory role in this response. A number of studies have alluded to a connection between ABA, potassium and proline accumulation. This raises the possibility that regulation of potassium levels by ABA may control proline accumulation and perhaps other responses in cells growing at low Ψ_w.

At least part of the requirement for high ABA concentrations in roots growing at low Ψ_w appears to be due to the inhibition of ethylene synthesis or action. Preliminary studies showed that 2,5 norbornadiene, a competitive inhibitor of ethylene action, partially restored root growth in fluridone-treated seedlings at a low Ψ_w of −0.6 MPa (W.G. Spollen and R.E. Sharp, unpublished data).

3.3.2 *Other hormones*

Much less is known about the involvement of hormones besides ABA in regulating root and shoot growth at low Ψ_w. It is interesting to note that certain concentrations of indoleacetic acid (IAA) applied to roots at high Ψ_w result in a similar spatial distribution of elongation to that caused by low Ψ_w (Evans *et al.*, 1990; Goodwin, 1972). Whether or not endogenous IAA is involved in the responses of root or shoot elongation to low Ψ_w remains to be addressed, however. Levels of gibberellins were decreased in the elongation zone of soybean hypocotyls when growth was inhibited at low Ψ_w (Bensen *et al.*, 1990), but a causal relationship has yet to be established. Ethylene has been invoked in the responses of root and shoot growth to various adverse environmental conditions, although information on its role in responses to low Ψ_w is scant (Abeles *et al.*, 1992). Finally, cytokinins, along with ABA, have been implicated as root 'signals' that modulate responses of shoot growth and stomatal aperture to soil drying (Zhang and Davies, 1991). This root to shoot communication is discussed elsewhere in this volume (Chapter 9).

3.4 Future directions

Progress has been made in the past few years in understanding the differential responses of root and shoot growth to water-limited conditions. With renewed recognition of the dynamic nature of cell wall yielding properties, increased emphasis should be given to elucidating the metabolic regulation of adjustments in cell wall yielding in response to water deficits. In particular, experiments are required to confirm the role of enhanced XET activity in increasing wall yielding in roots at low Ψ_w. This will require specifically altering XET activity or xyloglucan composition in the growing cell walls and assessing the resultant changes in wall yielding and growth responses. Other putative wall-loosening enzymes should also be studied in this regard.

A role has been established for endogenous ABA in the maintenance of primary root elongation and inhibition of shoot elongation in maize seedlings at low Ψ_w. The involvement of ABA in growth responses to low Ψ_w has been determined with spatial resolution, and these results have identified locations in the primary root and mesocotyl elongation zones where cell expansion is maintained or inhibited at low Ψ_w by endogenous ABA. Specific changes in gene expression that are associated with these responses have been identified (I.N. Saab, T.-H.D. Ho and R.E. Sharp, unpublished data), and future studies will characterize these genes and their products to gain an understanding of their possible function in growth regulation.

References

Abeles, F.B., Morgan, P.W. and Saltveit, M.E. (1992) *Ethylene in Plant Biology*, 2nd edn. Academic Press, San Diego.

Ahn, H.-J. and MacAdam, J.W. (1992) Localization of ferulic acid in the grass leaf blade elongation and secondary cell wall deposition zones (abstract). In: *Current Topics in Plant Biochemistry and Physiology* (eds D.D. Randall, R.E. Sharp, A.J. Novacky and D.G. Blevins), Vol 11. University of Missouri, Columbia, MO, p. 312.

Bensen, R.J., Beall, F.D., Mullet, J.E. and Morgan, P.W. (1990) Detection of endogenous gibberellins and their relationship to hypocotyl elongation in soybean seedlings. *Plant Physiol.*, **94**, 77–84.

Boyer, J.S. (1985) Water transport. *Ann. Rev. Plant. Physiol.*, **36**, 473–516.

Bozarth, C.S., Mullet, J.E. and Boyer, J.S. (1987) Cell wall proteins at low water potentials. *Plant Physiol.*, **85**, 261–267.

Creelman, R.A. and Mullet, J.E. (1991) Water deficit modulates gene expression in growing zones of soybean seedlings. Analysis of differentially expressed cDNAs, a new β-tubulin gene, and expression of genes encoding cell wall proteins. *Plant Mol. Biol.*, **17**, 591–608.

Evans, M.L., Kiss, H.G. and Ishikawa, H. (1990) Interaction of calcium and auxin in the regulation of root elongation. In: *Calcium in Plant Growth and Development* (eds R.T. Leonard and P.K. Hepler), Current Topics in Plant Physiology, Vol. 4. American Society of Plant Physiologists, Rockville, MD, pp. 168–175.

Fry, S.C. (1986) Cross-linking of matrix polymers in the growing cell walls of angiosperms. *Ann. Rev. Plant Physiol.*, **37**, 165–186.

Fry, S.C. (1989) Cellulases, hemicelluloses and auxin-stimulated growth: a possible relationship. *Physiol. Plant.*, **75**, 532–536.

Fry, S.C., Smith, R.C., Renwick, K.F., Martin, D.J., Hodge, S.K. and Matthews, K.J. (1992) Xyloglucan endotransglycosylase, a new wall-loosening enzyme activity from plants. *Biochem. J.*, **282**, 821–828.

Goodwin, R.H. (1972) Studies on roots. V. Effects of indoleacetic acid on the standard root growth pattern of *Phleum pratense. Bot. Gaz.*, **133**, 224–229.

Greacen, E.L. and Oh, J.S. (1972) Physics of root growth. *Nature*, **235**, 24–25.

Green, P.B., Erickson, R.O. and Buggy, J. (1971) Metabolic and physical control of cell elongation rate. *In vivo* studies of *Nitella. Plant Physiol.*, **47**, 423–430.

Hetherington, A.M. and Quatrano, R.S. (1991) Mechanisms of action of abscisic acid at the cellular level. *New Phytol.*, **119**, 9–32.

Hsiao, T.C. and Jing, J. (1987) Leaf and root expansive growth in response to water deficits. In: *Physiology of Cell Expansion During Plant Growth* (eds D.J. Cosgrove and D.P. Knievel). American Society of Plant Physiologists, Rockville, MD, pp. 180–192.

Ishida, K. and Katsumi, M. (1992) Effects of gibberellin and abscisic acid on the cortical micro-tubule orientation in hypocotyl cells of light-grown cucumber seedlings. *Int. J. Plant Sci.*, **153**, 155–163.

Itoh, K., Nakamura, Y., Kawata, H., Yamada, T., Ohta, E. and Sakata, M. (1987) Effect of osmotic stress on turgor pressure in mung bean root cells. *Plant Cell Physiol.*, **28**, 987–994.

Kutschera, U., Bergfeld, R. and Schopfer, P. (1987) Cooperation of epidermis and inner tissues in auxin-mediated growth of maize coleoptiles. *Planta*, **170**, 168–180.

Kuzmanoff, K.M. and Evans, M.L. (1981) Kinetics of adaptation of osmotic stress in lentil (*Lens culinaris* Med.) roots. *Plant Physiol.* **68**, 244–247.

Lockhart, J.A. (1965) An analysis of irreversible plant cell elongation. *J. Theor. Biol.* **8**, 264–275.

MacAdam, J.W., Sharp, R.E. and Nelson, C.J. (1992) Peroxidase activity in the leaf elongation zone of tall fescue. II. Spatial distribution of apoplastic peroxidase activity in genotypes dif-fering in length of the elongation zone. *Plant Physiol.*, **99**, 879–885.

Michelena, V.A. and Boyer, J.S. (1982) Complete turgor maintenance at low water potentials in the elongating region of maize leaves. *Plant Physiol.*, **69**, 1145–1149.

Nonami, H. and Boyer, J.S. (1989) Turgor and growth at low water potentials. *Plant Physiol.*, **89**, 798–804.

Nonami, H. and Boyer, J.S. (1990a) Primary events regulating stem growth at low water poten-tials. *Plant Physiol.*, **93**, 1601–1609.

Nonami, H. and Boyer, J.S. (1990b) Wall extensibility and cell wall hydraulic conductivity decrease in enlarging stem tissues at low water potentials. *Plant Physiol.*, **93**, 1610–1619.

Ordin, L. (1960) Effect of water stress on cell wall metabolism of *Avena* coleoptile tissue. *Plant Physiol.*, **35**, 443–450.

Paolillo, D.J. Jr (1989) Cell and axis elongation in etiolated soybean seedlings are altered by moisture stress. *Bot. Gaz.*, **150**, 101–107.

Passioura, J.B. and Fry, S.C. (1992) Turgor and cell expansion: beyond the Lockhart equation. *Aust. J. Plant Physiol.*, **19**, 565–576.

Pritchard, J., Wyn Jones, R.G. and Tomos, A.D. (1990) Measurement of yield threshold and cell wall extensibility of intact wheat roots under different ionic, osmotic and temperature treatments. *J. Exp. Bot.*, **41**, 669–675.

Pritchard, J., Wyn Jones, R.G. and Tomos, A.D. (1991) Turgor, growth and rheological gra-dients of wheat roots following osmotic stress. *J. Exp. Bot.*, **42**, 1043–1049.

Saab, I.N., Sharp, R.E., Pritchard, J. and Voetberg, G.S. (1990) Increased endogenous abscisic acid maintains primary root growth and inhibits shoot growth of maize seedlings at low water potentials. *Plant Physiol.*, **93**, 1329–1336.

Saab, I.N., Sharp, R.E. and Pritchard, J. (1992) Effect of inhibition of ABA accumulation on the spatial distribution of elongation in the primary root and mesocotyl of maize at low water potentials. *Plant Physiol.*, **99**, 26–33.

Sakurai, N., Tanaka, S. and Kuraishi, S. (1987) Changes in wall polysaccharides of squash (*Cur-curbita maxima* Duch.). I. Wall sugar composition and growth as affected by water stress. *Plant Cell Physiol.*, **28**, 1051–1058.

Schultz, H.R. and Matthews, M.A. (1988) Vegetative growth distribution during water deficits in *Vitis vinifera* L. *Aust. J. Plant Physiol.*, **15**, 641–656.

Sharp, R.E. and Davies, W.J. (1989) Regulation of growth and development of plants growing with a restricted supply of water. In: *Plants Under Stress* (eds H.G. Jones, T.J. Flowers and M.B. Jones). Society for Experimental Biology, Seminar Series 39. Cambridge University Press, Cambridge, pp. 71–94.

Sharp, R.E., Silk, W.K. and Hsiao, T.C. (1988) Growth of the maize primary root at low water potentials. I. Spatial distribution of expansive growth. *Plant Physiol.*, **87**, 50–57.

Sharp, R.E., Hsiao, T.C. and Silk, W.K. (1990) Growth of the maize primary root at low water potentials. II. Role of growth and deposition of hexose and potassium in osmotic adjustment. *Plant Physiol.*, **93**, 1337–1346.

Spollen, W.G. and Sharp, R.E. (1991) Spatial distribution of turgor and root growth at low water potentials. *Plant Physiol.*, **96**, 438–443.

Surowy, T.K. and Boyer, J.S. (1991) Low water potentials affect expression of genes encoding vegetative storage proteins and plasma membrane proton ATPase in soybean. *Plant Mol. Biol.*, **16**, 251–262.

Sweet, W.J., Morrison, J.C., Labavitch, J.M. and Matthews, M.A. (1990) Altered synthesis and composition of cell wall of grape (*Vitis vinifera* L.) leaves during expansion and growth-limiting water deficits. *Plant Cell Physiol.*, **31**, 407–414.

Van Volkenburgh, E. and Davies, W.J. (1983) Inhibition of light-stimulated leaf expansion by abscisic acid. *J. Exp. Bot.*, **34**, 835–845.

Voetberg, G.S. and Sharp, R.E. (1991) Growth of the maize primary root at low water potentials. III. Role of increased proline deposition in osmotic adjustment. *Plant Physiol.*, **96**, 1125–1130.

Wakabayashi, K., Sakurai, N. and Kuraishi, S. (1989) Role of the outer tissue in abscisic acid-mediated growth suppression of etiolated squash hypocotyl segments. *Physiol. Plant.*, **75**, 151–156.

Wakabayashi, K., Sakurai, N. and Kuraishi, S. (1991) Effects of abscisic acid on the synthesis of cell-wall polysaccharides in segments of etiolated squash hypocotyl. II. Levels of UDP-neutral sugars. *Plant Cell Physiol.*, **32**, 427–432.

Weaver, J.E. (1926) *Root Development of Field Crops*. McGraw-Hill, New York.

Westgate, M.E. and Boyer, J.S. (1985) Osmotic adjustment and the inhibition of leaf, root, stem and silk growth at low water potentials in maize. *Planta*, **164**, 540–549.

Zhang, J. and Davies, W.J. (1991) Root signals and the regulation of growth and development of plants in drying soil. *Ann. Rev. Plant Physiol. Plant Mol. Biol.*, **42**, 55–76.

Zhong, H. and Läuchli, A. (1988) Incorporation of [^{14}C]glucose into cell wall polysaccharides of cotton roots: effects of NaCl and $CaCl_2$. *Plant Physiol.*, **88**, 511–514.

4

Correlating biophysical and biochemical control of root cell expansion

Jeremy Pritchard and A. Deri Tomos

4.1 Introduction

In the study of the extension growth of plant cells we are faced with the problem of trying to integrate a range of different approaches. Biochemists describe the polymers present in the cell wall and their cross-links (e.g. Fry, 1986), microscopists the micro-anatomy (e.g. Hogetsu, 1986), and physiologists the physical behaviour of the cell walls (e.g. Boyer et al., 1985; Cosgrove, 1987; Pritchard et al., 1991). Each attempts to describe his or her understanding of 'expansion' to the others. In this chapter we describe some of our recent attempts to correlate these approaches.

Our starting point has been the Lockhart equation (Lockhart, 1965) which describes the biophysical parameters of expansion growth. Following most workers in the field, as well as our own measurements of hydraulic conductivity (Jones et al., 1988; Pritchard et al., 1987), we conclude that, to a first approximation, water flow into the expanding cell is not a limiting factor in root expansion. Thus the Lockhart equation reduces to the description of a Bingham substance (i.e. strain rate is linearly proportional to stress (P) above a yield stress value (yield threshold, Y)) (Ray et al., 1972). The proportionality constant (ϕ) is called the extensibility. Whatever the chemical or architectural basis of growth regulation, it *must* be manifested by modification of one or more of these parameters (Tomos et al., 1989);

$$\text{strain rate (growth)} = \phi(P - Y). \qquad 4.1$$

In principle, Y and ϕ can be estimated by imposing a range of turgor pressures on the expanding cell and measuring the growth rate (Pritchard et al., 1990a). This straightforward approach is complicated by the increasing evidence that neither Y nor ϕ is necessarily constant with time (Pritchard et al.,

1990a; Schmastig and Cosgrove, 1988) and that osmotic adjustment can counteract attempts to alter turgor at a rate faster than the experiments can be successfully performed (Pritchard *et al.*, 1990a).

In addition, growing organs are not homogeneous and the stresses which limit growth may be confined to only one tissue. For example, the epidermis is growth limiting in coleoptile systems (Kutshera *et al.*, 1987). Measurements of the parameters of the Lockhart equation must be made with reference to individual tissues. At finer resolution, only a portion of the wall may be load bearing and therefore relevant to a consideration of growth regulation. Many workers have observed that the mechanical properties of the cell wall are highly correlated with the orientation of the cellulose microfibrils. An orientation parallel to the axis of growth has often been considered to be growth limiting (Preston, 1982). The most complete exposition of this being the multi-net hypothesis (Roelofsen and Houwink, 1953).

Changes in the physical properties of the cell wall and any changes in its structure on a macro-level must be caused ultimately by alterations in its structure at the molecular level. These must be mediated by enzymes in the wall which can form or break cross-links between wall components (Fry, 1986). Any complete model of growth control must consider the role of such wall enzymes in relation to the physiological behaviour of the wall.

In this chapter we describe two conditions in which the rate of growth of maize roots is dynamically controlled. First, we describe the acceleration and deceleration of expansion that occurs as the cell passes from the apical meristem to the mature region of the root. Secondly, we describe some changes that occur to these same cells following long- and short-term osmotic stress.

4.2 Biophysical changes that control expansion rate

4.2.1 *Turgor pressure*

Control of cell expansion may be mediated through a change in the transport of solutes and water or through changes in cell-wall rheology. The former would result in a change in P, the latter would be associated with a change in Y and/or ϕ in Equation 4.1. Turgor pressure may be measured with a pressure probe. However, the accuracy of the measurements (approximately ± 0.04 MPa) may preclude the detection of small differences in P and Y. The scatter in a 'constant' P might mask small, but significant, changes in P. We have found this apparently to be the case for cereal leaves (Thomas *et al.*, 1988). For entire maize roots $(P - Y)$ is in the order of 0.2 MPa (Figure 4.1), certainly large enough for significant changes in P to be detected with the probe. Figure 4.2 illustrates the characteristics of the roots described in this report. The techniques used have been described previously (Pritchard *et al.*, 1991). Longitudinal changes in growth rate (Figure 4.2a) are not due to

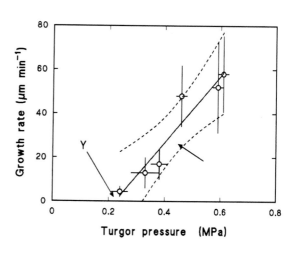

Figure 4.1. *Growth rate of entire maize roots as a function of cell turgor pressure. Roots were bathed in a series of mannitol solutions and growth and turgor were measured (modified from Pritchard et al., 1990b). Values of growth rates are the means of 8–17 measurements ($\pm$ SD). Values of turgor are the means of < 30 measurements ($\pm$ SD). Dotted lines show the 95% confidence interval for a linear regression on the means.*

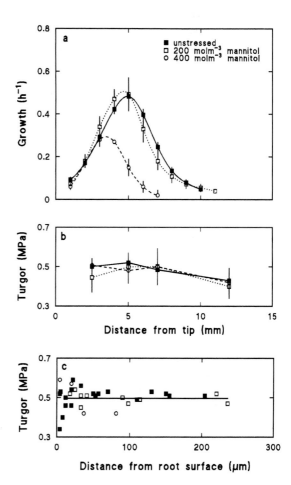

Figure 4.2. *Growth rate and turgor pressure of individual cells over the apical 12 mm of maize roots. Unstressed roots ($\blacksquare$); after 24 h in 200 mol m^{-3} mannitol (0.48 MPa) ($\square$); after 24 h in 400 mol m^{-3} mannitol (0.96 MPa) ($\bigcirc$). (a) Local longitudinal growth rate n > 15 $\pm$ SD. (b) Turgor pressure of epidermal and surface cortical cells measured longitudinally along the root tip, n = 5–25 $\pm$ SD. (c) Turgor pressure measured across the root radius at 7 mm from the root tip, each point is a single measurement.*

concomitant changes in turgor pressure (Figure 4.2b), implying that growth is not controlled by changes in water or solute transport. Growth was not a function of turgor in underlying tissue since turgor pressure was not changed across the radius of the root (Figure 4.2c).

In the (non-transpiring) plants studied there were no differences in turgor pressure in different cell layers at 7 mm from the top of unstressed roots (Figure 4.2c). Following 24 hours of 0.48 MPa osmotic stress, turgor pressure was generally uniform across the root cortex, with the exception of the epidermal layer, which tended to have lower values.

4.2.2 Wall rheology

In vivo *measurements*. The constancy of turgor pressure throughout the growing region (Figure 4.2) implies that the cell wall properties must change in order to generate the pattern in growth along the root. The walls must 'loosen' (by a decrease in Y or an increase in ϕ) along the region 0–5 mm from the root tip and harden in the 5–10 mm region until growth is finally stopped.

Growth rate and turgor pressure are tightly regulated following imposition of osmotic stress. After 24 hours in 200 mol m^{-3} mannitol (osmotic pressure of the medium $(\pi_0) = 0.48$ MPa) both parameters were the same as in unstressed roots (Figure 4.2a,b). Therefore, cell wall properties are unchanged by the treatment. After 24 hours at a higher stress (400 mol m^{-3}; $\pi_0 = 0.96$ MPa) growth and turgor pressure were similar over the apical 3 mm. Beyond this point, turgor pressure was unchanged but growth was greatly reduced (Figure 4.2a). The growing region became shorter following stress. This had been noted previously for wheat (Pritchard *et al.*, 1991) and maize (Sharp *et al.*, 1988) and indicates a hardening of the cell wall in the 3–10 mm zone following the higher osmotic stress.

Tensiometric measurements of whole-tissue rheology. The change in cell wall properties upon water stress inferred from the comparison of growth and turgor pressure was also detected by tensiometric measurements performed on methanol-killed tissue taken 5–10 mm from the root tip (Figure 4.3a). Both elastic and plastic deformation were measured using the 'Instron' technique as described by Cleland (1984). Following 24 hours in 400 mol m^{-3} mannitol, both the plastic and elastic components of extensibility were decreased (Figure 4.3a). There was no difference between the extensibilities of the unstressed roots and those grown in 200 mol m^{-3} mannitol. These observations confirm the conclusions drawn from the *in vivo* measurements.

Tensiometric measurements of isolated stele and cortex of unstressed roots. It is not known which tissues regulate the growth of the root (Tomos *et al.*, 1989). We used the tensiometric technique on roots in which the stele had been separated from the cortex.

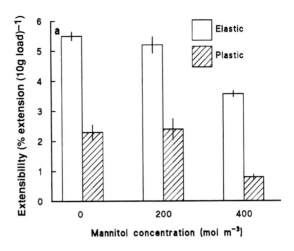

Figure 4.3. *Tissue elasticity and plasticity of the region 5– 10 mm from the root tip. Tissue was methanol killed and rehydrated prior to measurement with a tensiometer. (a) Intact tissue. Following 24 h in 0, 200 and 400 mol m^{-3} mannitol, n = 10–31 ± SD. (b) Tissue of unstressed roots separated into cortical sleeves and stele, n = 5–31 ± SD.*

The plastic deformation of the entire 5–10 mm portion of an intact root was 2.3% extension for a 10 g load (Figure 4.3b). The isolated cortical sleeve for the same region of the root had a plastic deformation of 1.4%. This was significantly less than that of the stele at 4.9% ($p < 0.001$). The elasticity (reversible increase in length) followed a similar pattern, being 5.5, 4.0 and 7.6% for the intact, cortical and stelar tissue, respectively ($p < 0.001$ in all cases).

Microscopic examination of resin-embedded sections of these isolated tissues indicated that the stele had split from the cortex mostly along the radial wall of the endodermis, leaving the outer wall of the endodermis with the cortical sleeve and the inner wall with the stele. This implies that the site of the rheological changes controlling extension is found in the cortex or in the outer wall of the endodermis rather than in the stele. Clearly, much more work needs to be performed relating the changes observed in intact tissue with the behaviour of the individual tissues.

Short-term changes in wall rheology. We have noted that 24 hours after experiencing moderate water stress ($200 \, \text{mol m}^{-3}$ mannitol; $0.48 \, \text{MPa}$) both rheology and turgor were the same as the control conditions. This was despite a series of significant changes in the interim period. Comparing growth rate and turgor pressure at each point along the growth profile allowed changes in the properties of the cell walls to be described during the osmotic adjustment that followed the stress.

The profile of local growth immediately following immersion in $0.48 \, \text{MPa}$ solution is shown in Figure 4.4a. Due to initial elastic shrinkage it was not possible to measure local (plastic) growth for the first 10 min after immersion, therefore the profile given here was measured 10–20 minutes following immersion. The growth rate was reduced significantly 4–10 mm from the tip, but was relatively unaffected nearer the apex (0–3 mm, Figure 4.4a). In contrast, turgor pressure was uniformly reduced over the entire apical 12 mm of the root (Figure 4.4b). From Equation 4.1, this implies that the smaller change in growth rate at the apex of stressed roots (see the insert in Figure 4.4a) was due to a rapid loosening of the cell wall. In contrast, the reduced turgor in the proximal regions of the growing zone was not accompanied by wall-loosening. However, both turgor and local growth rate eventually recovered over both proximal and apical regions following immersion in $200 \, \text{mol m}^{-3}$ ($\pi_0 = 0.48 \, \text{MPa}$) mannitol (Figure 4.2). This implies that the response of the walls of the apical 3 mm to water stress undergoes a cycle of an initial rapid loosening followed by a stiffening as turgor pressure recovers to its unstressed level. The relative constancy of growth rate demonstrates that wall stiffness and turgor pressure respond in concert.

4.3 Wall microfibril orientation

The rheological changes associated with extension growth may be due to changes in the cell wall 'macro' architecture. Cellulose microfibrils are a major component of the structure of the cell wall and clearly play a role in determining its mechanical properties (Preston, 1974). An orientation parallel to the direction of expansion is often considered to be growth limiting. We tested the hypothesis that microfibrillar orientation of the innermost layer of the cell wall may determine growth rate.

Microfibril orientation was measured on the inner face of cortical cell walls along the root axis of control roots (Figure 4.5) using a specially developed replica method (Pritchard, unpublished). The angle of microfibrils (where 0° is parallel to the root axis and 90° is transverse) decreased from 88° at 2.5 mm from the tip to 44° at 12 mm. Microfibril orientation cannot be a major component of cell wall rheology since, during the gradual change from transverse to parallel orientation (which should cause the walls to

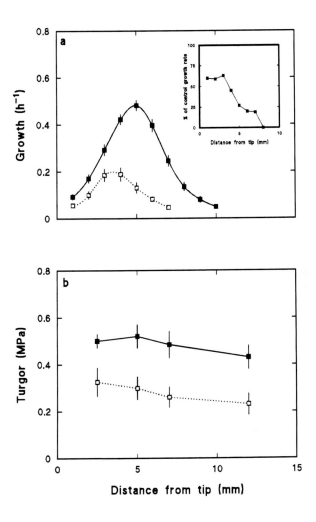

Figure 4.4. *Local growth and turgor pressure profiles 10–0 min before (■) and 10–20 min after (□) transfer from 0 mol m⁻³ (0 MPa) to 200 mol m⁻³ mannitol (0.48 MPa). (a) Local growth, n = 15 ± SD; insert shows the 'stressed' growth rate as a percentage of the control. (b) Turgor pressure, n = 5–15 ± SD.*

harden), the walls actually loosen over the apical 5 mm. However, basal to 5 mm the trend toward a parallel orientation is consistent with such a role.

Despite the lack of good correlation with developmental changes, the possibility remains that microfibrils could correlate with the changes in rheology following water stress. However, microfibril angle was unaffected by 24 h in either 200 or 400 mol m⁻³ mannitol (π_0 = 0.48 and 0.96 MPa) in comparison to unstressed roots. The angles at 7 mm from the tip (where wall properties

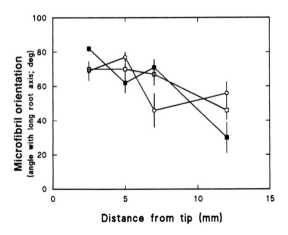

Figure 4.5. *Angle of cellulose microfibrils to the long axis of the root on the inner face of cortical cell walls over the apical 12 mm. Unstressed roots (■); 24 h in 200 mol m⁻³ mannitol (0.48 MPa) (□); 24 h in 400 mol m⁻³ mannitol (0.96 MPa) (○). (n = 6–26 ± SD).*

were altered in the 0.96 MPa treatment) were $66 \pm 4.6°$, $67 \pm 6.3°$ and $50 \pm 9.8°$ for the 0, 0.48 and 0.96 MPa treatments, respectively.

4.4 Wall enzyme distribution

4.4.1 *Xyloglucan endotransglycosylase*

A cross-linking enzyme in the cell wall. Differences in microfibril orientation could not satisfactorily account for the changes in rheology along the root. Alternatively, cross-links *between* microfibrils may determine wall properties. A possible candidate for this is a cross-link between cellulose polymers formed by xyloglucan hemicelluloses (Smith and Fry, 1991). Their length (150–1500 nm) could readily span the gap between adjacent microfibrils (about 30 nm) with each end hydrogen-bonded onto adjacent cellulose surfaces. Fry and colleagues (Fry *et al.*, 1992) have recently characterized from plant walls a xyloglucan endotransglycosylase (XET) activity that can cut such xyloglucan molecules. If that cross-link were load-bearing, XET activity could indeed result in wall loosening.

We measured the activity of XET along the root profile using the technique described by Fry *et al.* (1992) with radioactive substrate kindly provided by Dr Fry. Enzymic activity along unstressed roots increased from the root tip to a maximum of 3700 c.p.m. per mg fresh weight at 2.5 mm (Figure 4.6a), thereafter decreasing until at 12 mm activity was 1100 c.p.m. per mg fresh weight. Note that maximum growth was at 4.5 mm from the tip.

The increase and decrease in enzyme activity precede acceleration and deceleration of growth in a way that suggests cause and effect. Significant enzyme activity, however, remains in fully mature tissue in which growth has stopped. Clearly other factors are involved, at least in the cessation of growth.

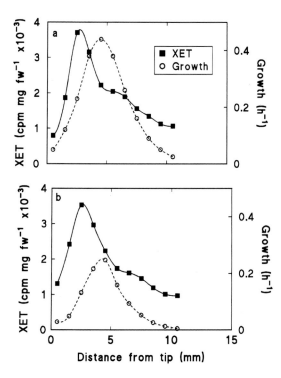

Figure 4.6. *Xyloglucan endo-transglycosylase activity (■) and growth (○) along the apical 12 mm. Each point represents four independent measurements. Roots were photographed for growth measurements immediately before sectioning for XET extraction. (a) Unstressed roots. (b) After immersion in 400 mol m⁻³ mannitol (0.96 MPa) for 24 h.*

XET activity following osmotic stress. The conclusion that other factors are involved is further strengthened by the observation that following osmotic stress at 0.96 MPa for 24 hours the longitudinal profile of XET activity was unchanged (Figure 4.6b) despite a hardening of the cell wall.

4.4.2 *Peroxidase*

Peroxidase enzymes may increase cross-linkage between cell wall components (Fry, 1987). An increase in peroxidase levels might, therefore, be involved in hardening the cell wall. The level of enzyme activity (sequentially extracted from the wall by weak buffer, high salt and digestion of cellulose) was unchanged along the growth region, showing no correlation with the hardening of the cell wall as growth ceased (Figure 4.7). However, peroxidase activity *did* increase over the whole growth profile in roots that had been exposed to 400 mol m⁻³ (0.96 MPa) mannitol (Figure 4.7) suggesting that the increase in cell wall peroxidase may be responsible for hardening the wall following stress.

Thus it remains unclear which factor hardens the wall in the proximal region of both stressed and unstressed roots, since no increase in peroxidase was measured in this region. A decrease in XET activity cannot be the sole cause, since non-growing tissue still has significant XET activity. Perhaps during development the wall structure becomes resistant to XET. Alternatively, a

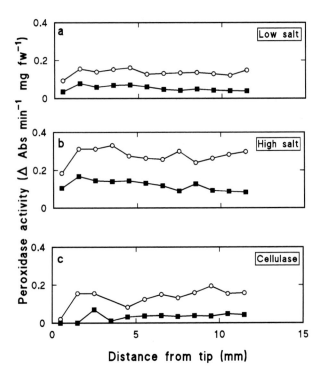

Figure 4.7. *Peroxidase activity along the apical 12 mm in unstressed roots (■) and roots grown in 400 mol m⁻³ mannitol (0.96 MPa) for 24 h (○). (a) Low-salt-extractable activity. (b) Additional activity extractable with 1 M NaCl. (c) Additional activity extractable with cellulase.*

different cross-linkage may predominate, so that the xyloglucan cross-linkages are no longer load-bearing.

It is not a prerequisite for a putative wall-hardening agent that it increases as growth ceases. Peroxidase activity may be responsible for the developmental wall hardening if it catalyses cross-links which gradually accumulate during development. In addition, there are a large number of peroxide iso-enzymes present in plants, any one of which may form cross-links between wall components. Any relevant developmental changes in this putative iso-enzyme might be masked by the large background level of peroxidase activity.

4.5 Osmotic characteristics of the expanding zone

4.5.1 *Osmotic pressure of control plants*

Although developmental and long-term changes in growth rate in response to stress are regulated by changes in wall rheology, the root displays strong osmotic adjustment and turgor maintenance. Growth cannot be divorced from membrane transport; cell turgor pressure is determined by solute transport influencing the osmotic pressure of the protoplast. The osmotic pressures of individual cells were therefore determined by microsampling and picolitre osmometry (Malone *et al.*, 1989) and were found to be constant along the apical 12 mm of unstressed roots (Figure 4.8a). In this context, isolated cells in a bathing medium would be expected to be in water potential equilibrium (water potential, $\Psi_w = P - \pi$). However, in hydroponically grown roots, vacuolar osmotic pressure exceeded the turgor by an average of 0.19 MPa over the whole growing zone. In this case Ψ_w for the cell was −0.19 MPa, while the water potential in the control medium (0.5 mol m^{-3} CaCl$_2$) was 0.004 MPa. A similar situation has been shown for root cells of maize, tomato, wheat and rice grown in nutrient solutions of different water potential (J. Pritchard, unpublished data).

The difference of −0.19 MPa between the water potential of cell and medium remained the same following the mannitol-induced water stress, as the cell osmotic pressure rose by the same amount as the osmotic pressure of the medium. This indicates 'full' osmotic adjustment of the protoplast to maintain turgor pressure.

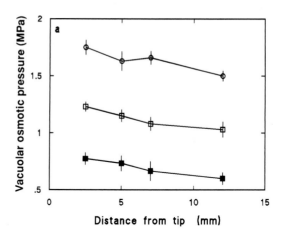

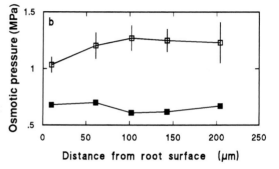

Figure 4.8. Osmotic pressure profiles of unstressed roots (■), roots stressed for 24 h in 200 mol m^{-3} mannitol (0.48 MPa, □), or for 24 h in 400 mol m^{-3} mannitol (0.96 MPa, ○, n => 12 ± SD). (a) Longitudinal profiles; (b) radial profiles at 7 mm.

The step in water potential between cell and the bathing medium requires explanation. A number of possible reasons for this may be proposed:

(1) Cell wall hydrostatic pressure could be lower than atmospheric due to transpiration tension. This seems unlikely since the transpiration rate was minimal in these plants. In addition, the root cell walls would be expected to be in hydrostatic equilibrium with the bathing solution surrounding them.

(2) If the entry of water into the cells were limiting growth, turgor would be less than the potential maximum (growth-induced water potential gradient) (Boyer, 1985; Cosgrove, 1987; Silk and Wagner, 1980). This is also unlikely since the half-times of hydraulic equilibrium are very short in these cells, indicating very low resistance to water flow from the medium. This explanation would also imply a greater discrepancy between turgor and osmotic pressure in the region where growth is most rapid, which (within experimental limits) is not observed.

(3) The average reflection coefficient (σ) of the cell solutes to the plasma membranes of the root cells may be significantly less than 1 (a value of 0.75 would be consistent with the observations).

(4) The osmotic pressure of the cell wall immediately adjacent to the plasma membrane is not zero, i.e. there is an unstirred layer within the wall. This could be stabilized by a flow of water across the wall and into the cell. If such a flow were driven by growth, the effect would be greatest in the fastest growing region, which has been precluded above. Water flow due to the transpiration stream across the cortex is a second source of such flow. This would contradict the view that the expanding zone of the root is relatively isolated from the transpiration steam (Sanderson *et al.*, 1988; Steudle and Frensch, 1989). Steudle (see Chapter 2) has recently proposed, however, a circulation of water and solutions across the root cortex—inward via the cell-to-cell pathway and out via the apoplast. The influence that such a scheme would have on the solute relations of the apoplast is unclear.

4.5.2 *Time-course of recovery from osmotic stress*

Following water stress, the rates of osmotic adjustment and of turgor regulation were not the same at all points along the root. On treatment with $200 \, mol \, m^{-3}$ mannitol (0.48 MPa) at 2.5 mm from the root tip, turgor was rapidly reduced from 0.5 MPa to around 0.27 MPa. Turgor pressure then recovered, reaching control values after about 2 hours, and thereafter remained constant (Figure 4.9a).

Turgor recovered in a similar manner at 7 mm from the tip, although the rate of recovery was slower than that at 2.5 mm (Figure 4.9b). Comparison of the rate of turgor recovery at $10 \, \mu m$ and $204 \, \mu m$ from the root surface showed no detectable difference in those cells adjacent to the stele ($204 \, \mu m$) and those at the root surface ($10 \, \mu m$) (Figure 4.9b).

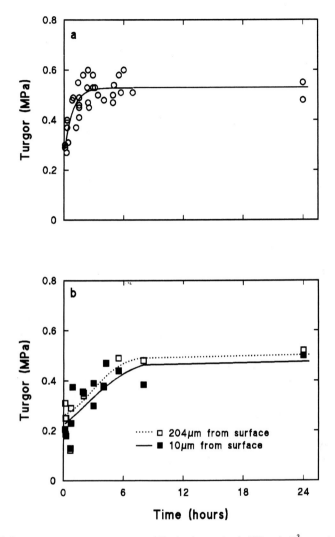

Figure 4.9. *Time-course of turgor recovery following immersion in 200 mol m^{-3} mannitol, each point is a single determination. (a) Cortical cells 2.5 mm from the root tip. (b) 7 mm from the root tip:* ■*, 10 µm from the root surface;* □*, 204 µm from the root surface.*

A similar slowing in the rate of turgor recovery at increasing distance from the root tip was reported for wheat roots (Pritchard *et al.*, 1991).

4.5.3 *Identification of osmotic solutes*

In order to identify the nature of the membrane transport activities responsible for osmotic adjustment and turgor maintenance, the inorganic composition of the vacuoles was assayed by X-ray microanalysis of samples extracted from individual cells, using the method of Malone *et al.* (1991).

The elemental composition was analysed along the root profile for the control and the two stress levels (Figure 4.10). K^+ and Cl^- were the major ions present, P and S were present at less than $10\,mol\,m^{-3}$ and Ca^{2+} was only present in trace amounts (data not shown).

K^+ levels decreased from $97 \pm 7.3\,mol\,m^{-3}$ at 2.5 mm to $60 \pm 8.0\,mol\,m^{-3}$ at 12 mm from the tip of unstressed roots (Figure 4.10a). This pattern was essentially unchanged following 24 h of 0.48 or 0.96 MPa stress. At 2.5 mm from the tip, K^+ was 88 ± 7.5 and $97 \pm 9\,mol\,m^{-3}$ in the 0.48 and 0.96 MPa treatments, respectively. At 12 mm from the tip this had decreased to 55 ± 6 and $63 \pm 10\,mol\,m^{-3}$ for the two treatments, respectively. The profile of K^+ concentration was broadly similar to that noted in another study of maize roots which only measured bulk tissue concentrations (Silk *et al.*, 1986).

Thus there was no significant increase in K^+ levels following stress, despite an increase in vacuolar osmotic pressure of 0.96 MPa. If all the K^+ was associated with a monovalent anion (e.g. Cl^- or NO_3^-) and the osmotic coefficient was 0.9 (Robinson and Stokes, 1965), potassium salts would only account for 57% of the osmotic pressure at 2.5 mm (307 mOsmolal) and 45% at 12 mm (240 mOsmolal) of unstressed roots. Most of the shortfall could be accounted for by glucose and fructose ($71\,mol\,m^{-3}$ and $31\,mol\,m^{-3}$, respectively; Figures 4.11a,b) along the growth profile. Measurements of bulk tissue suggested that neither proline nor sucrose could account for the shortfall (data not shown).

Chloride was the major anion in the vacuolar sap. Its concentrations in unstressed roots was essentially unchanged along the apical 12 mm (Figure 4.10b) with an average of $41.5 \pm 8.7\,mol\,m^{-3}$. Unlike K^+, Cl^- was lowest at 2.5 mm from the root tip. Following 24 h at $200\,mol\,m^{-3}$ mannitol, chloride concentration was effectively unchanged. At the highest stress treatment (0.96 MPa) Cl^- concentration increased to $68.8 \pm 14\,mol\,m^{-3}$, again the concentration was lower in the 2.5 mm section.

The increase of 0.48 and 0.96 MPa in vacuolar osmotic pressure following the two water stress levels (Figure 4.9a) was not due to accumulation of potassium salts. (Na^+ and Ca^{2+} salts did not contribute significantly.)

Osmotic adjustment is most likely due to an increase in neutral organic molecules. In unstressed roots the concentration of vacuolar glucose was unchanged along the growing zone but was increased by about 50 mOsmolal following 24 h in $200\,mol\,m^{-3}$ mannitol (Figure 4.11a). If glucose were soley responsible for the osmotic adjustment, an increase of around $200\,mol\,m^{-3}$ would have been expected. Fructose was also unchanged along the growth profile but did not increase following osmotic stress (Figure 4.11b). Other, as yet unidentified, solutes must be involved in osmotic adjustment.

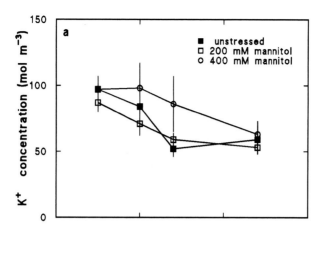

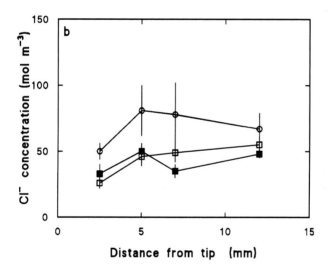

Figure 4.10. *Vacuolar K^+ and Cl^- concentrations along the apical 12 mm of unstressed roots (■), and roots after 24 h in 200 mol m^{-3} mannitol (0.48 MPa) (□) or 400 mol m^{-3} mannitol (0.96 MPa) (○ ; n = 10–16 ± SD). (a) K^+; (b) Cl^-.*

4.6 Conclusions

In this work we attempted to integrate anatomical and biochemical parameters with changes in wall properties. These changes were identified by comparing turgor and growth rate at high resolution along the growing zone.

The constancy of turgor pressure, both along and across the growing region, indicated that the variation in growth rate along unstressed roots is due to variation in cell wall properties. Specifically, the walls must loosen in

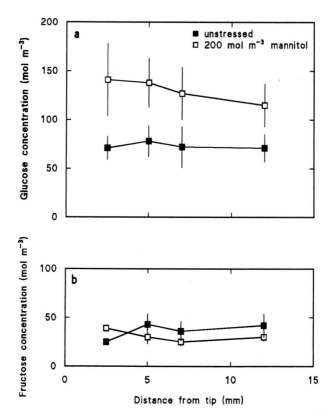

Figure 4.11. *Vacuolar glucose and fructose concentration along the apical 12 mm of unstressed roots* (■) *and roots following 24 h in 200 mol m^{-3} mannitol (0.48 MPa,* □*; n = 10–16 ± SD). (a) Glucose; (b) fructose.*

the region 0.5 mm from the root tip, allowing expansion to increase despite a constant turgor, and then become tighter, slowing and finally stopping growth. This has been noted previously in other studies on roots (Pritchard *et al.*, 1991; Spollen and Sharp, 1991).

Osmotic stress causes a transient drop in the turgor pressure, which induces the walls of the youngest cells to loosen even faster. This helps to maintain the local growth rate despite the lowered turgor. Turgor is rapidly brought back to the control levels by active solute transport, and the wall properties in this region return to their original value. Since neither turgor nor osmotic pressure are constant during this period, the mechanical properties of the wall must respond directly to *growth rate*. We currently have no hint as to the basis of this remarkable behaviour.

Following high levels of osmotic stress (0.96 MPa) growth is restricted by a hardened cell wall, despite a maintenance of turgor. Such stress-induced hardening has been noted in other tissues, for example in maize coleoptiles (Matsuda and Riazi, 1981) and in wheat roots (Pritchard *et al.*, 1991).

These changes in wall properties identified by comparisons of growth rate and turgor pressure must reflect changes in the structure of the cell wall or the interactions between its components.

The 'classical' reorientation of the innermost cellulose microfibrils to a more transverse orientation with expansion does occur in expanding maize roots. This has been shown previously by Hogetsu (1986) for pea roots. However, there is not a good correlation of the microfibril angle with the regions of acceleration and deceleration of growth. In addition, there is no change in microfibril orientation following the highest osmotic stress, even though there are large changes in cell wall properties. It appears that the orientation of the microfibrils is not as important in control of extension growth as the interactions between wall components.

The clearest correlation of the activity of a wall enzyme with growth was the increase in XET activity in cells approaching the maximum rate of expansion (0–3 mm from the tip). Similarly, XET activity declines prior to wall stiffening in the region 4–10 mm from the root tip. However, although XET may indeed be the primary regulator of wall properties in the apical region of the root, its influence is reduced during the later stages of development. This is seen in the mature (non-expanding) cells which still possess significant levels of XET activity. Presumably either the substrate of the enzyme is no longer present or the cross-links which it cuts have ceased to be load-bearing during development. Further evidence for a modification of wall properties by a factor other than XET was seen following the highest level of osmotic stress, when there was significant wall stiffening along the growing region but XET activity was unchanged in comparison to unstressed roots.

Clearly, a number of parameters play a role in determining growth rate. Of the candidates studied here, XET activity drops from its maximum and microfibril orientation becomes more longitudinal. The potentially cross-linking activity of peroxidase does not increase, but this does not preclude a role in wall hardening, as its effect may well be cumulative. A wall loosening or tightening factor may not simply correlate with the growth profile, which may be the result of several antagonistic or synergistic factors.

Studies of tissues other than roots have suggested that growth control may be restricted to a particular layer of cells or even part of a layer (Kutshera *et al.*, 1987). It is important to know the nature of the load-bearing layer since relevant biochemical changes occurring in this layer may be difficult to see against the background of the whole tissue.

For the first time, tensiometric measurements allowed us to rule out the stele as the site of growth control in maize roots. The *outer* wall of the endodermis is a possible candidate, although tissues further from the central axis may be more likely.

This work has brought to light several issues of the water and solute relations of maize root growth, both under control and stressed conditions. Curiously, the wall water potential along the entire expanding zone does not appear to be in equilibrium with the bathing medium. Several suggestions have been presented to explain this; two are more likely. These are an unusually low reflection coefficient of 0.75 and the presence of unstirred layers of solutes within the walls against the plasma membranes of the cells. Upon water stress the cells adjusted their osmotic pressure in order to not only maintain turgor constant, but also to maintain the same water potential gradient between their walls and the adjacent bathing medium. This observation is consistent with both a constant reflection coefficient and a constant apparent osmotic pressure in the walls. The constancy of pressure and water potential gradient is similar to the situation in the leaves of salt-stressed *Suaeda maritima* (Clipson *et al.*, 1985). More recently, Zimmermann *et al.* (1992) demonstrated a similar situation for the roots of *Aster tripolium*, where a water potential step exists between bathing medium and root cortex. These authors have no explanation to offer for this phenomenon. Clearly, whatever its basis, this behaviour has considerable significance in stress tolerance.

Osmotic adjustment resulting in total turgor recovery occurs throughout the growing zone. Its rate is highest towards the apex. The molecular nature of this adjustment is not fully clear. Neither K^+ nor Cl^- transport is involved, but *glucose* import is accelerated by water stress. However, this sugar only accounts for some 30% of the osmotic adjustment. We have yet to identify the solute responsible for most of the osmotic adjustment.

In addition to externally applied water stress, the expanding cells also experience considerable variations in growth rate. This has a direct bearing on solute transport, since as cell expansion accelerates so must the net uptake of solutes (and water) to maintain turgor. Unlike the response to stress, the highest rate of uptake is not found towards the apical end of the growing zone. The demand for solutes is a function of growth rate. The mechanism by which solute uptake and cell growth are kept in step is not understood, although it is probable that turgor pressure itself has a central role.

Solute demand at a constant growth rate is a function of internal osmotic pressure (Greenway and Munns, 1983). At the maximum growth point (5 mm from the tip) the constant growth rate demands an increased solute uptake of 60% as internal osmotic pressure rises from 0.71 MPa to 1.15 MPa.

The expanding zone of the maize root shows strong dynamic control of both growth and water relations. Both wall mechanics and membrane transport respond during development and to stress. The activities of XET and peroxidase appear to play a role in the regulation of wall properties, although other activities must be involved. The changes relevant to growth regulation occur outside the stele. Continuing study will aim to further characterize the load-bearing layers in the root and to examine the biochemical response of their walls.

Acknowledgements

We wish to thank Drs Stephen Fry and Richard Hetherington for their collaboration with the enzymatic measurements. This work is currently being supported by a joint AFRC grant to ADT and Stephen Fry.

References

Boyer, J.S. (1985) Water transport. *Ann. Rev. Plant Physiol.*, **36**, 473–516.

Boyer, J.S., Cavalieri, A.J. and Schulze, E.D. (1985) Control of the rate of cell enlargement: excision, wall relaxation and growth induced water potentials. *Planta*, **163**, 527–543.

Cleland, R.E. (1984) The Instron technique as a measure of immediate past wall extensibility. *Planta*, **160**, 514–520.

Clipson, N.J.W., Tomos, A.D. and Flowers, T.J. (1985) Salt tolerance in the halophyte *Suaeda maritima* L. Dum. The maintenance of turgor pressure and water potential gradients in plants growing at different salinities. *Planta* **165**, 392–396.

Cosgrove, D.J. (1987) Wall relaxation and the driving forces for growth. *Plant Physiol.*, **84**, 561–564.

Fry, S.C. (1986) Cross linking of matrix polymers in the growing walls of angiosperms. *Ann. Rev. Plant Physiol.*, **37**, 165–186.

Fry, S.C. (1987) Formation of iso-dityrosine by peroxidase isozymes. *J. Exp. Bot.*, **38**, 853–862.

Fry, S.C., Smith, R.C., Renwick, K.F., Martin, D.J., Hodge, S.K. and Matthews, D.J. (1992) Xyloglucan-endotransglycosylase, a new wall-loosening enzyme-activity from plants. *Biochem. J.*, **282**, 821–828.

Greenway, H. and Munns, R. (1983) Interactions between growth, uptake of Cl^- and Na^+, and water relations of plants in saline environments, II Highly vacuolated cells. *Plant Cell Environ.*, **6**, 575–589.

Hogetsu, T. (1986) Orientation of wall microfibril deposition in root cells of *Pisum sativum* L. var. Alaska. *Plant Cell Physiol.*, **27**, 947–951.

Jones, H., Leigh, R.A., Wyn Jones, R.G. and Tomos, A.D. (1988) The integration of whole-root and cellular hydraulic conductivities in cereal roots. *Planta*, **174**, 1–7.

Kutschera, U., Bergfeld, R. and Schopfer, P. (1987) Co-operation of epidermis and inner tissues in auxin-mediated growth of maize coleoptiles. *Planta*, **170**, 168–180.

Lockhart, J.A. (1965) An analysis of irreversible plant cell elongation. *J. Theor. Biol.*, **8**, 264–275.

Malone, M., Leigh, R.A. and Tomos, A.D. (1989) Extraction and analysis of sap from individual wheat leaf cells: The effect of sampling speed on the osmotic pressure of extracted sap. *Plant Cell Environ.* **12**, 919–926.

Malone, M., Leigh, R.A. and Tomos, A.D. (1991) Concentrations of vacuolar inorganic ions in individual cells of intact wheat leaf epidermis. *J. Exp. Bot.*, **42**, 385–389.

Matsuda, Y. and Riazi, A. (1981) Stress induced osmotic adjustment in growing regions of barley leaves. *Plant Physiol.* **68**, 571–576.

Preston, R.D. (1974) *The Physical Biology of Plant Cell Walls*. Chapman & Hall, London.

Preston, R.D. (1982) The case for multinet growth in growing walls of plant cells.

Pritchard, J., Tomos, A.D. and Wyn Jones, R.G. (1987) Control of wheat root elongation growth I. Effects of ions on growth rate, wall rheology and cell water relations. *J. Exp. Bot.*, **38**, 948–959.

Pritchard, J., Wyn Jones, R.G. and Tomos, A.D. (1990a) Measurement of yield threshold and cell wall extensibility of intact wheat roots under different ionic, osmotic and temperature treatments. *J. Exp. Bot.*, **41**, 669–675.

Pritchard, J., Barlow, P.S., Adam, J.S. and Tomos, A.D. (1990b) Biophysics of the growth of maize roots by lowered temperature. *Plant Physiol.*, **93**, 222–230.

Pritchard, J., Wyn Jones, R.G. and Tomos, A.D. (1991) Turgor, growth and rheological gradients of wheat roots following osmotic stress. *J. Exp. Bot.*, **42**, 1043–1049.

Ray, P.M., Green, P.B. and Cleland, R. (1972) Role of turgor in plant cell growth. *Nature*, **239**, 163–164.

Robinson, R.A. and Stokes, R.H. (1965) *Electrolyte Solutions, the Measurement and Interpretation of Conductivity*. Butterworths, London.

Roelofsen, P.A. and Houwink, A.L. (1953) Architecture and growth of the primary wall in some plant hairs and in the *Phycomyces* sporangiophore. *Acta Botanica Neerlandica*, **2**, 218–225.

Sanderson, J., Whitbread, F.C. and Clarkson, D.J. (1988) Persistent xylem cross-walls reduce the axial conductivity in the apical 20 cm of barley seminal root axes: implications for the driving forces for water movement. *Plant Cell Environ.*, **11**, 247–256.

Schmastig, J.G. and Cosgrove, D.J. (1988) Growth inhibition, turgor maintenance and changes in yield threshold after cessation of solute import in pea epicotlys. *Plant Physiol.*, **88**, 1240–1245.

Sharp, R.E., Silk, W.K. and Hsiao, T.C. (1988) Growth of the primary maize root at low water potentials. I. Spatial distribution of expansive growth. *Plant Physiol.*, **87**, 50–57.

Silk, W.K., Hsiao, T.C., Diederhofen, U. and Matson, C. (1986) Spatial distribution of potassium, solutes, and their deposition rates in the growth zone of the primary corn root. *Plant Physiol.*, **82**, 853–858.

Silk, W.K. and Wagner, K.K. (1980) Growth sustaining water potential distribution in the primary corn root. *Plant Physiol.*, **66**, 859–863.

Smith, R.C. and Fry, S.C. (1991) Endotransglycosylation of xyloglucans in plant cell suspension cultures. *Biochem. J.*, **279**, 529–535.

Spollen, W.G. and Sharp, R.E. (1991) Spatial distribution of turgor and root growth at low water potentials. *Plant Physiol.*, **96**, 438–443.

Steudle, E. and Frensch, J. (1989) Osmotic responses of maize roots. Water and solute relations. *Planta*, **177**, 281–295.

Thomas, A., Tomos, A.D., Stoddart, J.L., Thomas, H. and Pollock, C.J. (1988) Cell expansion rate, temperature and turgor pressure in growing leaves of *Lolium temulentum* L. *New Phytol.* **112**, 1–5.

Tomos, A.D., Malone, M. and Pritchard, J. (1989) The biophysics of differential growth. *Environ. Exp. Bot.*, **29**, 7–25.

Zimmermann, U., Rygol, J., Balling, A. and Klock, G. (1992) Radial turgor and osmotic pressure profiles in intact and excised roots of *Aster tripolium* – pressure probe measurements and nuclear magnetic resonance-imaging analysis. *Plant Physiol.*, **99**, 186–196.

Pressure probe methods to measure transpiration in single cells

Joseph K.E. Ortega

5.1 Introduction

The rate of plant and fungal cell enlargement depends on two simultaneous and interrelated physical processes, the *net* rate of water uptake and the rate of cell wall extension. Transpiration is an important process that impacts both water transport and enlargement (growth) of plant and fungal cells. In the past, pressure probe methods have been developed to determine the biophysical and biomechanical properties of plant and fungal cells that control water transport (Husken *et al.*, 1978; Wendler and Zimmermann, 1985; Zimmermann and Steudle, 1978) and cell wall extension (Cosgrove, 1985; Ortega *et al.*, 1989, 1991). This chapter discusses and compares two pressure probe methods that have recently been developed to measure transpiration rates from single plant and fungal cells.

5.2 General theoretical foundations

Pressure probe methods require that there is a known relationship between the turgor pressure and other parameters which represent relevant biophysical and biomechanical properties of plant and fungal cells. Governing equations which establish such a relationship for many plant and fungal cells have been termed the 'augmented growth equations' (Ortega, 1985, 1990; Ortega *et al.*, 1988). The basic form of these governing equations was introduced by Lockhart (1965) and they were termed the 'growth equations' by Taiz (1984). Subsequently, each of the original two growth equations was augmented with an additional term to extend its utility (Ortega, 1985; Ortega *et al.*, 1988). In general, the augmented growth equations describe two

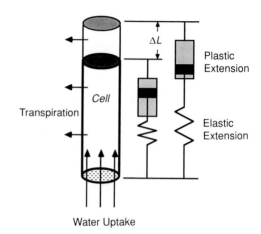

Figure 5.1. *A schematic illustration of the two simultaneous and interrelated physical processes that control the rate of plant and fungal cell enlargement; the net rate of water uptake, and the rate of cell wall extension. The net rate of water uptake is the difference between the rate of water uptake and the transpiration rate. The rate of cell wall extension is the sum of plastic (dashpot) and elastic (spring) extension rates.*

simultaneous and interrelated physical processes that control the relative rate of plant and fungal cell enlargement (Figure 5.1), the *net* relative rate of water uptake (Equation 5.1) and the relative rate of extension of the cell wall chamber (Equation 5.2):

$$v_w = (dV_w/dt)/V_w = L(\sigma\Delta\pi - P) - T, \qquad 5.1$$

{*net* relative rate of water uptake} = {relative water uptake rate}

$$- \text{\{relative transpiration rate\}}$$

$$v_c = (dV_c/dt)/V_c = \phi(P - Y) + (dP/dt)/\epsilon, \qquad 5.2$$

{relative cell wall extension rate} = {plastic}

$$+ \text{\{\textit{elastic}\} [relative extension rates]}$$

where v is the relative rate of change in volume, V is the volume, L is the relative hydraulic conductance of the cell membrane, σ is the solute reflection coefficient, $\Delta\pi$ is the osmotic pressure difference, P is the turgor pressure, ϕ is the relative irreversible wall extensibility, Y is the yield threshold (sometimes referred to as the critical turgor pressure, P_c), ϵ is the volumetric elastic modulus, and the subscripts 'w' and 'c' correspond to 'water' and 'cell wall', respectively. It should be noted that the volume of the cell contents (mostly water) must equal the volume of the cell wall chamber ($V_w = V_c$) and that the respective relative rates of change in volume must also be equal ($v_w = v_c$).

Of the many pressure probe methods that have been established over the past two decades, most of them may be categorized as either 'pressure relaxation' methods or 'pressure clamp' methods, depending on whether the turgor pressure decays from one value to another (relaxation), or is maintained (clamped) at a fixed magnitude.

5.3 Pressure relaxation method

5.3.1 *Theory*

The turgor pressure of a plant or fungal cell will decay when its water source is removed. If the cell is growing, then the turgor pressure decay rate depends on both the plastic extension rate and the transpiration rate. The governing equation for the turgor pressure decay rate $(-dP/dt)$ is obtained by combining Equations 5.1 and 5.2 with the elimination of v (since $v_w = v_c$):

$$-(dP/dt) = \epsilon[\phi(P - Y) + T],\qquad 5.3$$

{turgor pressure decay rate} = {ϵ} × {relative plastic extension rate

+ relative transpiration rate}

where the term $L(\sigma\Delta\pi - P)$ is zero, because the cell is isolated from its water source. It should be noted that the governing equation for *in vivo* stress relaxation, $dP/dt = -\epsilon\phi(P - Y)$ (Cosgrove, 1985; Ortega, 1985), is recovered from Equation 5.3 when transpiration is also eliminated, i.e. $T = 0$. If Equation 5.3 is solved for T, Equation 5.4 is obtained:

$$T = -(dP/dt)/\epsilon - \phi(P - Y),\qquad 5.4$$

{relative transpiration rate} = {turgor pressure decay rate/ϵ}

− {relative plastic extension rate}.

It is apparent that, in theory, the relative transpiration rate may be determined when ϵ, ϕ, and Y are known and the turgor pressure decay rate, $-(dP/dt)$, is measured. In practice, it would be difficult to establish a pressure probe method using Equation 5.4, because the term, $\phi(P - Y)$, is only valid when P is greater than Y. Therefore, the range of turgor pressure decay in which Equation 5.4 can be used may be as small as 0.02 MPa (Green *et al.*, 1971).

A simpler relationship is obtained for the case of a *non-growing* cell because the term, $\phi(P - Y)$, is zero for the non-growing cell and Equation 5.4 reduces to Equation 5.5 (Ortega *et al.*, 1988):

$$T = -(dP/dt)/\epsilon,\qquad 5.5$$

{relative transpiration rate} = {turgor pressure decay rate}

÷ {volumetric elastic modulus}.

Therefore, T may be determined by measuring the turgor pressure decay rate, when ϵ is known. In general, the volumetric elastic modulus is a non-linear function of the turgor pressure (Husken *et al.*, 1978; Zimmerman and Steudle, 1978); see Figure 5.2, open circles. The non-linear relationship between

Figure 5.2. The volumetric
elastic modulus, ϵ, as a function of
the turgor pressure, P, for a typical
plant cell (data points, ○). This
non-linear function may be
approximated by Equation 5.6 (see
text) over the entire range of P, or
by a line of slope, b, over the low
range of P and a horizontal line over
the high range of P.

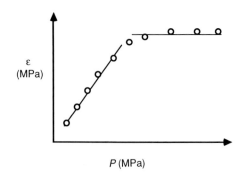

ϵ and P may be approximated by the following equation (Ortega, 1990; Ortega et al., 1988):

$$\epsilon(P) = \epsilon_m - [(\epsilon_m - \epsilon_0) \exp{(-kP)}], \qquad 5.6$$

where ϵ_m is the maximum value, ϵ_0 is the value at zero turgor pressure, and k is a constant which is determined from the experimental results. This relationship has the properties that at low pressures ϵ increases linearly with P (so at low pressures ϵ can be approximated by the relationship, $\epsilon = \epsilon_0 + bP$, where b is the slope of the line; see Figure 5.2, solid line), and at higher pressures ϵ becomes constant ($\epsilon = \epsilon_m$; see Figure 5.2, solid line). Therefore, depending on the range of turgor pressure decay, the following equations may be used to determine T for non-growing cells:

$$T = -(dP/dt)/\epsilon_m, \qquad \{\text{high } P\} \qquad 5.7a$$

$$T = -(dP/dt)/(\epsilon_0 + bP), \qquad \{\text{low } P\} \qquad 5.7b$$

$$T = -(dP/dt)/\{\epsilon_m - [(\epsilon_m - \epsilon_0) \exp{(-kP)}]\}, \quad \{\text{all } P\} \qquad 5.7c$$

5.3.2 *Application*

The pressure relaxation method has been used to determine transpiration rates of single-celled sporangiophores of *Phycomyces blakesleeanus*. Transpiration is especially important to these aerial fungal cells, since they can transpire up to 85% of their water uptake (Bergman *et al.*, 1969). Also, it has been suggested that transpiration rates mediate some of the sporangiophores' sensory responses (Cerda-Olmedo and Lipson, 1987).

Pressure relaxation experiments were conducted on 'plucked', non-growing (stage III) sporangiophores of *Phycomyces* (Ortega *et al.*, 1988). Plucked sporangiophores (sporangiophores which are carefully removed, at the base, from the mycelium) were used so that they could be isolated from their water source at a prearranged time in the experimental protocol. In general, plucked sporangiophores with their bases in water continue to grow, develop and respond to sensory stimuli, although their growth rate is slightly reduced (Bergman *et al.*, 1969).

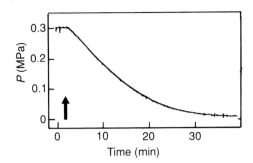

Figure 5.3. *The turgor pressure trace* (P *vs.* t) *of a 'pressure relaxation' experiment, from which the transpiration rate is determined* (*see text*).

In general, the experimental protocol was as follows. A stage III sporangiophore was plucked and transferred to a support chamber (this holds the sporangiophore vertical, with its base immersed in water) where it equilibrated for 10–20 min, and during which time the length was measured to ensure that it was not growing. Then the sporangiophore was impaled with the microcapillary tip of the pressure probe, and the turgor pressure was measured and recorded on a stripchart recorder for the remainder of the experiment. After the turgor pressure stabilized, the water level was lowered in the support chamber to isolate the sporangiophore from its water source, at which time the turgor pressure began to decay.

Figure 5.3 plots the turgor pressure, P (the pressure trace is taken directly from the chart recorder), as a function of time (the upward-pointing arrow indicates the time when the sporangiophore was isolated from its water source). Some features of the turgor pressure decay curve are worthy of discussion. The early portion of the pressure decay (approximately 3–8 min) is nearly linear. According to theory, a linear turgor pressure decay would be expected, because at high turgor pressures $\epsilon = \epsilon_m$ (for these sporangiophores, $\epsilon \approx \epsilon_m$, when $P \geq 0.2\,\text{MPa}$). Therefore, when $P > 0.2\,\text{MPa}$, Equation 5.7a may be used and solved to obtain a linear turgor pressure decay: $P(t) = P_i - \epsilon_m Tt$, where P_i is the initial pressure. At lower turgor pressures $(P < 0.2\,\text{MPa})$, the pressure decay curve appears to become exponential. This is consistent with the turgor pressure behaviour predicted by Equation 5.7b, in which ϵ increases linearly with P $(\epsilon = \epsilon_0 + bP)$. When Equation 5.7b is solved for $P(t)$, an exponential decay is obtained: $P(t) = (P_i + \epsilon_0/b) \exp(-bTt) - (\epsilon_0/b)$, where P_i is again the initial pressure.

The relative transpiration rate (T) and volumetric transpiration rate (VT) of plucked stage III sporangiophores were determined from pressure relaxation experiments (using Equation 5.7a) to be: $T = 0.0049 \pm 0.0011\,\text{min}^{-1}$ $(n = 12)$, and $VT = 3.2 \pm 1.2\,\text{nl min}^{-1}$ $(n = 12)$; $25\% < \text{RH} < 60\%$. The volumetric elastic modulus (maximum value) was determined with other experiments (J.K.E. Ortega, M.E. Smith, S.A. Bell, A.J. Erazo and M.A. Espinosa, unpublished results) to be $\epsilon_m = 2.0 \pm 1.5\,\text{MPa}$ $(n = 19)$. However, the value $\epsilon_m = 2.75\,\text{MPa}$ was used to determine T. The reason that the value of $\epsilon_m = 2.75\,\text{MPa}$ was used, instead of the mean value of $\epsilon_m = 2.0\,\text{MPa}$, is discussed in Section 5.5.

5.4 Pressure clamp method

5.4.1 *Theory*

Another pressure probe method (pressure clamp) was recently developed to measure relative and volumetric transpiration rates of single plant and fungal cells (Ortega *et al.*, 1992). The pressure clamp method has two major advantages over the pressure relaxation method; first, the cell may be growing or non-growing, and secondly, ϵ is not needed for the determination of either the relative transpiration rate or the volumetric transpiration rate.

The first augmented growth equation (Ortega *et al.*, 1988; Equation 5.1) provides the theoretical foundation for the pressure clamp method:

$$v_w = (dV_w/dt)/V_w = L(\sigma\Delta\pi - P) - T,$$

$\{net$ relative rate of water uptake$\} = \{$relative water uptake rate$\}$

$- \{$relative transpiration rate$\}.$

Assuming (as did Lockhart, 1965) that during cell enlargement the rate of change of cell contents is due primarily to *net* water uptake, then the net relative rate of water uptake, $(dV_w/dt)/V_w$, is equal to the relative growth rate, $[(dV/dt)/V]_g$. It follows that Equation 5.1 may be rewritten in terms of the relative growth rate, $[(dV/dt)/V]_g$, which may be expressed as the difference between the relative water uptake rate, $[(dV/dt)/V]_i$, and the relative transpiration rate, $[(dV/dt)/V]_T$:

$$[(dV/dt)/V]_g = [(dV/dt)/V]_i - [(dV/dt)/V]_T, \qquad 5.8$$

$\{$relative growth rate$\} = \{$relative water uptake rate$\}$

$- \{$relative transpiration rate$\}$

where $[(dV/dt)/V]_i = L(\sigma\Delta\pi - P)$, and $[(dV/dt)/V]_T = T$ (see Equation 5.1). A 'volumetric' equation may be obtained by multiplying Equation 5.8 by the cell volume, V:

$$[dV/dt]_g = [dV/dt]_i - [dV/dt]_T, \qquad 5.9$$

$\{$volumetric growth rate$\} = \{$volumetric water uptake rate$\}$

$- \{$volumetric transpiration rate$\}.$

Solving Equation 5.9 for the volumetric transpiration rate, $[dV/dt]_T$,

Equation 5.10 is obtained:

$$[dV/dt]_T = [dV/dt]_i - [dV/dt]_g, \qquad 5.10$$

{volumetric transpiration rate} = {volumetric water uptake rate}

− {volumetric growth rate}.

It is apparent that Equation 5.10 may be used to determine the magnitude of the volumetric transpiration rate when the volumetric growth rate and the volumetric rate of water uptake are determined.

5.4.2 *Application*

The pressure clamp method was demonstrated on the sporangiophores of *Phycomyces blakesleeanus* for both growing (stage IV) and non-growing (stage III) cells (Ortega *et al.*, 1992). Stage IV sporangiophores grow predominantly in length and the mean cross-sectional area remains relatively constant. Thus, the volumetric growth rate of stage IV sporangiophores can be determined by measuring the longitudinal growth rate, dl/dt, and multiplying this value by the cross-sectional area of the sporangiophore. It is noted that the volumetric growth rate, $[dV/dt]_g$, is zero for non-growing, stage III sporangiophores. Therefore, using Equation 5.10, the following relationship is obtained for non-growing cells; $[dV/dt]_T = [dV/dt]_i$.

The volumetric rate of water uptake, $[dV/dt]_i$, was determined by a pressure clamp method using the pressure probe (Ortega *et al.*, 1992). When a plucked sporangiophore is removed from its water source, the turgor pressure decays as transpiration occurs (see Figure 5.3), unless the volume of water lost by transpiration is replaced with biologically inert silicone oil (using the pressure probe) at the same rate. Thus, the magnitude of $[dV/dt]_i$ can be measured by clamping the turgor pressure at its equilibrium value with the pressure probe and measuring the volumetric rate at which the oil is injected into the sporangiophore. As previously shown (Cosgrove *et al.*, 1987), the oil injected into the sporangiophore stalk is easily visible and can be measured directly with the use of a stereomicroscope.

Pressure clamp experiments were conducted using the following experimental protocol. A plucked sporangiophore was transferred to a support chamber (this holds the sporangiophore vertical with its base immersed in water) where it was allowed to equilibrate for 10–30 min. Following the equilibration period, growth rate measurements were initiated and continued for the remainder of the experiment. When a 5–10 min period of steady growth rate was observed, the sporangiophore was impaled by the microcapillary tip of the pressure probe to measure the turgor pressure. Then turgor pressure and growth rate were simultaneously measured for another 5 min, to ensure that both were constant. Following this monitoring period, the sporangiophore was isolated from its water source (by lowering the water level in

Figure 5.4. *The results of a pressure clamp experiment. The turgor pressure (a), volumetric rate of silicone oil uptake (b; upper curve, □-□), volumetric growth rate (b; lower curve, ●-●) and volumetric transpiration rate (c) are plotted against the same time scale. The small arrows on the time scale and pointing upward (a and b) indicate the time when the sporangiophore was impaled with the microcapillary tip of the pressure probe to measure the equilibrium turgor pressure. The wide arrows pointing downward (a, b and c) indicate the time when the water level was lowered below the base of the sporangiophore, and the beginning of the period when silicone oil was continuously injected into the sporangiophore's vacuole to maintain (clamp) the turgor pressure at its equilibrium value.*

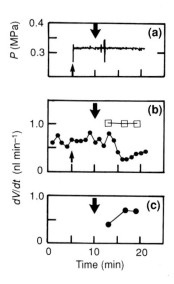

the support chamber) and the sporangiophore's turgor pressure was maintained (clamped) at its previously measured equilibrium value by injecting inert silicone oil into the cell vacuole with the pressure probe. The volumetric rate at which oil was injected into the sporangiophore's vacuole to maintain a constant turgor pressure, $[dV/dt]_i$, was measured. Thus, knowing the oil injection rate, $[dV/dt]_i$, and the volumetric growth rate, $[dV/dt]_g$, the volumetric transpiration rate, $[dV/dt]_T$, could be calculated using Equation 5.10. The volumetric growth rate of non-growing, stage III sporangiophores is zero; thus $[dV/dt]_T = [dV/dt]_i$ for these cells.

The results of a typical pressure clamp experiment conducted on a stage IV sporangiophore are presented in Figure 5.4. The turgor pressure, P (Figure 5.4a; pressure trace from the chart recorder), the volumetric rate of oil uptake (Figure 5.4b; upper curve) after the water source is removed, the volumetric growth rate of the sporangiophore (Figure 5.4b; lower curve) and the volumetric transpiration rate (Figure 5.4c) are plotted against the same time scale. It should be noted that different time intervals are used to calculate the values represented by the data points for the volumetric rate of oil uptake; the volumetric transpiration rate is calculated as the volume of oil uptake occurring during a time interval *minus* the 'total' volumetric growth occurring during the same time interval, and *divided* by the time interval.

The mean values obtained for volumetric transpiration rate, relative transpiration rate and other relevant parameters for non-growing stage III and growing stage IV sporangiophores are presented and compared in Table 5.1. The results indicate that volumetric transpiration rates and relative

Table 5.1. *Results of pressure clamp experiments*

Parameter	(units)	Stage III Mean ± SD ($n = 18$)	Stage IV Mean ± SD ($n = 18$)
$[dV/dt]_T$	$(nl\,min^{-1})$	2.94 ± 0.80	1.41 ± 1.13
$[(dV/dt)/V]_T$	(min^{-1})	0.0042 ± 0.0019	0.0020 ± 0.0015
RH	(%)	51.4 ± 4.5	53.8 ± 10.0
dl/dt	$(\mu m\,min^{-1})$	0.0	19.7 ± 11.8

transpiration rates of stage III sporangiophores are slightly larger than the respective rates of stage IV sporangiophores.

5.5 Combined pressure clamp and pressure relaxation experiments

Some experiments were conducted in which both pressure clamp and pressure relaxation methods were used on the same stage III sporangiophore (Ortega *et al.*, 1992). A similar protocol, as previously described, was used to conduct the pressure clamp experiment, but when this experiment was completed, the

Figure 5.5. *The results of a combined pressure clamp and pressure relaxation experiment. The turgor pressure (a), volumetric rate of silicone oil uptake (b; upper curve, □-□), volumetric growth rate (b; lower curve, ●-●) and volumetric transpiration rate (c) are plotted against the same time scale. The small arrows on the time scale and pointing upward (a and b) indicate the time when the sporangiophore was impaled with the microcapillary tip of the pressure probe to measure the equilibrium turgor pressure. The wide arrows pointing downward (a, b and c) indicate the time when the water level was lowered below the base of the sporangiophore, and the beginning of the period when silicone oil was continuously injected into the sporangiophore's vacuole to maintain (clamp), the turgor pressure at its equilibrium value (pressure clamp experiment). The narrow arrows pointing downward (a, b and c) indicate the time when the injection of silicone oil was stopped, which allows the turgor pressure to decay (pressure relaxation experiment).*

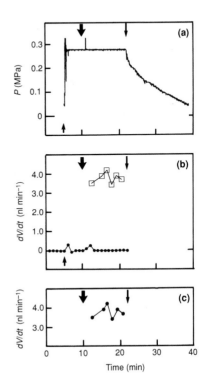

Table 5.2. *Results of combined pressure clamp and pressure relaxation experiments*

Expt	Pressure clamp VT (nl min^{-1})	T (min^{-1})	Pressure relaxation ($\epsilon = 2.0$ MPa) VT (nl min^{-1})	T (min^{-1})	(1.9 MPa $< \epsilon <$ 3.3 MPa) VT (nl min^{-1})	T (min^{-1})	ϵ (MPa)
1	3.93	0.0049	3.84	0.0048	3.93	0.0049	2.0
2	3.38	0.0069	3.90	0.0080	3.38	0.0069	2.3
3	3.61	0.0056	5.67	0.0088	3.61	0.0056	3.1
4	2.71	0.0037	4.43	0.0060	2.71	0.0037	3.2
5	3.02	0.0038	4.70	0.0060	3.02	0.0038	3.1
					$\epsilon_{av} = 2.75$ (MPa)		

injection of oil (which is required to clamp the turgor pressure at a constant magnitude) was stopped, and the turgor pressure was allowed to decay (pressure relaxation experiment). Figure 5.5 presents the results of one of these combined experiments. The turgor pressure, P (Figure 5.5a; pressure trace from the chart recorder), the volumetric rate of oil uptake (Figure 5.5b; upper curve) after the water source is removed, the volumetric growth rate of the sporangiophore (Figure 5.5b; lower curve) and the volumetric transpiration rate (Figure 5.5c) are plotted against the same time scale. The small upward-pointing arrow (on the time scale of Figures 5.5a and b) indicates the time when the sporangiophore was impaled to measure the turgor pressure. The large downward-pointing arrow indicates the time when the sporangiophore was isolated from its water source, thus initiating the pressure clamp portion of the experiment. The small downward-pointing arrow indicates the time when the oil injection (which is necessary to clamp the turgor pressure) is stopped, thus initiating the pressure relaxation portion of the experiment. During the pressure clamp portion of the experiment (the period between the downward-pointing arrows), it can be seen that the volumetric growth rate (Figure 5.5b; lower curve) is essentially zero, and that the volumetric transpiration rate (Figure 5.5c) is equal to the volumetric rate of oil uptake (Figure 5.5b; upper curve) for this non-growing cell. In this way, both pressure clamp and pressure relaxation experiments were conducted on the same sporangiophore, back to back, under identical environmental conditions.

Table 5.2 presents the results of five combined experiments. The results of the pressure clamp method are compared to 'two' sets of results ($\epsilon = 2.0$ MPa, and 1.9 MPa $< \epsilon <$ 3.3 MPa) obtained from the pressure relaxation method. Equation 5.7a was used to determine relative transpiration rates for pressure relaxation experiments. Thus, it is apparent that the magnitude of ϵ is needed to calculate $[(dV/dt)/V]_T$ (where $T = [(dV/dt)/V]_T$). Other investigations (not presented here) were conducted to determine ϵ for stage III sporangiophores (J.K.E. Ortega, M.E. Smith, S.A. Bell, A.J. Erazo and M.A. Espinosa, unpublished results) and the following mean value was obtained:

$\epsilon = 2.0 \pm 1.5$ MPa $(n = 19)$. Using the mean value for ϵ $(\epsilon = 2.0$ MPa), together with corresponding turgor pressure decay curves to determine $[(dV/dt)/V]_T$, produces the first set of values for transpiration rates. It can be seen, using this first set of values, that while the respective values for volumetric transpiration rates (VT) and relative transpiration rates (T) obtained from the pressure clamp and pressure relaxation methods are similar, the values from the pressure relaxation method are slightly larger. The following mean values were obtained for this first set of 'calculated' transpiration rates (from the five pressure relaxation experiments): $[dV/dt]_T = 4.51 \pm 0.74$ nl min^{-1} $(n = 5)$, and $[(dV/dt)/V]_T = 0.0067 \pm 0.0016$ min^{-1} $(n = 5)$. These mean values are slightly larger than those obtained from the pressure clamp method for the same five experiments: $[dV/dt]_T = 3.33 \pm 0.48$ nl min^{-1} $(n = 5)$ and $[(dV/dt)/V]_T = 0.0050 \pm 0.0013$ min^{-1} $(n = 5)$. It is both interesting and important to note that identical respective values for volumetric transpiration rates and relative transpiration rates (obtained from individual pressure clamp experiments) can be obtained from corresponding individual pressure relaxation experiments, if different values of ϵ (within the range of measured values for ϵ: 1.9 MPa $< \epsilon <$ 3.3 MPa, where the mean value is $\epsilon = 2.75 \pm 0.59$ MPa, $n = 5$) are used for the respective calculation (Table 5.2, second set of values). Furthermore, when the value of $\epsilon = 2.75$ is used for all the calculations for the pressure relaxation method, the results obtained from the five pressure relaxation experiments $(n = 5; [dV/dt]_T = 3.28 \pm 0.54$ nl min^{-1}, and $[(dV/dt)/V]_T = 0.0049 \pm 0.0012$ min$^{-1})$ are statistically identical to those obtained from corresponding (five) pressure clamp experiments.

5.6 Control experiments

Control experiments were also conducted in an environment of approximately 100% RH, to suppress transpiration. Stage IV sporangiophores were tested in a small, sealed chamber with free-standing water at the base of the chamber. This chamber apparatus was constructed to produce an environment of nearly 100% RH. The chamber apparatus was nearly identical in design to the chamber apparatus used previously to produce an environment of approximately 100% RH for *in vivo* stress relaxation experiments (Ortega et al., 1989). After the water source was removed from the sporangiophore in this environment of approximately 100% RH, the volumetric rate of oil uptake was nearly equal to the volumetric growth rate of the sporangiophore, and the difference (the volumetric transpiration rate) was essentially zero (Ortega et al., 1992).

5.7 Discussion

Although the augmented growth equations provide the theoretical foundation for both pressure relaxation and pressure clamp methods, the pressure clamp

Table 5.3. *Results of pressure clamp and pressure relaxation experiments*

	Pressure clamp Mean ± SD ($n = 18$)		Pressure relaxation Mean ± SD ($n = 12$) ($\epsilon = 2.75$ MPa)	
	VT (nl min^{-1})	T (min^{-1})	VT (nl min^{-1})	T (min^{-1})
Stage III	2.94 ± 0.80 (RH = 51.4 ± 4.5%)	0.0042 ± 0.0019	3.15 ± 1.20 (RH = 48.0 ± 9.2%)	0.0049 ± 0.0011
Stage IV	1.41 ± 1.13 (RH = 53.8 ± 10.0%)	0.0020 ± 0.0015	–	–

method has the theoretical advantage of being independent of the volumetric elastic modulus, ϵ. As it was demonstrated with experiments in which both methods were used on the same non-growing sporangiophore, and in identical environmental conditions, similar values for transpiration rate and relative transpiration rates are obtained when the appropriate value for ϵ is used. The determination of an accurate value for ϵ is not always possible because of inherent variability due to natural differences in cells and because of technical difficulties in establishing appropriate methods to determine ϵ for different plant and fungal cells. For the experiments in which both pressure clamp and pressure relaxation methods are conducted on the same stage III sporangiophore, good agreement is obtained between these two methods when the measured mean value of $\epsilon = 2.0$ MPa is used. However, statistically identical results are obtained when the value of $\epsilon = 2.75$ MPa (which is within the range of values of ϵ measured) is used. When the value of $\epsilon = 2.75$ MPa is used to analyse the results of all pressure relaxation experiments, very good agreement between the pressure clamp method and the pressure relaxation method is obtained for respective values of volumetric transpiration rates and relative transpiration rates of stage III sporangiophores (see Table 5.3).

Another advantage of the pressure clamp method is that it can be used on *growing* single plant and fungal cells, while the pressure relaxation method can only be used for non-growing single plant and fungal cells. The transpiration rates determined for growing, stage IV sporangiophores by the pressure clamp method are in the same range as those obtained by Foster (Bergman *et al.*, 1969), although slightly larger. However, Foster were conducted at 75% RH (Bergman *et al.*, 1969), and the pressure clamp experiments reported here were conducted in less humid conditions (see Table 5.3). This difference in relative humidity between the two investigations may account for the slightly larger transpiration rates obtained for stage IV sporangiophores using the pressure clamp method.

It is thought that the two pressure probe methods described in this chapter need not be restricted to single-celled plants and fungi, but in time these methods may be extended to study plant cells in tissue. For example, it may be possible to adapt the experimental protocol used to conduct *in vivo* stress relaxation experiments on growing plant tissue to measure ϕ (Cosgrove, 1985) in order to conduct pressure relaxation experiments on non-growing plant tissue, so that transpiration rates can be measured. In general, the non-growing tissue to be studied could be isolated from its water source, and the turgor pressure (as a function of time) of the cells in the tissue could be measured with the pressure probe. If the volumetric elastic modulus of these cells is also measured with the pressure probe, then the relative transpiration rate may be determined by use of Equations 5.7a, b or c, depending on which range of the turgor pressure decay curve is used to make the determination. The volumetric transpiration rate could be obtained by multiplying the relative transpiration rate by the volume of the plant tissue.

Interestingly, and perhaps importantly, the pressure relaxation method, which was used to measure transpiration rates, may be adapted to determine the 'average' volumetric elastic modulus of cells in plant tissue. If the 'total' transpiration rate of the plant tissue is measured independently, then a pressure relaxation experiment can be conducted to determine the 'average' turgor pressure decay rate of the cells in the tissue (much like Cosgrove's experiments on *in vivo* stress relaxation (1985)), and Equation 5.11 may be used to determine ϵ as a function of P:

$$\epsilon_{av} = (-dP/dt)_{av}/T, \qquad\qquad 5.11$$

$\{$volumetric elastic modulus$\} = \{$turgor pressure decay rate$\}$

$\div \{$relative transpiration rate$\}$.

Because of inherent variability due to natural differences in cells, and because of technical difficulties in measuring ϵ in all locations of plant tissue, perhaps this potential method might complement current methods used to determine the average volumetric elastic modulus of plant tissue.

Acknowledgements

We acknowledge the contributions of K. Manica and R. Keanini to some of the early pressure relaxation experiments, of S. Bell and A. Erazo to the pressure clamp experiments, and of M. Smith and M. Espinosa to measurements of the volumetric elastic modulus. The work reported in this chapter was supported by US National Science Foundation Grants DCB-8514902, DCB-8801717 and IBN-9103760 to JKEO.

References

Bergman, K., Burke, P.V., Cerda-Olmedo, E., David, C.N., Delbruck, M., Foster, K.W., Goodell, E.W., Heisenberg, M., Meissner, G., Zalokar, M., Dennison, D.S. and Shropshire, W. Jr (1969) *Phycomyces. Bacteriol. Rev.*, **33**, 99–157.

Cerda-Olmedo, E. and Lipson, E.D. (eds) (1987) *Phycomyces*. Cold Spring Harbor Laboratory, Cold Spring Harbor, New York.

Cosgrove, D.J. (1985) Cell wall yield properties of growing tissue: evaluation by *in vivo* stress relaxation. *Plant Physiol.*, **78**, 347–356.

Cosgrove, D.J., Ortega, J.K.E. and Shropshire, W. Jr (1987) Pressure probe study of the water relations of *Phycomyces blakesleeanus* sporangiophores. *Biophys. J.*, **51**, 413–423.

Green, P.B., Erickson, R.O. and Buggy, J. (1971) Metabolic and physical control of cell elongation rate: *in vivo* studies in *Nitella*. *Plant Physiol.*, **47**, 423–430.

Husken, K., Steudle, E. and Zimmermann, U. (1978) Pressure probe technique for measuring water relations of cells in higher plants. *Plant Physiol.*, **61**, 158–163.

Lockhart, J.A. (1965) An analysis of irreversible plant cell elongation. *J. Theor. Biol.*, **8**, 264–275.

Ortega, J.K.E. (1985) Augmented growth equation for cell wall expansion. *Plant Physiol.*, **79**, 318–320.

Ortega, J.K.E. (1990) Governing equations for plant cell growth. *Physiol. Plant.*, **79**, 116–121.

Ortega, J.K.E., Keanini, R.G. and Manica, K.J. (1988) Pressure probe technique to study transpiration in *Phycomyces* sporangiophores. *Plant Physiol.*, **87**, 11–14.

Ortega, J.K.E., Zehr, E.G. and Keanini, R.G. (1989) *In vivo* creep and stress relaxation experiments to determine the wall extensibility and yield threshold for the sporangiophores of *Phycomyces*. *Biophys. J.*, **56**, 465–475.

Ortega, J.K.E., Smith, M.E., Erazo, A.J., Espinosa, M.A., Bell, S.A. and Zehr, E.G. (1991) A comparison of cell wall yielding properties for two developmental stages of *Phycomyces* sporangiophores: determination by *in vivo* creep experiments. *Planta*, **183**, 613–619.

Ortega, J.K.E., Bell, S.A. and Erazo, A.J. (1992) Pressure clamp method to measure transpiration in growing single plant cells: demonstration with sporangiophores of *Phycomyces*. *Plant Physiol.*, **100**, 1036–1041.

Taiz, L. (1984) Plant cell expansion: regulation of cell wall mechanical properties. *Ann. Rev. Plant Physiol.*, **35**, 585–657.

Wendler, S. and Zimmermann, U. (1985) Determination of the hydraulic conductivity of *Lamprothamnium* by use of the pressure clamp. *Planta*, **164**, 241–245.

Zimmermann, U. and Steudle, E. (1978) Physical aspects of water relations of plant cells. *Adv. Bot. Res.*, **6**, 45–117.

Xylem pressure and transport in higher plants and tall trees

U. Zimmermann, R. Benkert, H. Schneider, J. Rygol, J.J. Zhu and G. Zimmermann

6.1 Introduction

The problem of the ascent of water in higher plants and tall trees has attracted the attention of botanists for more than a century. Many different theories have been put forward in the past. The cohesion theory introduced by Dixon and Joly (1894) is nowadays generally accepted by plant physiologists. This theory postulates that transpiration pulls water from the roots to the leaves to heights greater than can be achieved by use of a vacuum pump. This implies that the water in the xylem vessels must be under considerable tension (negative pressure)—at least in plants and trees taller than 10 m (the height of a water column that can counterbalance the pressure of the atmosphere).

From the time of the introduction of the cohesion concept for water ascent, much effort was expended on the measurement of the postulated high negative pressure values in the xylem. The Scholander bomb technique (Scholander *et al.*, 1965), the 'root pressurization' technique (Passioura and Munns, 1984) and psychrometry (Turner *et al.*, 1984) are currently the standard methods for exploring the water status in both laboratory and field studies. These techniques have yielded negative pressure values down to $-12\,\text{MPa}$ (Lange *et al.*, 1975). However, these methods are all indirect.

The recent technical achievement by Balling and Zimmermann (1990)—the direct measurement of negative hydrostatic pressures in the xylem of whole plants and trees—has renewed a rigorous and controversial debate about the mechanisms of water ascent. The xylem pressure probe has given results which are clearly at variance with those obtained by indirect methods and which conflict with the assumptions that plant physiologists use when discussing water ascent in higher plants and tall trees.

In this chapter the authors will discuss the principle of the xylem pressure probe technique, the physics involved in the development of negative pressures in liquids, and will re-examine some paradigms in the light of new evidence.

6.2 Pressure probe technique

One of the most useful tools for studies of water relations at the cell and tissue level is the cell turgor pressure probe introduced by Zimmermann and co-workers more than 20 years ago (Zimmermann *et al.*, 1969; also Oparka *et al.*, 1991; Tomos, 1988; Zimmermann and Steudle, 1978). The principle of the xylem pressure probe is similar to that of the cell turgor pressure probe. In both probes the pressure is transmitted through a liquid-filled microcapillary to a pressure transducer mounted in a small perspex chamber (Figure 6.1). Injection of volume (pressure) pulses can be performed by (manual or automatic) displacement of the metal rod via a micrometer screw. For the determination of turgor pressure (and of the hydraulic conductivity of the cell membrane, as well as of the elastic modulus of the cell wall) the microcapillary is filled with silicone oil (Figure 6.1, inset). When the pipette tip has been inserted into a turgid plant cell, oil will be displaced from the tip by cell sap under turgor pressure. Appropriate

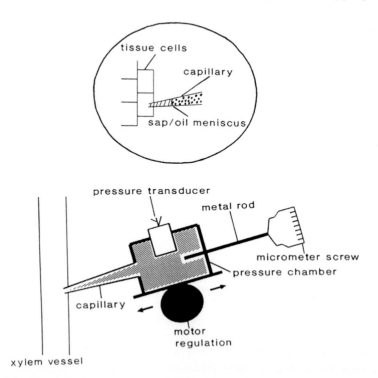

Figure 6.1. *Schematic diagrams of the xylem pressure probe and (inset) cell turgor pressure probe. For a detailed explanation, see the text.*

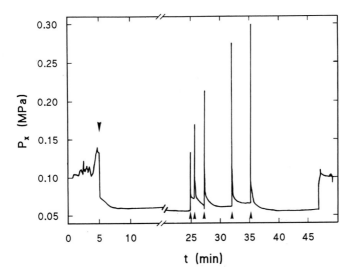

Figure 6.2. *Pressure recording measured during the insertion of the microcapillary of the xylem pressure probe into a vessel of a 1-year-old willow twig (excised in air and kept in water). The puncture of the vessel is indicated by a pressure drop to subatmospheric (downward arrow). After injection of volume pulses of various amplitudes (upward arrows) the xylem pressure relaxed to the original value within a very short time (only a part of the pulse series is shown). Similar pressure values were obtained from in situ measurements. However, it must be stated that the injection of volume pulses into vessels in the tree was made difficult due to mechanical oscillations of the twigs in the absence of a vibration-free table.*

displacement of the metal rod allows the oil/sap boundary to return to the reference line.

For measurements of subatmospheric pressures, as in the xylem vessels (and in or around secretory cells, see below) of plants, the probe is filled with degassed water. For insertion, the microcapillary is advanced slowly into the tissue. Penetration of a xylem vessel is detected as a rapid decrease of pressure. At this instant, the advance of the microcapillary is stopped automatically (or manually). A pressure recording measured during the insertion of the pipette tip of the xylem pressure probe into a water-conducting vessel of a 1-year-old willow twig (excised in air and kept in water) is shown in Figure 6.2. Puncture of the vessel is evident as a decrease of pressure from >0.1 MPa to subatmospheric or negative values (downward arrow). After equilibration between the vessel and the probe, volume increments of different amplitude are injected into the vessel (upward arrows) by appropriate displacement of the metal rod (Figure 6.1). Independent of the magnitude of the change imposed, the xylem pressure recovers almost completely within a very short time. This response pattern could not have been given by a vapour-filled vessel. It also excludes the possibility that the tension originally present in the xylem vessel is irreversibly changed, on initial penetration of the vessel, by injection of water from the microcapillary (Benkert *et al.*, 1991).

Since pressures which are subatmospheric or negative can exist temporarily under some circumstances in other regions of a plant than in the xylem vessels (see below), it is advisable to use additional methods to confirm that it is the xylem that is being measured. One possibility is to identify the region of xylem vessels in a plant by comparison of the insertion depth of the microcapillary with an anatomical cross-section. The penetration depth can be measured easily during each experiment by using a commercially available digital read-out (digimatic indicator) which is attached to the automatically driven support of the micromanipulator. This system allows the penetration depth of the microcapillary to be read with an accuracy of about $10\,\mu m$ after calibration.

An alternative approach for identification of a penetrated vessel is to use dye injection. In this method the microcapillary is filled with a dye (eosin, fluorescein, Indian ink, etc.). When a xylem vessel under tension is penetrated, the dye is sucked in. Cross-sections taken near the penetration point clearly show that only the punctured vessel is stained (Figure 6.3). The use of Indian ink has the advantage that the dye is not transported by the transpiration stream (because of its colloidal properties), whereas fluorescein can be used for the determination of the flow velocity because of its low molecular weight (Benkert et al., 1991).

6.3 Negative pressures in liquids

The ability of a liquid to sustain a state of tension (negative pressure) arises from the enormous attractive forces acting between the molecules in a condensed phase. A liquid under a negative pressure cannot be considered as thermodynamically stable, since its free energy is diminished if it is allowed to break up by the process of cavitation (with the formation of water vapour and gas bubbles). However, such a state can be relatively stable (if the pressures are not too negative) so that the usual thermodynamic treatment of liquids can be applied to the region of negative pressures.

Let us consider an osmotic two-compartment system in which two phases, x and c, containing different concentrations of solutes (e.g. the xylem and the surrounding cells) are separated via a solute-impermeable membrane. The thermodynamic analysis of this system shows (Balling and Zimmermann, 1990; Mauro, 1965) that the following equation holds for the equilibrium (or a quasi-stationary) state:

$$P_x - \sigma_x \pi_x = P_c - \sigma_c \pi_c \qquad 6.1$$

where $P_{x,c}$ = hydrostatic pressures, $\pi_{x,c}$ = osmotic pressures and $\sigma_{x,c}$ = average reflection coefficients of the solutes in the two compartments. $\sigma\pi$ is called the effective osmotic pressure.

Equation 6.1 allows estimations from an osmotic standpoint of the magnitude of the negative pressure which can be developed in the xylem lumen of a plant.

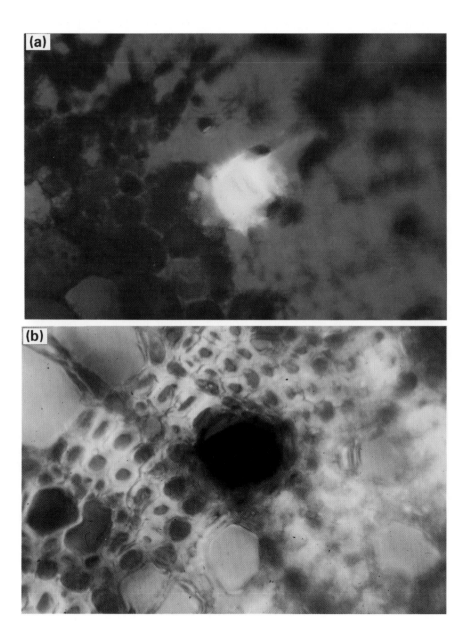

Figure 6.3. *Injection of dye into a large vessel using the xylem pressure probe. A twig of willow (a) or beech (b) was excised and the cut end placed in water. The capillary tip of the xylem pressure probe was loaded with a solution of fluorescein (a) or Indian ink (b) and inserted into a xylem vessel of the twig. After about 1 minute the transverse sections shown here were taken at 2 mm above the point of insertion. Magnification: (a) ×590; (b) ×825.*

Let us assume as a first approximation that the effective osmotic pressure in one of the compartments (e.g. the xylem) is negligible. According to Equation 6.1 the hydrostatic pressure, P_x, in the water phase is then positive for the case $P_c > \sigma_c \pi_c$ and negative for the opposite case, $P_c < \sigma_c \pi_c$. The first case is realized in the classical Pfeffer osmometer and the second in the Hepp-type osmometer. An osmotic two-compartment system can switch between the two states depending on the ratio of the hydrostatic pressure to the osmotic pressure in the solute phase.

6.3.1 *Hepp-type osmometer*

In the (modified) Hepp-type osmometer (Balling *et al.*, 1988; Hepp, 1936) the chamber containing the water phase is provided by a glass capillary of small diameter. The capillary, mounted vertically, is sealed at its lower end by a pressure transducer (which transforms a pressure signal into a proportional voltage signal) and at its upper end by a solute-impermeable membrane (backed up by a metal grid). This membrane separates the water phase in the capillary from the solute reservoir above (which is open to the atmosphere). In the Hepp-type bio-osmometer of Balling *et al.* (1988) the lower end of the capillary is sealed to the isolated vascular bundle of a plant (e.g. *Plantago*) and the pressure probe is inserted before or during the experiments into one of the vessels (Figure 6.4, inset). Addition of sufficient osmoticum (e.g. polyethyleneglycol; PEG 6000) to the reservoir leads to a drop of the pressure in the capillary below zero (Figure 6.4) provided that the osmotic pressure in the solute phase exceeds 1 atmosphere (about 40 mOsmol).

Using this set-up, absolute negative pressures of about -0.6 MPa can be attained for some time in the glass capillary and about -0.25 MPa (Balling *et al.*, 1988) when the vascular bundle is attached. Since a state of tension is metastable (see above) cavitation can set in, particularly if the negative pressure falls below about -0.2 MPa. When cavitation occurs (see, for example, Figure 6.5a) the pressure very quickly becomes that of the water vapour ($+0.0023$ MPa at $20\,^{\circ}$C). If there is a leak in the system, the pressure relaxes to atmospheric (Figure 6.5b).

Assuming that the xylem/parenchymal tissue cell system can be considered as a Hepp-type osmometer, it is evident that a tension can exist in the water of the vessels when the intracellular effective osmotic pressure (or the effective osmotic pressure difference between the cells and the apo-plastic space) is larger than the turgor pressure. Such a case can be envisaged for a transpiring plant when the cell turgor drops considerably because of water loss and the water supply from the roots is rate-limiting. The magnitude of the tension which will develop with progressive transpiration during the day depends on the effective osmotic pressure within the xylem vessels (see Equation 6.1 and below) and on the water exchange time between the xylem and the tissue compartment in relation to the transpiration rate.

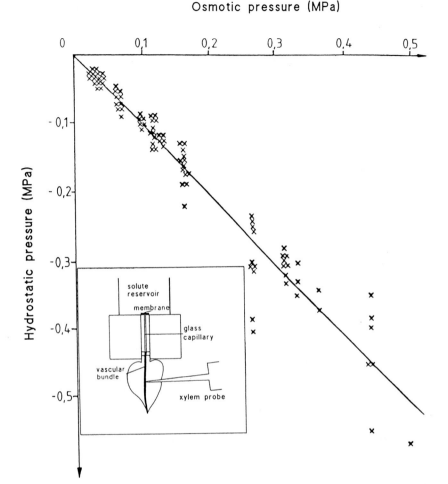

Figure 6.4. *Schematic diagram of the Hepp-type bio-osmometer (inset) and a plot of the relative negative pressure (tension) created in the xylem vessels of* Plantago *(and the glass capillary) versus the osmotic pressure of the polyethylene glycol solution. For a detailed description, see the text. (Redrawn from Balling* et al., *1988.)*

6.3.2 *Viscous solutions*

Equation 6.1 also predicts that under some circumstances tensions can (temporarily) be created in a solvent phase which is in contact with a viscous solution. Consider a water-filled microcapillary with a small outer tip diameter (say, 5–15 μm as used in the xylem pressure probe) which is inserted into a highly viscous solution (e.g. 2% DNA). The viscosity of the solution and the small tip diameter create a diffusion barrier for the long-chain polymers (by analogy to the PEG-impermeable membrane in the Hepp-type

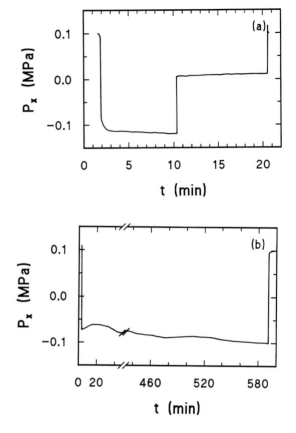

Figure 6.5. *Pressure recording in the case of cavitation (a) and of a leak (b), once a negative (or sub-atmospheric) pressure (P$_x$) had been monitored in a xylem vessel of* Nicotiana tabacum.

osmometer). Equilibrium between the two compartments can only be achieved—at least initially—by diffusion of water molecules from the micro-capillary into the viscous solution. This will induce a tension in the water phase. The duration of the tension state depends on the ratio of the chain length of the macromolecules to the inner tip diameter of the micro-capillary. As shown in Figure 6.6 (curve a) transient absolute negative press-ures of -0.007 MPa could easily be produced using DNA solutions. Cavita-tion was a rare event because of the viscosity of the solution.

Stable recordings of negative pressures in viscous solutions were only observed with seeds of *Lepidium sativum* suspended in water (Figure 6.6, curve b). (Note that the *Marangoni* streaming (see below) may also play a role in the development of subatmospheric and slightly negative pressures under these conditions if the outer surface of the microcapillary is covered by a thin film of water.)

These results may be of great relevance for the discussion of whether there exist regions (apart from the water-conducting vessels) in the plant tissue in which subatmospheric or even negative pressures can be developed—at least temporarily. Potential candidates are gland cells which secrete polysac-

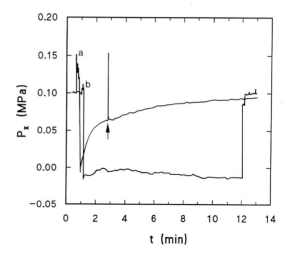

Figure 6.6. *Recordings of sub-atmospheric or negative pressures after insertion of the microcapillary of the xylem pressure probe into an aqueous 2% DNA solution (curve a; the arrow indicates the injection of a volume pulse), and into a suspension of seeds of* Lepidium sativum *(curve b).*

charides, i.e. highly viscous solutions. Indeed, recent measurements on *Sparmannia africana* (unpublished data) showed that in secretory cells, in lysigenous cavities (formed by the disintegration of cells) and/or in their environment the probe recorded absolute values below atmospheric (positive) and negative values down to −0.195 MPa. Such absolute negative pressure values could also be observed in small water droplets placed on the surface of the stem in those areas where glands were located about 100 μm deep in the tissue (data not shown). The occurrence of subatmospheric or negative pressures in the water droplet generally required pre-injury of the tissue surface by using a needle or the micropipette, presumably in order to enhance secretion. The negative pressures (created by the viscous aqueous secretory products) within the pressure probe usually relaxed with a similar time constant to that measured in the DNA solution experiment.

From these data we can conclude that accumulation of high molecular weight secretory products may also create subatmospheric or even negative pressures some distance away from the region of secretion (e.g. in the apoplastic spaces), if the water supply from the xylem vessels (located more deeply in the tissue) is not sufficient.

6.3.3 *Microcapillary structures*

Pressures that are slightly subatmospheric can be measured within the pressure probe if the microcapillary is exposed to air for about 30 minutes. The minimum absolute pressure (arising from the evaporation of water from the tip) was recorded to be +0.085 MPa (outer diameter about 10 μm). Since the possibility exists that an air-filled interspace is penetrated during insertion of the xylem pressure probe, only values below this threshold value were included in the analysis of the xylem pressure data.

From a theoretical standpoint, larger tensions than those predicted by Equation 6.1 can be expected at air–water interfaces in a capillary network, e.g. in the microporous structure of the plants, because this equation only

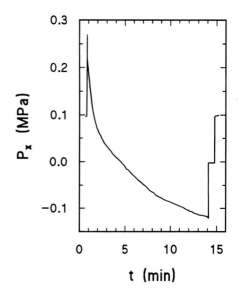

Figure 6.7. *Typical pressure recording from the capillary network of tissue paper made with the xylem pressure probe. For a detailed discussion, see the text.*

describes the osmotic equilibrium between the xylem and tissue cell compartment.

In order to test the hypothesis that significant negative pressures can be developed in the capillary network of the walls, we used tissue paper (or parcel string of natural material) as a model system (Zimmermann *et al.*, 1993a). Such material should provide a reasonable approximation to the plant cell wall. Results (Figure 6.7) verified that negative (absolute) capillary pressures (down to about −0.32 MPa) could be measured in such material, that the measured pressure was dependent on the water status and on the micropore structure of the material, and that recordings were stable for some time (depending on the seal between the microcapillary tip and the tissue paper or the string). The continuous increase in tension in such material (until cavitation occurs) can be explained by assuming that the material will be gradually drying out and its water content declining. However, such an effect would also be expected even if the material was not drying out, and its water content was constant. This is because the mass flow of water to the neighbouring tissue from the matrix directly under the tip is slow compared to the hydraulic capacity of the probe itself. According to the Hagen–Poiseuille equation, the flow velocity is proportional to the second power of the capillary radius and thus will be extremely low in such small capillaries. Furthermore, Marangoni convection, which is an additional force for water flow in the tissue, can occur (see below).

The considerations and the experimental evidence given above clearly demonstrate that water can be kept temporarily under tension and that several reasons can be envisaged for the origin of negative pressures in the water-conducting vessels or other regions of higher plants and tall trees.

6.4 Xylem pressure

The values recorded by the probe in the xylem of well-watered intact herbaceous plants (*Nicotiana rustica, N. tabacum, Zea mays* and *Lycopersicon esculentum*, etc.) were significantly less negative than those measured by the Scholander bomb technique (Balling and Zimmermann, 1990; Benkert *et al.*, 1991; and unpublished data). Even though negative values down to about -0.3 MPa could be recorded by use of the pressure probe, the average values were positive, but subatmospheric (e.g. $+0.03$ MPa for *N. rustica*).

Combined experiments, using the bomb and the probe on excised leaves of *N. rustica* showed that the xylem pressure did not respond to external gas pressure until a threshold value of the bomb pressure of about 0.25 MPa was reached. Above this value, collapse of the xylem tension and equilibration with atmospheric pressure was observed. The collapse of xylem tension coincided with the value at which water appeared at the cut end. When the pressure transducer was directly sealed to the cut end protruding through the seal of the bomb, the pressure response in the tissue was 1 : 1, provided that the cut end of the petiole had been pre-infiltrated with water under vacuum. Without pre-infiltration, a pressure response in the tissue could be recorded, but it was less than 1 : 1 up to a bomb overpressure of about 0.25 MPa. Beyond this threshold value the pressure in the tissue increased in a 1 : 1 ratio.

From these and other data (Balling and Zimmermann, 1990) it is evident that external application of gas pressure did not lead to a pressure equilibration between the xylem vessels and their surroundings, presumably because of the air within the leaves (Zimmermann *et al.*, 1993a).

The discrepancy between the readings of the Scholander bomb and the pressure probe becomes particularly obvious when willow twigs on the tree or excised twigs were examined. The tree had been subjected to prolonged drought (Zimmermann *et al.*, 1993a). When 1-year-old twigs cut in air 1 m above the ground were used, the average balancing bomb overpressure was about 1.1 MPa, which agreed well with published values in the literature (Zimmermann and Brown, 1980). The corresponding bomb pressure for twigs which had been taken at 5.8 m height was slightly greater than 1.2 MPa.

Renner (1925) and other authors of his time demonstrated that rapid drying of the tissue at the cut end of twigs excised in air can have a tremendous effect on the magnitude of the estimated negative pressure values. In view of this, some twigs were cut under water prior to pressure bomb measurements. The bomb overpressure required for water exudation from these twigs was reduced to 0.2–0.5 MPa. A similar reduction in bomb overpressure was also observed when leaves or twigs of sugarcane, birch, poplar or coffee trees were cut under water instead of in air, but not when excised leaves of well-watered tobacco plants were used.

In contrast to the Scholander bomb results, positive, but below-atmospheric pressures were recorded *in situ* using the xylem pressure probe in willow twigs on the tree. The average absolute value was $+0.063$ MPa and independent of the height (1 m or 5.8 m) at which the measurements were performed. This value agreed well with that measured by Renner (1925) for willow twigs during early spring before sprouting of the leaves, using the vacuum pump technique. Measurements on excised twigs yielded identical results (Figure 6.2) independent of whether the twigs had been cut in air or under water.

Recent *in situ* measurements with the xylem pressure probe in the xylem of leaves of beech and tropical trees (*Anacardium excelsum*, Panama and *Argyrodendron peralatum*, Queensland, Australia) during the rainy and dry seasons yielded results similar to those from willow twigs. The pressure in the lumen of most of the large vessels was, on average, in the positive, sub-atmospheric range (+0.02 to +0.06 MPa) and was independent of height up to 35 m (Zimmermann *et al.*, 1993a,b). In contrast, the 'xylem pressure' as measured by the Scholander bomb technique was of the order of 2–3 MPa for leafy twigs cut from trees in air during the dry season. High-resolution NMR imaging showed that the leaves of such twigs contained large amounts of air-filled spaces (Zimmermann *et al.*, 1993a). The bomb readings were, however, considerably lower (0.3 MPa or less) when well-hydrated leafy twigs of a transpiring tropical tree (*A. excelsum*) were taken during the rainy season. Interestingly, in the latter case a vertical 'tension' gradient up to a height of 35 m could not be detected by the Scholander bomb technique. An absolute tension of 0.3 MPa is not sufficient to overcome gravitational and frictional resistances in order to supply the transpiring leaves at a height of 35 m with water from the roots. This and the absence of a vertical 'tension' gradient are not consistent with the cohesion theory. Similarly, Tobiessen *et al.* (1971) concluded, on the basis of their measurements in a giant *Sequoia* tree, that the pressure bomb technique and/or the cohesion theory must be called into question.

These, as well as other, unpublished results (see also Zimmermann *et al.*, 1993a) show once again that the large amount of air in twigs of trees subjected to drought apparently prevents equilibration between external gas pressure and the pressure of the xylem vessels. This, together with the drying process at the cut end, requires excess bomb pressure in order to yield liquid at the cut end of the excised leaf, with the consequence that the tension in the xylem vessels is considerably overestimated.

6.5 Transpiration-induced changes in xylem pressure

Xylem pressure probe measurements in some rapidly transpiring plants (e.g. sunflower and sugarcane) and twigs on transpiring tall trees (beech, tropical trees) have shown that pressure in the xylem vessels follows a characteristic pattern. At day-break regions of subatmospheric pressures could not be found despite many insertions of the pressure probe into various parts of plants and twigs. This indicated that the pressure in the vessels was at atmospheric or higher. As the day progressed, temperature increased and humidity decreased. This caused an increase in the rate of transpiration, which was reflected in lower xylem pressures. The development of tension occurred discontinuously, that is the pressure in the xylem often recovered slightly between successive pressure drops (Figure 6.8). About 3–4 hours after daybreak the pressures in the vessels reached zero or even became negative. During outdoor measurements cavitation usually occurred when slightly negative pressures were established, because the wind speed increased perceptibly around this time (Figure 6.8). Under laboratory conditions measurements on plants exposed to an air-stream were much more stable. As in the field, the average pressure in the vessels continued to decrease over time. The increase in tension was accompanied with fluctuations of the pressure, sometimes even between the positive and negative ranges (Zimmermann *et al.*, 1993b). Under these conditions negative pressures of about −0.2 MPa could frequently be recorded before the measurement ended in cavitation followed by a leak.

Insertion of the probe into vessels of the same plants or twigs on the tree around noon revealed pressure values which were in the positive, subatmospheric range. This was evidence that cavitation occurred more frequently at this time of the day when the absolute tension in the vessels exceeded about 0.03 MPa.

In contrast, under constant environmental conditions in the laboratory such diurnal patterns in the development of tension in the vessels were not observed. When well-watered plants (such as tobacco) were used, slightly

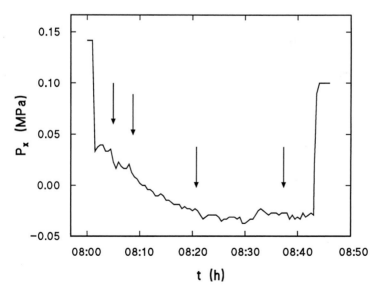

Figure 6.8. *Pattern of xylem pressure recorded* in situ *in a xylem vessel of a well-hydrated transpiring leaf of a tropical tree* (A. excelsum) *at 35 m height. Arrows indicate the lines of occurrence of a short breeze. For a detailed discussion, see the text.* (*Redrawn from Zimmermann* et al., *1993b.*)

positive or negative absolute pressures could be recorded over the whole day (starting about 2 hours after illumination).

6.6 Probe measurements in microporous structures of plants

If a large vessel was penetrated, the final pressure was established within a very short time (Figure 6.2). In cases in which the probe was placed into xylem walls or in the region of very small microcapillaries (less than about 1 μm) a slow response was observed, as expected from the tissue paper and string experiments (Figure 6.7). In the tissues of well-hydrated plants, the final pressure recorded from the slow-kinetic response within the vascular tissue may also reflect xylem tension. However, after a severe drought or when the humidity is low and/or the temperature high, the microcapillary regions may contain very little water. Under these conditions, insertion of the probe may provide a significant additional source of water for the microcapillary region, so that a gradually increasing tension of considerable magnitude (about 0.2 MPa) is created in the probe until cavitation occurs. Consistent with the model experiments (Figure 6.7), after cavitation and leak, negative pressures in these microcapillary regions were often re-established after the injection of water pulses from the probe into this region (Figure 6.9).

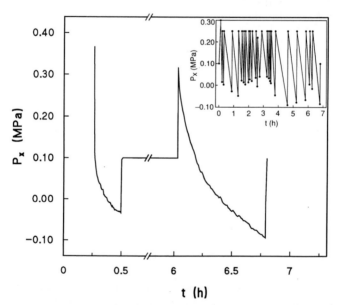

Figure 6.9. Recording of a slow type of response in a leaf of Argyrodendron peralatum. The xylem pressure probe was advanced into the microcapillary region (close to the midrib of the excised and dehydrated leaf). The tension increased continuously until cavitation, followed by a leak, occurred. Repeated injection of volume pulses (as shown schematically in the inset) induced slow transient returns of the pressure to negative values, followed by a leak giving recovery to atmospheric pressure. Similar recordings were obtained under in situ conditions on twigs and leaves of various species.

6.7 Flow velocity measurements

Evidence that water in the xylem of twigs was only under subatmospheric (positive) pressure was also obtained from flow velocity measurements. As with tobacco plants (Benkert *et al.*, 1991) flow velocity was determined in excised twigs of willow by injection of fluorescein into the vessel via the xylem pressure probe, and by subsequent fluorescence microscopy of the cross-sections made at regular intervals above the injection point (Figure 6.3a). The most distant cross-section in which fluorescein could be detected gave the distance travelled by the dye within 1 minute, and thus the velocity.

Under normal transpiration rates (about 0.5–$0.6 \, \text{ml h}^{-1}$) the average velocity was determined to be $2 \, \text{mm s}^{-1}$ and $6 \, \text{mm s}^{-1}$ when the willow twigs were cut in air and under water, respectively. The flow velocities were of the same order as those measured by the heat pulse technique (Zimmermann and Brown, 1980).

In one set of experiments, the leaves of the willow twig were excised under water, sealed and the distal end connected to a vacuum line. Using the vacuum pump, the xylem pressure could be reduced to $+0.04 \, \text{MPa}$. Consistent with the increase in xylem tension observed in twigs after replacement of the leaves, the flow velocity increased on average to $7 \, \text{mm s}^{-1}$ and $10 \, \text{mm s}^{-1}$, after the leaves were severed in air or water, respectively.

In the twig/vacuum system, unlike the leafy twig, the flow velocity can be calculated from the Hagen–Poiseuille law, if the vessels are assumed to be cylindrical:

$$v = r^2 \Delta P / 8 \eta L \qquad\qquad 6.2$$

where $r =$ the average radius of the large vessels (in which the fluorescein was transported, about $20 \, \mu\text{m}$), $L =$ length of the twig (about $200 \, \text{mm}$), $\Delta P =$ pressure difference between the xylem vessels $(+0.04 \, \text{MPa})$ and vacuum, and $\eta =$ the viscosity of water $(0.001 \, \text{g mm}^{-1} \text{s}^{-1})$. The calculations yielded a value of about $10 \, \text{mm s}^{-1}$, which agrees well with that determined experimentally. Since the flow velocity in the leafy twig (excised under water) was only about half of that in the twig/vacuum system, the pressure difference along the former twigs must have been smaller by a similar proportion. It can therefore be concluded that the average pressure at the sites where the xylem ends in the leaves must be about $+0.04 \, \text{MPa}$.

It is interesting to note that an introduction of a restriction at the base of the willow twig (by clamping, as done by Renner) did not lead to a significant change in the flow velocity. This was partly at variance to the findings on intact, well-hydrated tobacco plants. Experiments under various conditions (vacuum pump attached to the decapitated stem with or without leaves and/ or restrictions) showed that the vacuum pump always created higher flow velocities than the leaves of the intact plants (Figure 6.10). This was similar

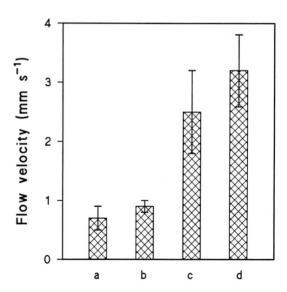

Figure 6.10. *Flow velocity of tobacco plant xylem measured by the fluorescein dye injection method. (a) Flow velocity measured in an intact stem having seven leaves on average. (b) Flow velocity determined after removal of the leaves and sealing of the cut ends. (c) Flow velocity measured after attachment of a vacuum line to the decapitated end of the stem (with leaves); and (d) after the further introduction of a restriction. Each bar represents 3–5 measurements.*

to the results with willow twigs. In the absence of a clamp, the average velocity increased from $0.8 \, \text{mm s}^{-1}$ in the control plants to about $2.5 \, \text{mm s}^{-1}$ when the vacuum pump was attached to the top of the stem. However, in the presence of a restriction, a further increase of the flow velocity to about $3 \, \text{mm s}^{-1}$ was observed.

These data demonstrate that the leaves are not necessarily more effective than a vacuum pump at inducing xylem tension. In his interpretation, Renner (1925) assumed that the effective cross-sectional area for water flow remained unchanged when the twig was clamped and the leaves were replaced by a single vacuum line. He calculated the unknown tension values in the xylem from the ratio of the water volume flow through a leafy twig to that through the twig/vacuum system, by assuming that the vacuum line creates an absolute underpressure of $+0.01 \, \text{MPa}$ in the vessels (this value is apparently too low, see above). According to the Hagen–Poiseuille equation, the velocity (but not the rate of the volume uptake by the twig) is directly proportional to the driving force. Therefore, total flow in the twig/vacuum system could be lower than that in the intact twig despite an increase in the flow velocity, if the number of conducting xylem vessels decreases due to the experimental manipulation.

From this work it is clear that Renner's results and interpretations should not be taken as unequivocal evidence of the existence of large tensions in the xylem vessels of higher plants and trees. Despite this, we should point out that the tension values (minimum $-0.9 \, \text{MPa}$) calculated in the past from the vacuum/twig experiment were physically much more reasonable than those recorded by the Scholander bomb technique (particularly if the systematic error in Renner's calculation, due to the use of a tension value which is too low, is taken into account).

6.8 Turgor pressure profiles in the parenchyma tissue

Equation 6.1 demonstrates that the xylem pressure must be linked to the turgor pressure and to the osmotic pressure of the parenchymal tissue. It is well
known from work on roots of halo- and glycophytes (using the cell turgor
pressure probe) that standing turgor and osmotic pressure gradients can
exist between cells of adjacent layers (Rygol and Zimmermann, 1990; Rygol
et al., 1993; Zimmermann et al., 1992). The turgor and osmotic pressure profiles collapsed to a uniform (intermediate) value when the roots were excised
or when the relative humidity of the leaf environment was adjusted to nearly
100%. The existence of radial turgor pressure gradients was also confirmed by
Meshcheryakov et al. (1992) for hypocotyls. These data throw into question all
the work that has been performed so far on excised roots.

Radial turgor pressure gradients also exist in the outer cell layers of twigs,
such as willow twigs. In most experiments the turgor pressure increased progressively from the outermost to the third subjacent parenchymal layer (Figure
6.11). The turgor pressure also collapsed in twigs deprived of their leaves,
with all cells assuming an intermediate value of about 0.1 MPa.

Radial turgor pressure gradients through the parenchymal tissue could also
be recorded *in situ* in beech twigs at heights of up to 17 m. In a few cases no
turgor pressure differences were found between the various layers of cortical
parenchyma. However, turgor pressure in the epidermal cells was always significantly lower than in the cortical cells.

Slight differences between measurements on the northern and southern
sides of the tree were observed (Figure 6.12). There was no consistent relationship between the magnitude of the turgor pressure gradient and the environmental conditions, or the height on the tree. Data from all heights were
therefore averaged for Figure 6.12. Pressure gradients were also found during
measurements in the laboratory, on twigs excised under water.

The demonstration in the willow and beech tissues of stable turgor pressure
profiles casts further doubt on the reliability of indirect methods for the estimation of the water-relation parameters of whole tissues. For example, from
the relationship between chamber pressure and exudate volume, a value for
'whole-tissue turgor pressure' can be estimated with the Scholander bomb
(Scholander et al., 1965). However, where the tissue exhibits marked and
stable turgor pressure gradients, such average values of tissue turgor will be
largely meaningless.

In addition, the rapid redistribution of solutes after excision of the twigs
also suggests that obtaining a representative sample of xylem sap may involve
hitherto unrecognized technical difficulties. Therefore, it is not unlikely that
the solute content of the xylem sap of the large vessels is much higher than
is generally supposed (see Equation 6.1 and below).

6.9 Conclusions

In the light of the results obtained by the pressure probe technique, Passioura

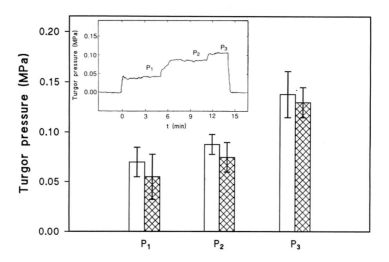

Figure 6.11. *Radial turgor pressure gradients in parenchyma cells of willow twigs. The twigs were excised under water and their cut ends were kept in water for 2h (open columns) or 5h (hatched columns). P_1 refers to the outermost (parenchymal) cell layer while P_2 and P_3 are successively deeper layers of cells. Bars show means ($\pm SD$) of four separate experiments. The inset shows a typical individual recording during stepwise insertion of the pressure probe.*

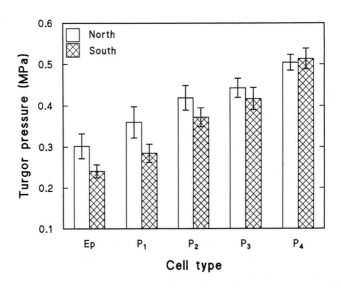

Figure 6.12. *Radial turgor pressure gradients in the epidermal (Ep) and parenchymal (P) cells of beech twigs. These measurements were made in situ, on twigs at various heights up a beech tree, by stepwise insertion of the cell pressure probe into successively deeper cell layers (as in Figure 6.11). Data from the various heights showed no consistent differences and they have therefore been combined (see text). Open columns show data from the northern side of the tree and hatched columns from the southern. Each bar is a mean ($\pm SD$) for 30–120 cells.*

(1991; see also Smith, 1991) asked very recently if plant physiologists were faced with an impasse in water relations. The answer is 'certainly not' if we accept that the conclusions drawn from bomb experiments (and the other indirect methods, see Zimmermann *et al.*, 1993a) are misleading.

The data obtained on tall trees by using the xylem pressure probe are apparently in contrast to the cohesion theory, that water is pulled from the roots to the foliage of a tree through continuous 'vulnerable pipelines' (Milburn, 1979).

We believe that nature has developed different strategies to supply tall trees with water necessary for the compensation of transpirational water loss:

(1) The microcapillary system is an essential reservoir for the water demand of a plant (Sachs, 1887). The experiments in which water volume increments were injected into the microcapillary region by use of the probe have shown clearly that the water transport properties of dried microporous elements can be easily regenerated provided that water is available. Capillary-induced water uptake is, however, a slow process and may not meet the immediate water demand of a transpiring plant during the day (Tyree *et al.*, 1991).

(2) A second potential water reservoir is the turgescent living cells which are hydraulically connected to the microcapillary regions and to the large vessels. The long-term *in situ* xylem pressure measurements (Figure 6.8) have provided convincing evidence that the development of tension in the large vessels is delayed by water flow from the surrounding cells (as depicted by Equation 6.1).

(3) The main pathway for water transport is the large xylem vessels which will function as conductance amplifiers. The average tension measured in the vessels of the twigs and leaves up to a height of 35 m by using the probe is not consistent with the assumption of a tension gradient in a standing continuous vertical water column. We, therefore, have to postulate (in agreement with similar ideas of Sachs, 1887) that the large vessels are segmented in axial compartments of modest length bounded by (continuous) solute-reflecting barriers both in axial and radial directions. In other words, the large axial tension gradients postulated by the cohesion theory would be replaced by axial solute gradients within the large vessels. Axial barriers in the xylem have not been reported, but they may be associated with tyloses, resins, polysaccharides or other blockages (Morse, 1990; Sachs, 1887). In addition, concentric solute-reflecting sheaths have recently been found around the vascular bundles of maize (Canny, 1990) and sugarcane (Jacobsen *et al.*, 1992; Welbaum *et al.*, 1992). The hypothesis of vertically segmented xylem compartments bounded by solute-reflecting barriers requires that the effective osmotic pressure in the vessels should be much (at least locally, see Zimmermann *et al.*, 1993a) higher than is generally supposed (for remarkable exceptions, see Smith and Lüttge, 1985). This is not unlikely in view of the technical problems involved in the collection of xylem sap and of the turgor pressure results reported here (see above).

If compartmentation of the xylem exists, and water is raised in smaller vertical steps than postulated by the cohesion theory, water flow to the foliage can also be achieved by Marangoni convection. Marangoni convection arises if a gradient in interfacial tension exists along a fluid–fluid or fluid–gas interface (Langbein and Heide, 1986; Young et al., 1959). In contrast to free convection, Marangoni convection is gravity-independent and can, therefore, also operate under microgravity.

Xylem probe measurements and other experiments (Laschimke, 1989; Zimmermann and Brown, 1980) have provided strong evidence for the presence of tiny air bubbles adhering to the inner walls of the xylem elements. This would provide the fluid–gas interface required for Marangoni streaming. Tension gradients along such an interface can be created easily by local solute gradients in each vessel, by temperature gradients or by surface-active compounds (such as proteins, see Biles and Abeles, 1991). The theory of Marangoni streaming shows (Zimmermann et al., 1993b) that the flow velocity increases with increasing capillary diameter. For vessels of 10 cm length (typical for diffuse-porous species, see Zimmermann and Jeje, 1981) and diameters of 20–100 μm the flow velocity is in the order of $0.2 \, \text{mm} \, \text{s}^{-1}$ and $1 \, \text{mm} \, \text{s}^{-1}$, respectively, if the tension gradient is assumed to be $10^{-3} \, \text{N} \, \text{m}^{-1}$. These values are in the range measured both by the heat pulse and xylem pressure probe technique.

At first sight, the assumption of Marangoni streaming along a water–air interface in the large vessels of the xylem would not be consistent with the observation that negative pressures could be developed in the morning hours. However, in model experiments it can easily be shown that adhered gas bubbles are swept along with water flow. Cavitation at relatively low negative pressures and the formation of tiny gas bubbles would then be absolutely necessary conditions to re-create the Marangoni streaming for replacing the water loss. According to this scenario, the occurrence of cavitation may not necessarily always lead to a catastrophic dysfunction of the xylem (Tyree and Sperry, 1989), but may represent a survival strategy for replacement of transpirational water loss.

Acknowledgements

We are very grateful to Dr W.M. Arnold and Dr M. Malone for critical discussion of the experiments. This work was supported by grants of the Deutsche Forschungsgemeinschaft (SFB 251) and by the NMR-Graduiertenkolleg (DFG, Prof. Haase) to UZ.

References

Balling, A. and Zimmermann, U. (1990) Comparative measurements of the xylem pressure of *Nicotiana* plants by means of the pressure bomb and pressure probe. *Planta*, **182**, 325–338.

Balling, A., Zimmermann, U., Büchner, K.-H. and Lange, O.L. (1988) Direct measurement of negative pressure in artificial-biological systems. *Naturwissenschaften*, **75**, 409–411.

Benkert, R., Balling, A. and Zimmermann, U. (1991) Direct measurements of the pressure and flow in the xylem vessels of *Nicotiana tabacum* and their dependence on flow resistance and transpiration rate. *Bot. Acta*, **104**, 423–432.

Biles, C.L. and Abeles, F.B. (1991) Xylem sap proteins. *Plant Physiol.*, **96**, 597–601.

Canny, M.J. (1990) What becomes of the transpiration stream? *New Phytol.*, **114**, 341–368.

Dixon, H.H. and Joly, J. (1894) On the ascent of sap. *Phil. Trans. R.˙Soc. Lond. B*, **186**, 563–576.

Hepp, O. (1936) Ein neues Onkometer zur Bestimmung des kolloid-osmotischen Druckes mit gesteigerter Meßgenauigkeit und vereinfachter Handhabung. *Z. Ges. Exp. Med.*, **99**, 709–719.

Jacobsen, K.R., Fisher, D.G., Maretzki, A. and Moore, P.H. (1992) Anatomy of sugarcane stem in relation to phloem unloading and sucrose storage. *Bot. Acta*, **105**, 70–80.

Langbein, D. and Heide, W. (1986) Study of convective mechanisms under microgravity conditions. *Adv. Space Res.*, **6**, 5–17.

Lange, O.L., Schulze, E.-D., Evenari, M., Kappen, L. and Buschbom, U. (1975) The temperature-related photosynthetic capacity of plants under desert conditions. *Oecologia*, **18**, 45–53.

Laschimke, R. (1989) Investigation of the wetting behaviour of natural lignin – a contribution to the cohesion theory of water transport in plants. *Thermochimica Acta*, **151**, 35–56.

Mauro, A. (1965) Osmotic flow in a rigid porous membrane. *Science*, **149**, 867–869.

Meshcheryakov, A., Steudle, E. and Komor, E. (1992) Gradients of turgor, osmotic pressure and water potential in the cortex of the hypocotyl of growing *Ricinus* seedlings. *Plant Physiol.*, **98**, 840–852.

Milburn, J.A. (1979) *Water Flow in Plants*. Longman Press, London.

Morse, S.R. (1990) Water balance in *Hemizonia luzulifolia*: the role of extracellular polysaccharides. *Plant Cell Environ.*, **13**, 39–48.

Oparka, K.J., Murphy, R., Derrick, P.M., Prior, D.A.M. and Smith, J.A.C. (1991) Modification of the pressure probe technique permits controlled intracellular microinjection of fluorescent probes. *J. Cell Sci.*, **98**, 539–544.

Passioura, J.B. (1991) An impasse in plant water relations? *Bot. Acta*, **104**, 405–411.

Passioura, J.B. and Munns, R. (1984) Hydraulic resistance of plants. II. Effects of rooting medium, and time of day, in barley and lupin. *Aust. J. Plant Physiol.*, **11**, 341–350.

Renner, O. (1925) Zum Nachweis negativer Drucke im Gefäßwasser bewurzelter Holzgewächse. *Flora*, **119**, 402–408.

Rygol, J. and Zimmermann, U. (1990) Radial and axial turgor pressure measurements in individual root cells of *Mesembryanthemum crystallinum* grown under various saline conditions. *Plant Cell Environ.*, **13**, 15–26.

Rygol, J., Pritchard, J., Zhu, J.J., Tomos, D. and Zimmermann, U. (1993) Transpiration induces radial turgor pressure gradients in wheat and maize roots. *Plant Physiol.*, in press.

Sachs, J. (1887) *Vorlesungen über Pflanzen-Physiologie*. Verlag Wilhelm Engelmann, Leipzig.

Scholander, P.F., Hammel, H.T., Bradstreet, E.D. and Hemmingsen, E.A. (1965) Sap pressure in vascular plants. *Science*, **148**, 339–346.

Smith, J.A.C. (1991) Ion transport and the transpiration stream. *Bot. Acta*, **104**, 416–421.

Smith, J.A.C. and Lüttge, U. (1985) Day–night changes in leaf water relations associated with the rhythm of crassulacean acid metabolism in *Kalanchoe daigremontiana*. *Planta*, **163**, 272–282.

Tobiessen, P., Rundel, P.W. and Stecker, R.E. (1971) Water potential gradient in a tall *Sequoiadendron*. *Plant Physiol.*, **48**, 303–304.

Tomos, A.D. (1988) Cellular water relations of plants. In: *Water Science Reviews*, Vol. 3 (ed. F. Franks). Cambridge University Press, Cambridge, pp. 186–277.

Turner, N.C., Spurway, R.A. and Schulze, E.-D. (1984) Comparison of water potentials measured by *in situ* psychrometry and pressure chamber in morphologically different species. *Plant Physiol.*, **74**, 316–319.

Tyree, M.T. and Sperry, J.S. (1989) Vulnerability of xylem to cavitation and embolism. *Ann. Rev. Plant Physiol. Plant. Mol. Biol.*, **40**, 19–38.

Tyree, M.T., Snyderman, D.A., Wilmot, T.R. and Machado, J.-L. (1991) Water relations and hydraulic architecture of a tropical tree (*Schefflera morototoni*). *Plant Physiol.*, **96**, 1105–1113.

Welbaum, G.E., Meinzer, F.C., Grayson, R.L. and Thornham, K.T. (1992) Evidence for and consequences of a barrier to solute diffusion between the apoplast and vascular bundles in sugarcane stalk tissue. *Aust. J. Plant. Physiol.*, **19**, 611–624.

Young, N.O., Goldstein, J.S. and Block, M.J. (1959) The motion of bubbles in a vertical temperature gradient. *J. Fluid Mech.*, **6**, 350–356.

Zimmermann, M.H. and Brown, C.L. (1980) *Trees – Structure and Function.* Springer-Verlag, Berlin.

Zimmermann, M.H. and Jeje, A.A. (1981) Vessel-length distribution in stems of some American woody plants. *Can. J. Bot.*, **59**, 1882–1892.

Zimmermann, U. and Steudle, E. (1978) Physical aspects of water relations of plant cells. *Adv. Bot. Res.*, **6,** 45–117.

Zimmermann, U., Räde, H. and Steudle, E. (1969) Kontinuierliche Druckmessung in Pflanzenzellen. *Naturwissenschaften*, **56**, 634.

Zimmermann, U., Rygol, J., Balling, A., Klöck, G., Metzler, A. and Haase, A. (1992) Radial turgor and osmotic pressure profiles in intact and excised roots of *Aster tripolium*: Pressure probe measurements and NMR-Imaging analysis. *Plant Physiol.*, **99**, 186–196.

Zimmermann, U., Benkert, R., Malone, M., Schneider, H., Rygol, J., Kuchenbrod, E., Haase, A. and Zimmermann, G. (1993a) Water ascent in trees: Direct *in situ* pressure probe measurements in twigs of willow and beech. *Plant Physiol.*, submitted.

Zimmermann, U., Benkert, R., Schneider, H., Meinzer, F., Goldstein, G., Kuchenbrod, E. and Haase, A. (1993b) Direct *in situ* measurements of xylem pressure in a tall tropical tree during the rainy season. *Plant Cell Environ.*, submitted.

Consequences of xylem cavitation for plant water deficits

John Grace

7.1 Introduction

The cohesion theory of water transport in plants has always been controversial. It seems to have been first expounded by Bohm, who demonstrated the power of a conifer shoot to lift columns of mercury beyond the normal barometric height (Böhm, 1893). The theory was elaborated by Askenasy (Askenasy, 1895), although in most texts on plant water relations it is attributed to Dixon and Joly (1896).

Although the theory is now widely accepted, there remain problems of understanding how the water transport system in plants can continue to operate, apparently reliably, after being subject to a wide range of physical conditions, including extreme tensions and freezing. According to the theory, water is drawn through the myriad of interconnected capillary tubes ('conduits') that connect the leaves to the soil. The driving force is not the pumping action of the roots, but the sucking action of leaves after they have lost water to the atmosphere. This sucking action is more powerful than the force a mechanical pump could ever produce, as it arises from the chemical potential of water in the presence of solutes. Water has a high tensile strength, so when the 'chains' of water molecules are drawn towards the leaves they hold together against the frictional drag imposed by the walls of the conduits. This attraction between water molecules is crucial to the safe working of the transport system. Actually, according to its phase diagram, liquid water might be expected to change state to vapour at low pressure (Franks, 1983). In the cohesion theory the water inside the xylem is not just at a low pressure, but at a negative pressure, or tension. Thus it may be expected to boil. Remarkably, liquid water is being transported in a metastable state.

The notion that water in the stems of tall plants is under tension is strongly supported by modern measurements, notably those made with the pressure chamber but also those made with thermocouple psychrometers attached to the stem (Dixon and Tyree, 1984). Typical tensions, for coniferous trees at midday, are 0.5–2 MPa (Whitehead and Jarvis, 1981).

The tensile strength of pure, de-aerated, water contained in glass tubes has been shown to be as high as 30 MPa (Briggs, 1950). Although this is much less than the theoretical values, calculated from the strength of the hydrogen bonds that attract water molecules to each other, it is nevertheless sufficient to keep water intact in plants, where maximal tensions rarely exceed 4 MPa. Thus it would seem at first sight that all is well. However, much recent work has demonstrated that breakages of water columns occur at much lower tensions than those predicted from the classical work using pure water. This happens especially if there is much dissolved gas in the water, or some small bubbles adhering to the walls of the tube or lodged inside crevices. In the latter case, the bubble expands when the tension reaches a critical point. This critical tension does not depend on the tensile strength of the liquid, but rather on the radius of the bubble. In the vascular system of plants, there is probably a third type of breakage. Air may be sucked into one of the tubes from an adjacent tube which has become air filled, as the tubes are interconnected by small pores. Bubbles of small radius require high tensions to make them grow, and it is this growth of the bubble which gives rise to the breakage of the column.

In this chapter, I will comment on the extent to which breakage occurs in vascular plants and, following breakage, the ease with which the refilling may follow. Finally, I will speculate on the importance of this process in the life of the plant.

7.2. Demonstration that breakage of water occurs

7.2.1 Fluctuations in xylem water content

Circumstantial evidence that water columns do not remain intact comes from quite early measurements of the water content in the xylem of trees over the course of the season. Even now, this work is not well known and deserves attention. Some of the most thorough measurements were performed on Canadian forest trees, motivated apparently by the need of the forest industry to understand the seasonal changes in sinkage of logs (Gibbs, 1930, 1958). It was shown that the wood of young trees of *Betula populifolia* and other angiosperm trees underwent an annual cycle of water content, varying from 100% of dry weight in the early spring, to as little as 60% in the late summer (Figure 7.1). Gymnosperm trees also underwent seasonal cycles, though of a different kind. The interpretation of such fluctuations is not straightforward. Small fluctuations in the water content of xylem are not necessarily indicative of breakage of water columns, because xylem does contain some living cells,

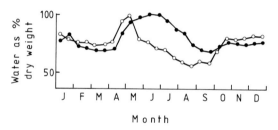

Figure 7.1. *Seasonal changes in the water content of* Betula populifolia *according to Gibbs (1958). Tops and middles of the stem are indicated by closed and open circles, respectively.*

the ray parenchyma. These cells presumably undergo changes in water content and water potential in much the same way as any other live cell in the plant. The percentage of total volume comprising ray parenchyma is, however, small. Measurements of *Pinus sylvestris* and *Malus sylvestris* (Dixon, 1914; Doley, 1974) suggest a typical range of 2–10%. By comparison with other living cells, it is unlikely that these small living volumes undergo changes in water content exceeding 25%. Thus, it is inconceivable that changes in the water content of the parenchyma cells over their physiological range could account for the massive seasonal changes in bulk water content that have been observed. Another variable component of water content is the 'capillary water' contained inside cracks between conduits, as well as already cavitated conduits, as discussed by Zimmermann (1983), and we will return to this later.

The usual conclusion to be drawn from Gibbs' work, as well as similar more modern studies (Wang *et al.*, 1992) is that the water content of forest trees is the result of an annual cycle of emptying and refilling. More surprising have been similar observations that suggest appreciable fluctuations in water contents of xylem over periods of weeks (Chalk and Bigg, 1956; Roberts, 1976; Waring and Running, 1978; Waring *et al.*, 1979) or even hours (Brough *et al.*, 1986). If fluctuations in the water content of xylem are to be interpreted as variations in the fraction of capillary tubes that are filled as opposed to empty, then breakage and repair of water columns must be commonplace.

The concept of ready fracture of stretched water columns in transpiring trees at relatively small negative pressures was presented by Greenidge (1955). Greenidge made extensive use of the injection of dye into trees to trace the pattern of transport in the trunk. When stressed by drought, the dye patterns were altered. He was able to infer a high percentage of gas-filled conducting tissue in trees which had been root-pruned to stress the plant.

7.2.2 *Acoustic emissions*

The subject was advanced considerably in the 1960s by Milburn's demonstration of acoustic emissions (AEs) from plants. The experimental set-up was extremely simple, comprising a record-player pick-up arm attached by a piece of wire to the petiole of a detached leaf of *Ricinus*, the castor-oil plant, and connected to an amplifier (Milburn, 1966; Milburn and Johnson, 1966). The whole set-up was in a sound-proof booth to avoid amplification of extraneous sound in the environment. The important observations were: as the leaf dehydrated, audible clicks were made; if transpiration was stopped experimentally the clicks stopped also. Milburn interpreted the clicks as breakages in water columns, pointing out that the use of sound to detect breakage of water columns was not new. It had been used in the original experiments to determine the tensile strength of water (Berthelot, 1850).

Later it was discovered that acoustic events could be detected at ultrasonic frequency using commercial sensors designed for detecting cracking in engineering structures (Tyree and Dixon, 1983). These sensors detect in the range 0.1–1 MHz. Ultrasonic frequencies are more convenient than audio-frequency because they are not propagated very far in air, and so the problems of interference from extraneous noise is negligible. Thus, the detector need not be housed in a sound-proof booth, and may be taken into the natural environment as field-portable equipment. When ultrasound sensors are attached to the stem of a water-stressed plant they typically produce far more events than have been reported from sensors operating in the audio-frequency range (Ritman and Milburn, 1988).

The process whereby water 'breaks' is termed cavitation. It can be observed on a large scale in the wake of a ship's propellor, as a result of the sudden tension immediately behind the blades. It is also observed when attempting to suck water by a pump through a pipe to a height exceeding 10 m, a feat which is achieved by tall plants but not by water engineers. Cavitation was observed experimentally by Bertholet (1850) by cooling, and therefore contracting, degassed water that completely filled a thick-walled glass tube. In plant tissue it has been demonstrated directly by microscopy in dehydrating spores and fern sporangia (Milburn, 1970; Renner, 1915) and in suitably prepared vascular tissue of higher plants (Crafts, 1939; Milburn and McLaughlin, 1974). Non-biologists reading this should realize that the vascular system consists of dead, stiff-walled capillary tubes that are interconnected by pores. These tubes are formed from the walls of cells that were alive during their expansion phase, but began to serve their function as water conduits only after death. Whether cavitation can occur also in the living cells of higher plants is a matter of some controversy (Oertli, 1993). It occurs in the blood of divers when they surface from depth too rapidly, but for rather different reasons.

We can imagine several ways by which the water columns might become disrupted under tension:

(1) a cavity in the water may arise *de novo* (as it presumably does in a well-prepared Bertholet tube);
(2) a bubble may form from the growth of an existing microscopic bubble on the wall of the tube; and
(3) air may be drawn into the xylem from the outside environment, as a result of leakiness of the tissues.

In the case of *de novo* cavitation, it is expected that the bubble-like cavity is initially near vacuum, soon becoming saturated with water vapour (at the saturated vapour pressure of water) and finally containing gases diffusing from the surrounding air. Thus, cavitation leads to a bubble of air, referred to as an embolism by plant physiologists because it is supposed to stop water transport in any tracheid in which it occurs.

The physics of bubbles, necessary to understand the relationship between the bubble size and whether the bubble will grow or contract at any given pressure, has been presented elsewhere (Epstein and Plesset, 1950; Oertli, 1971; Skaar, 1972). There is a simple, and useful, relationship between the radius of the bubble, r, and the pressure, p, that would cause the bubble to neither expand nor contract:

$$p = \frac{2s}{r}$$

where s is the surface tension of the liquid $(N\,m^{-1})$.

There is good evidence that the acoustic signals from plants come directly from cavitations. The notion that one cavitation causes one signal is attractive, and if true, would enable a quantitative approach towards assessing cavitation in the field. When small segments of conifer wood are allowed to dry on the bench, the number of AEs produced over the entire drying period is found to be less than or equal to the number of tracheids the piece of wood contains (Sandford and Grace, 1985; Tyree et al., 1984a). It is held that they would be the same but for the less-than-perfect counting efficiency. Counting is limited ultimately by the signal-to-noise ratio of the equipment, as the counter depends on detecting a voltage signal against a background of electronic noise emanating from the sensor itself. Improved sensors and optimization of the counting procedure provide an almost perfect agreement between counts and numbers of tracheids (Tyree et al., 1984a).

Another line of evidence, supporting the contention that AEs arise from cavitation, is the good agreement found between the total embolism and the accumulated AEs found by Dixon et al. (1984). These authors used a beam of γ-radiation to assess the water content of the stem of *Thuja* while the tree was deprived of water (Edwards and Jarvis, 1983). Attached to the same stem, very close to the path of the γ-beam, was an ultrasound sensor. The accumulated AEs were linearly related to the stem density (Figure 7.2). Unfortunately, the crucial calculation of the volume of embolism associated with one AE could not be made because of uncertainty in the listening distance of the sensor.

There have been many speculations about the origins of acoustic emissions. It has been suggested that they could represent aspirating pits, or creaks in the cell walls as they come under strain. It is hard to test such hypotheses critically. Tyree et al. (1984b) attempted to show that the sounds are not related to tissue drought *per se* but to the tensions that normally accompany tissue drought. This was demonstrated by dehydrating a shoot in a pressure chamber with the petiole protruding from the chamber to the atmosphere. An ultrasonic

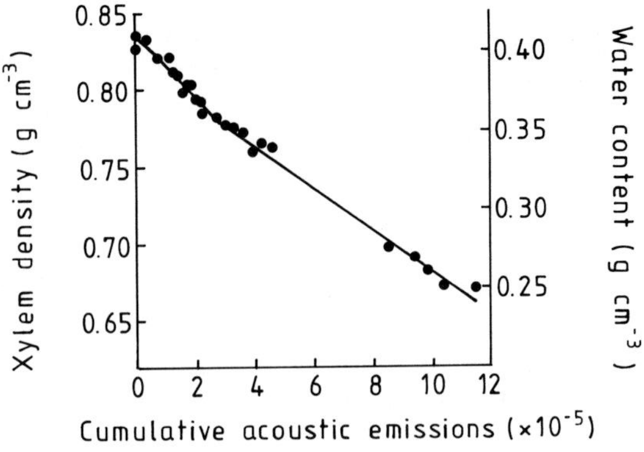

Figure 7.2. *The relationship between the density of the xylem, its water content and accumulated acoustic emissions in* Thuja occidentalis. *(From Dixon* et al. *1984.)*

sensor was attached to the stem inside the pressure chamber. By pressurizing the chamber, water was squeezed from the system and exuded at the protruding shoot. However, during this mode of dehydration, when the xylem is at a positive pressure inside the chamber, there were very few acoustic signals. When the chamber was retured to atmospheric pressure to bring the xylem under tension, there were many acoustic events.

Direct, or near-direct, observations of dry regions of xylem have been obtained. Embolized regions in the centres of large park-grown trees have been demonstrated with X-rays (Habermehl, 1982) and γ-rays (Zainalabidin and Crowther, personal communication). In both cases, the cross-section was scanned at several angles, and a map of water content was inferred using computer tomography. Water in plant tissue may also be visualized using nuclear magnetic resonance (NMR), although this is far from being a practical field technique. However, it is capable of relatively high resolution (10–50 μm) so that individual cells may be discerned, and it seems possible to distinguish water-filled cells from empty cells (Ratcliffe, 1991).

Ingenious methods of dye injection (Greenidge, 1955; Lo Gullo and Salleo, 1991; Milburn, 1979; Sperry, 1993; Zimmermann, 1983) have been used very convincingly to demonstrate that a proportion of the xylem conduits are air-filled.

In our own laboratory we have found that cavitation may be demonstrated using scanning electron microscopy on small stems which have been freeze fractured after plunging into liquid nitrogen to 'fix' the water.

7.3 The general features of cavitation

Acoustic emissions do not usually occur at night. They commonly start in the morning when a threshold water potential is reached, often in the range −1.0 to −1.4 MPa. Tensions of this magnitude are typical of nearly all terrestrial vascular plants during normal growth. Thus, it should not be supposed that cavitation is a rare event, nor one that is confined to droughted plants.

After the threshold, the cavitation rate increases with tension, measured as

Figure 7.3. *Diurnal variation in xylem water potential, stomatal conductance, transpiration rate and acoustic emissions (AEs) in young* Pinus sylvestris. *Measurements were made on (a) 9 May, (b) 31 May and (c) 12 June 1985. Control plants (○) were always well watered and drought plants (●) were not watered between 10 May and 3 June. Where data points are too close to be shown separately, they are shown as pied (◐). Bars are standard errors. (From Peña and Grace, 1986.)*

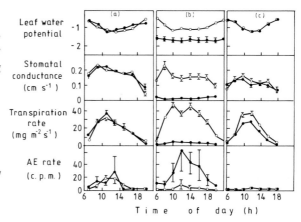

water potential. It appears that different species and different parts of the water transport system have their own characteristic 'vulnerability curve'. Acoustic emissions often peak in the early afternoon when tensions are maximal. In coniferous wood, rates of 10 per second are common in intact plants, and 100 per second in pieces of peeled wood left to dry on the bench. In angiosperm wood, lower rates are normal, but they can still be detected, even in herbaceous material.

There may be important seasonal changes in the vulnerability to cavitation. One reason for this may be the extent to which gas is dissolved in the sap. In the early summer, high temperatures and respiration rates may cause particularly high CO_2 concentrations in the sap, which may increase the probability of cavitation. Actually, there are rather few measurements of the gas composition of sap, but two studies suggest that it is CO_2-enriched air (Carrodus and Triffett, 1975; Sperry *et al.*, 1987). In saplings of *Pinus sylvestris* it was possible to demonstrate high rates of cavitation early in May, but not so by midsummer, suggesting that most of the cavitation that can occur easily does so in the early summer (Peña and Grace, 1986; Figure 7.3).

It is widely held that large vessels are more likely to cavitate than small ones, with tracheids least at risk. The reason for this view is not very clear, except that if cavitation is a spatially random event it will happen more often in a large volume than a small one. It may, of course, not be spatially random but associated with bubble growth from walls. If this is the case, large vessels would still cavitate more often, as they possess more surface area of wall, but large-vesselled species will contain less wall area in a given volume of xylem tissue. It is possible that the vulnerability to cavitation in many cases depends on the structure of the pit membranes, as suggested by Crombie *et al.* (1985). Actual data on the cavitation of vessels of different size have been obtained by perfusing wood segments with stain. Those into which the stain does not perfuse are deemed to have cavitated. Such data for *Ceratonia* show clearly that large vessels are the most vulnerable (Figure 7.4). Among vascular plants

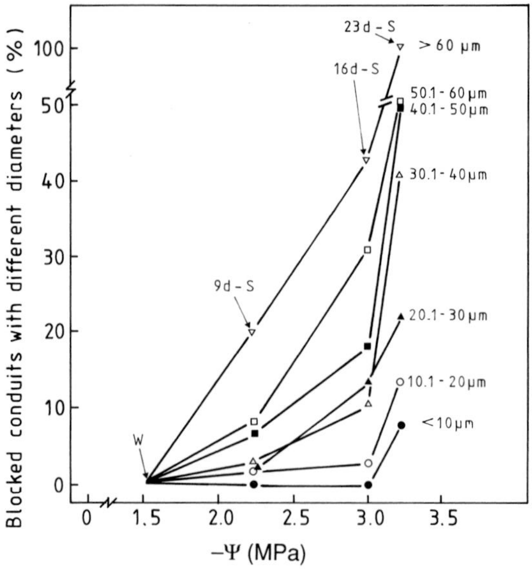

Figure 7.4. *Relationship between water potential and embolism in stems of* Ceratonia. *The data are presented as vulnerability curves for vessels of particular size classes. Measurements were made on plants which had been water stressed for 9 days, 16 days and 23 days after watering (W). (From Lo Gullo and Salleo, 1991.)*

there is a strong relationship between vessel or tracheid size and the availability of water in the habitat. Species that tolerate dry habitats, or regions where water is seasonally unavailable because of low temperatures, tend to have small conducting cells. Presumably this reduces the chance of embolism spreading.

7.4 Safety aspects

To withstand great tensions, the conduit walls are very stiff, and the taller the plant, the greater is the allocation of biomass to wall material. One of the selection pressures to drive this evolutionary tendency is simply the need for mechanical strength in the stem and branches that support the canopy. However, at a microscopic level it is apparent that much of the morphology of xylem tissue has nothing to do with strength *per se* but more to do with the need to prevent the spread of any gas that may be present as a bubble (Carlquist, 1975; Hebant, 1977; Zimmermann, 1983). The individual cells that make up the conducting pathway in conifers are rather small, typically 3–6 mm in length and 20 μm in radius. In a fully hydrated stem, water flows from one cell to another through the pores that occur in their common wall. These pores have a structure which works as a valve (Figure 7.5). They possess a central plug, or torus, suspended by radial microfibrils. In well-hydrated xylem the water flows through this pore around the torus (in between the microfibrils). If cavitation occurs in one of the cells, the gas bubble does not spread, because of the surface tension forces at the meniscus pressing against the microfibrils. These radial microfibrils are separated by gaps of about 0.1 μm and so the tensions that the meniscus can sustain before

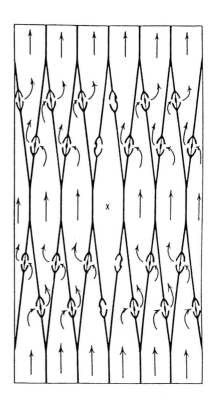

Figure 7.5. *Schematic diagram to show the valve action of bordered pits in gymnosperm wood. The pits on the walls of the embolized tracheid (×) have aspirated, preventing air-seeding of adjacent tracheids. (From Zimmermann, 1983.)*

it expands through the gap are typically 1.5 MPa. Mild tensions can therefore be resisted. At larger tensions, the torus may be displaced, being forced up against one of the walls and blocking the flow altogether, as a valve. This would effectively protect surrounding cells from spreading of the embolus. There may be cases where disease prevents the mechanism from working properly. For example, when xylem is attacked by parasitic fungi, the fungal hyphae secrete enzymes that digest the intricate pore mechanism, and large regions of air-filled xylem occur (Coutts, 1976, 1977). The pit mechanism may become dysfunctional without fungal infection, particularly in older vessels (Sperry *et al.*, 1991).

In gymnosperm tissue as a whole, water may flow laterally as well as vertically, so the presence of embolized tracheids does not prevent transport. The movement of water radially around saw-cuts is easily demonstrated with dyes (Greenidge, 1955). In this case, water will simply flow along gradients of water potential, taking the line of least resistance.

Apart from the pit mechanism, there appears to be another major safety device that operates on the scale of the whole plant. Zimmermann (1983) suggested that the tips of branches have 'designed leaks' in the tracheids, into which air may enter so that cavitation occurs easily. Moreover, the vessels may become narrower in the terminal regions of the hydraulic pathway, so that the resistance to flow increases sharply, and, during transpiration, the tensions are thus considerably greater in these areas than elsewhere. Both of these arrangements would mean that in times of drought, cavitation would occur first in the petioles and twigs rather than in the main stem. The petioles and twigs are expendable. The main stem is thus protected. If it were not so, water stress would cause a loss in hydraulic conductivity, which would itself cause more cavitation and consequent further loss in conductivity, or 'run-away cavitation' (Tyree and Ewers, 1991). But hydraulic segmentation provides a possibility of damage-limitation, through leaf and twig shedding. There is much evidence for this hypothesis, from measurements of vessel sizes in hardwoods (Ikeda and Suzaki, 1984), the occurrence of vessel ends at nodes rather than internodes (Salleo *et al.*, 1984), and the distribution of anomalous vessels near the bases of branches in oaks (Lev-Yadun and Aloni, 1990). These vessels lead nowhere in particular and may even be spherical.

7.5 Cavitation and hydraulic permeability

The volumetric flow of water, Q, through an individual tube of length L depends on the applied pressure differential between ends, p, and the fourth power of the radius, r, according to the Poiseuille equation:

$$Q = \frac{\pi r^4 \Delta p}{8\eta L}.$$

For a bundle of n long tubes of varying radius, the relationship becomes

$$Q = \frac{\pi \Delta p}{8\eta L} \sum_1^n r_i^4.$$

In most cases the measured permeability is less than that calculated on the basis of the Poiseuille equation. For example, Petty (1978) found that the measured value in *Betula* was only 34–38% of the calculated value, and attributed the discrepancy to the resistance to flow imposed by the perforation plates between vessels, and the fact that some vessels terminated inside the tested samples. The equation is likely to work best for xylem that consists mainly of vessels. In other tissues, for instance in conifers, the pit-pairs between adjacent tracheids consistute a considerable resistance to flow and an additional term should be added (e.g. Bolton and Petty, 1975).

From the foregoing, it may be expected that cavitation in the main stem might necessarily reduce the hydraulic permeability, as it will prevent water flow in some fraction of the tubes and may affect the tubes with the largest radius first. An estimate of permeability can be obtained *in situ* from measurements of whole-plant transpiration made simultaneously with leaf and soil water potentials, as discussed by Whitehead and Jarvis (1981). The estimate is made by plotting the water potentials obtained over the course of a day, against the corresponding rates of transpiration, and deriving the permeability from Darcy's law:

$$Q = \frac{kA\Delta p}{L\eta}$$

where A is the cross-sectional area and k is the permeability, with units m^2. By substituting this equation for Q in the previous equation, it is clear that

$$k = \frac{\pi \sum_{1}^{n} r_i^4}{8A}.$$

In practice, the plot of transpiration rate against water potential shows considerable hysteresis, which has been attributed to the withdrawal of water from storage sites. Thus, more complex models, involving capacitance, may have to be used to interpret the trends. Peña and Grace (1986) showed how the slope of the relationship between water potential and transpiration rate changed after Scots pine saplings had been subjected to a cycle of drought during which cavitation occurred. From the acoustic emission rate they estimated that about one-third of the total tracheid population had cavitated, and observed a corresponding decline in stem density and hydraulic conductance.

The second method of measurement utilizes a stem segment, cut from the plant (either a twig or a section of main stem). Water at a known pressure is allowed to flow through the stem, and is collected at the other end. Once again, the measurement depends on the relationship between flow and potential, and problems of capacitance may arise. There are many details of procedure that are important to observe. For example, the water should be filtered and de-aerated, and care must be taken to prevent blockage of the cut surface by resin secretion and debris of saw-cuts (Edwards and Jarvis, 1982; Lo Gullo and Salleo, 1991).

In many such perfusion experiments it has been found that the permeability gradually increases over the course of a few hours. This is interpreted as the restoration of maximum permeability as gas in the stem section is dissolved by the perfusing water. There is good evidence that the increase really is the result of dissolving gas, from perfusions with dyes (Sperry *et al.*, 1988). Others have found, or assumed, a linear relationship (Lo Gullo and Salleo, 1991; Sperry *et al.*, 1988). In fact, it is becoming widespread practice to estimate percentage cavitation from percentage reduction in maximum permeability; this is not advisable in view of the markedly non-linear relationship found by Edwards and Jarvis (1982). There are strong grounds for expecting a non-linear relationship: the conduits vary in size and, as we have seen, the larger ones are thought likely to cavitate first. There may be a real difference between gymnosperm and angiosperm wood, the latter often being a closer approximation to a bundle of long tubes of uniform radius, so that loss of conducting cross-sectional area as revealed by dyes is more directly related to loss of permeability. The situation in gymnosperm wood is less ideal, as the tracheids are short and lateral transport may be easier. There have been suggestions of 'spare capacity' in the amount of conducting tissue (Jones and Sutherland, 1991).

7.6 Cavitation and water storage

Cavitation is normally viewed as a 'bad thing', but there may be times when it is positively beneficial. Many woody tissues can be regarded as water stores, and cavitation is the means of drawing on the store. The question of how, and even where, water is stored in trees is still somewhat controversial. Large trees are known to undergo diurnal shrinkage and swelling in concert with the diurnal pattern of water potential. Most of the shrinking is in the elastic, living tissues of the bark. Similar shrinking will also occur in the leaves and other organs as their turgor falls. Measurements of the shrinking of the xylem alone, achieved with sensitive displacement transducers, show that xylem does shrink but to a much smaller extent. As the water comes under tension, the walls tend to be drawn in, much as a drinking straw that is sucked too enthusiastically, but the elastic modulus of the lignin–cellulose walls of vessels and tracheids is large and the shrinkage is small.

Another source of water stored in living tissues is the xylem parenchyma, already referred to above. In most temperature trees it is a minor component, but it may be quite significant in some tropical species.

In the xylem itself, the 'capillary water' may be important. Zimmermann (1983) drew attention to the possibility of capillary water between cells, but it is always difficult to see such intercellular spaces. Probably much more significant is the water in the ends of already cavitated conduits, which may vary directly with the water potential of the surrounding tissue. Finally, the release of water by cavitation offers a substantial supply of water, but at the cost of hydraulic permeability.

Estimates of how much water may be drawn from the store over periods in the summer vary considerably. Waring et al. (1979) used the seasonal changes in relative water content of different parts of a *Pinus sylvestris* tree as the basis for their estimates. In one 2-week period, there was a reduction of 27% in the relative water content of sapwood, which corresponds to the evaporation of 2.5 and 5.1 mm of water for low and high population densities, respectively. The water availability was as much as 21.2 mm in the high-density stand, of which 64% was from the sapwood and less than 5% in the phloem, cambium and foliage. These estimates are in contrast to those of Roberts (1976), perhaps because the latter measurements were made at the end of a dry summer, when sapwood water content was already very low. The maximum decline Roberts observed was only 10%.

The view that cavitation releases water from tracheids and makes it available to other tissues, particularly the leaves, is supported by the data of Dixon et al. (1984).

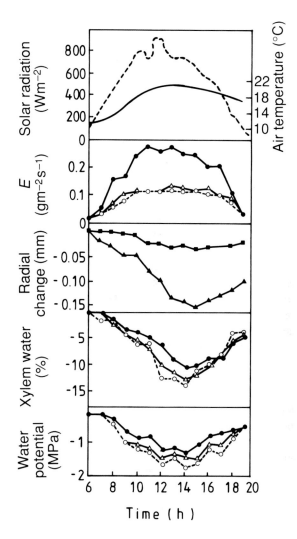

Figure 7.6. *Evaporation and transpiration rates (E) and water status of apple trees during a summer's day. Treatments were: (●) irrigated by mist, (△) irrigated by watering the soil, (○) unirrigated. Climatological variables in top window: solar radiation (broken line), air temperature (solid line). Measurements of stem shrinkage with a displacement transducer (third window): xylem shrinkage (×), total shrinkage of stem tissues (+). Note the decline and recovery of xylem water content over the course of the day. (From Brough et al., 1986.)*

These authors, working with *Thuja* saplings that were deprived of water, showed that after a burst of cavitation water potentials in the leaves increased.

Recently, Tyree and Yang (1990) introduced a way to distinguish between the mechanisms of water release from storage. Stem segments were allowed to dry out while recording their water potentials (with stem hygrometers), their relative water contents and the acoustic emission rate. It was possible to discern phases. In the first phase of dehydration, from 0 to -0.2 MPa, considerable amounts of water were lost but no cavitation occurred. This was attributed to loss of capillary water. The capacitance, corresponding to this phase, was 0.4 kg water per dm^3 of tissue per MPa. The second phase had a lower capacitance $(0.02\,kg\,dm^{-3}\,MPa^{-1})$. The final phase was associated with much cavitation and high capacitance $(0.07–0.22\,kg\,dm^{-3}\,MPa^{-1})$. The cavitation-released fraction is presumably important in times of severe water stress. This analysis underlines the importance of capillary water, and helps to explain the diurnal fluctuations in water content that Brough *et al.* (1986) observed, but does not help in identifying whether the capillary water is present in the cracks between cells or in the cells where embolism has already occurred (Figure 7.6).

7.7 The repair process

The refilling process is scarcely discussed in major reviews (Pickard, 1981; Tyree and Sperry, 1989; Zimmermann, 1983). It is widely held that refilling will occur only when the gas inside each tracheid comes under a positive pressure, causing it to dissolve (Milburn, 1979; Zimmermann, 1983). In herbaceous plants this may be achieved on a daily basis, as pre-dawn root pressure commonly creates positive pressure in the xylem. Thus, many herbaceous weeds and grasses guttate, and when cut in the early morning they exude sap. This situation does not generally hold for trees. Published diurnal trends in water potentials of trees usually show that pre-dawn water potentials do not reach zero (Whitehead and Jarvis, 1981). Yet if acoustic emissions represent cavitations, the rates observed every day are so large that they could not possibly occur unless refilling were possible and widespread.

It has been suggested that the ray parenchyma of the xylem may have a role in refilling. Anatomical studies of the stem show that in conifers, each tracheid is in contact with the medullary ray tissue (Carlquist, 1975). The early idea was that a 'vital' process was necessary, and this is still used to explain water transport phenomena by some authors (Braun, 1983; Wodzicki and Brown, 1970). One hypothesis is that water is pumped into tracheids, thus causing a localized positive pressure. Of course, plant cells are not known to pump water directly, so it may be proposed that they secrete ions into the residual water of the tracheids. Then, water would flow from surrounding cells to the embolized cell, putting the gas bubble under pressure. There are certain *a priori* objections to this hypothesis. One is that the tracheid does not have a semipermeable membrane, so as soon as pressure builds up the water will tend to be squeezed out of the cell. Tracheid walls are surprisingly permeable to water (Petty and Palin, 1983). It remains possible that a living process operates, and the data of Wodzicki and Brown (1970), where permeability was measured before and after killing the stem, provide some support for the hypothesis. Recently, attempts were made to repeat Wodzicki and Brown's work with a more rigorous control of experimental conditions (Borghetti *et al.*, 1991).

In this work, experiments on stem segments of *Pinus sylvestris* as well as whole plants were carried out. It was shown that stem segments, which had been partially dehydrated, were quite capable of lifting water and completely rehydrating under a low tension. The tension was achieved by connecting the stem segment to a tube supplying degassed filtered water, and raising the segment above the level of the water supply. More recently, Sobrado *et al.* (1992) increased the imposed tension to the equivalent of 4 m of water, i.e. 0.04 MPa. Note that such experiments are *not* the equivalent of those of Bohm (1893) in which columns of mercury were lifted. They are different in that the uptake of water in the experiments of Borghetti *et al.* (1991) and Sobrado *et al.* (1992) represents uptake into partially dehydrated stem segments in which much cavitation has occurred. Because the xylem rehydrates, we interpret the uptake as refilling, at least *some* of which must be refilling of embolized tracheids. So the experiments are important demonstrations that refilling can occur when the water is present at tensions, albeit low tensions. (At larger tension, it becomes difficult to maintain the water column: bubbles, evidently from the xylem, are sucked into the water supply and the column breaks.)

In this experimental system, it was possible to show that the rate of uptake, and the final extent of uptake, was not affected by prior perfusion with the respiratory poison sodium azide. It was also shown that the azide had indeed killed the ray parenchyma.

A similar experiment was also conducted on intact trees, with essentially the same result (Borghetti *et al.*, 1991). Refilling apparently took about 1 day, and the water potentials of the leaves remained well below zero. Again, refilling was insensitive to azide, and the conclusion was that the living cells did not influence the refilling process.

Subsequent experiments by Edwards *et al.* (1993) using the same system have shown that refilling is greatly facilitated by the *flow* of de-aerated water through the stem section, presumably to remove the dissolved gases.

Borghetti *et al.* (1991) considered several other possible refilling mechanisms. The most likely was considered to be that of capillarity. The residual water trapped in the ends of embolized tracheids is assumed to form a meniscus of radius r which will tend to pull water in when the water potential of the bulk tissue rises to near zero. In doing so, it would put the gases in the bubble under pressure (even though bulk water potentials are below zero) and they would dissolve. This mechanism works as long as the radii are small enough to generate the required 'pull' and in the work of Borghetti *et al.* (1991) they may have been. One technical problem is that few people have measured accurately the water potentials of stems (as opposed to leaves). We have few data to show what happens to stem water potentials at pre-dawn (Whitehead and Jarvis, 1981).

In the case of angiosperm wood, the mechanism outlined above does not work as the vessels have much larger radii. Sperry *et al.* (1987) examined springtime refilling in the wild grapevine. Such plants have large vessels that are gas-filled by late winter. Refilling is achieved by root pressure. As the sap rises in the vessels, much of the gas is simply squeezed out of the system under pressure. Very high root pressures have been recorded, for example, in palms, enough to raise water to 10 m, about the size of a mature palm tree (Davis, 1961).

A related question of great interest is the removal of embolisms that occur as a result of freezing in winter. Sucoff (1969) mounted pieces of *Pinus* wood in silicone oil and observed freezing under the light microscope. The expansion of water associated with the transition from liquid to ice caused water to be expelled from each freezing lumen, so frozen tracheids contained an (estimated) 91% of their original water. After freezing was complete, small bubbles could be seen inside the ice within many of the tracheids. On thawing, most of the bubbles redissolved, leaving only a few tracheids embolized. Sucoff's interpretation was very ingenious. On thawing, the system of interconnected tracheids was considered to be under tension (as a result of having lost water). As a result of the tension, the *largest* bubble in an interconnected set of tracheids expands rapidly, releasing the tension, so the small bubbles redissolve. In this theory the pits between the tracheids remain open, and water can move freely between cells. Afterwards, just a few tracheids remain embolized, and permeability is not appreciably impaired.

An alternative theory is that the pits do indeed aspirate because of the pressure rise inside each tracheid at ice formation (Hammel, 1967). Each tracheid is thus sealed off, so that no water can leave the cell on freezing. Consequently, on thawing, any bubbles present will dissolve as the pressure will remain high until all the ice has melted.

Robson and Petty (1993) reviewed evidence for and against these conflicting theories, and found that neither are fully consistent with observations. They produced a new theory, in which the capillaries of the cell wall matrix have a role. They pointed out that the freezing front is expected to move radially inwards. As the pits are on the radial walls, they are not very well positioned as pathways for outflow of water. So the internal pressure is more likely to force water into the tracheid walls. After freezing, the water in the walls (which may not freeze despite the subzero temperature) migrates into the frozen lumen as a result of differences in chemical potential. On thawing, the pressure inside the system remains high, and any bubbles present dissolve.

7.8 Ecological significance

The significance of breakage of the water columns in plants is becoming clearer. In fast-growing, short-lived plants having a relatively small number of large vessels, cavitation during periods of drought may be a major disaster if refilling by root pressure cannot occur during the following night. In this case, tensions on the following day will be even higher as the conducting pathway is correspondingly less, leading to still more cavitation. In long-lived species (tall trees and woody shrubs) the water transport system is designed for more safety, with many small tracheids instead of few large vessels, and a relatively large fraction of plant tissue devoted to water transport. The stems of large trees have several functions, and it may be that the biomechanical requirements for strength and elastic modulus happen to coincide with the design for water storage and safe water transport.

Refilling in tall trees is still an open question, as several datasets suggest that it is a regular phenomenon during the summer months, even though water potentials remain negative throughout the diurnal cycle. The capacity to refill may be crucial in determining the recovery of tree species following drought stress.

Acknowledgements

The author wishes to acknowledge financial support for cavitation research from the Natural Environment Research Council.

References

Askenasy, E. (1895) Uber das saftsteigen. *Bot. Zentral.*, **62**, 237–238.

Berthelot, M. (1850) Sur quelques phenomenens de dilatation forcée des liquides. *Ann. Chemie Physique, 3e Serie*, **30**, 232–237.

Böhm, J. (1893) Capillarität und Softsteigen. *Ber. deut. bot. Ges.*, **11**, 203–212.

Bolton, A.J. and Petty, J.A. (1975) Structural components influencing the permeability of ponded and unponded Sitka spruce. *J. Microscopy*, **104**, 3–46.

Borghetti, M., Edwards, W.R.N., Grace, J., Jarvis, P.G. and Raschi, A. (1991) The refilling of embolised xylem in *Pinus sylvestris* L. *Plant Cell Environ.*, **14**, 357–369.

Braun, H.J. (1983) Zur Dynamic des Wassertransportes in Baumen. *Ber. deut. bot. Ges.* **96**, 29–47.

Briggs, L.J. (1950) Limiting negative pressure of water. *J. Appl. Phys.*, **21**, 721–722.

Brough, D.W., Jones, H.G. and Grace, J. (1986) Diurnal changes in water content of the stems of apple trees, as influenced by irrigation. *Plant Cell Environ.*, **9**, 1–7.

Carlquist, S. (1975) *Ecological Strategies of Xylem Evolution.* University of California Press, Berkeley.

Carrodus, B.B. and Triffett, A.C.K. (1975) Analysis of the composition of respiratory gases in woody stems by mass spectroscopy. *New Phytol.*, **74**, 243–246.

Chalk, L. and Bigg, J.M. (1958) The distribution of moisture in the living stem of Sitka spruce and Douglas Fir. *Forestry*, **29**, 5–21.

Coutts, M.P. (1976) The formation of dry zones in the sapwood of conifers i. Induction of drying in standing trees and logs by *Fomes annosus* and extracts of infected wood. *Eur. J. Forest Pathol.*, **6**, 372–381.

Coutts, M.P. (1977) The formation of dry zones in the sapwood of conifers ii. The role of living cells in the release of water. *Eur. J. Forest Pathol.*, **7**, 6–12.

Crafts, A.S. (1939) Solute transport in plants. *Science*, **90**, 337–338.

Crombie, D.S., Hipkins, M.F. and Milburn, J.F. (1985) Gas penetration of pit membranes in the xylem of *Rhododendron* as the cause of acoustically detectable sap cavitations. *Aust. J. Plant Physiol.*, **12**, 445–453.

Davis, T.A. (1961) High root pressures in palms. *Nature*, **192**, 277–278.

Dixon, H.H. (1914) *Transpiration and the Ascent of Sap in Plants.* Macmillan, London.

Dixon, H.H. and Joly, J. (1896) On the ascent of sap. *R. Soc. Phil. Trans.*, **B186**, 563–576.

Dixon, M.A. and Tyree, M.T. (1984) A new stem hygrometer corrected for temperature gradients and calibrated against the pressure bomb. *Plant Cell Environ.*, **7**, 693–697.

Dixon, M.A., Grace, J. and Tyree, M.T. (1984) Concurrent measurements of stem density, leaf water potential and cavitation on a shoot of *Thuja occidentalis* L. *Plant Cell Environ.*, **7**, 615–618.

Doley, D. (1974) Alternatives to the assessment of latewood and earlywood on dicotyledonous trees: a study of structural variation in growth rings of apple (*Malus pumila* Mill.). *New Phytol.*, **73**, 157–171.

Edwards, W.R.N. and Jarvis, P.G. (1982) Relations between water content, potential and permeability in stems of conifers. *Plant Cell Environ.*, **5**, 271–277.

Edwards, W.R.N. and Jarvis, P.G. (1983) A method for measuring radial differences in water content of intact tree stems by attenuation of gamma radiation. *Plant Cell Environ.*, **6**, 225–260.

Edwards, W.R.N., Jarvis, P.G., Grace, J. and Moncrieff, J.B. (1993) Reversing cavitation in tracheids of *Pinus sylvestris* L. under negative water potentials. *Plant Cell Environ.*, in press.

Epstein, P.S. and Plesset, M.S. (1950) On the stability of gas bubbles in liquid-gas solutions. *J. Appl. Phys.*, **18**, 1505–1509.

Franks, F. (1983) *Water*. Royal Society of Chemistry, London.

Gibbs, R.D. (1930) Sinkage studies. ii The seasonal distribution of water and gas in trees. *Can. J. Res.*, **2**, 425–439.

Gibbs, R.D. (1958) Patterns in the seasonal water content of trees. In: *The Physiology of Forest Trees* (ed. K.V. Thimann). Ronald Press, New York, pp. 43–69.

Greenidge, K.N.H. (1955) Studies in the physiology of forest trees. ii Experimental studies of fracture of stretched water columns in transpiring trees. *Am. J. Bot.*, **42**, 28–37.

Habermehl, A. (1982) A new non-destructive method for determining internal wood condition and decay in living trees. ii. Results and further developments. *Arboricultural J.*, **6**, 121–130.

Hammel, H.T. (1967) Freezing of xylem sap without cavitation. *Plant Physiol.*, **42**, 55–66.

Hebant, C. (1977) *Conducting Tissues in Bryophytes*. J. Cramer, Vaduz.

Ikeda, T. and Suzaki, T. (1984) Distribution of xylem resistance to water flow in stems and branches of hardwood species. *J. Jap. Forestry Soc.*, **66**, 229–236.

Jones, H.G. and Sutherland, R.A. (1991) Stomatal control of xylem embolism. *Plant Cell Environ.*, **14**, 607–612.

Lev-Yadun, S. and Aloni, R. (1990) Vascular differentiation in branch junctions of trees: circular patterns and functional significance. *Trees*, **4**, 49–54.

Lo Gullo, M.A. and Salleo, S. (1991) Three different methods for measuring xylem cavitation and embolism: a comparison. *Ann. Bot.*, **67**, 417–424.

Milburn, J.A. (1966) The conduction of sap. i Water conduction and cavitation in water-stressed leaves. *Planta*, **65**, 34–42.

Milburn, J.A. (1970) Cavitation and osmotic potentials of *Sordaria* ascospores. *New Phytol.*, **69**, 133–141.

Milburn, J.A. (1979) *Water Flow in Plants*. Longman, London.

Milburn, J.A. and Johnson, R.P.C. (1966) The conduction of sap. ii Detection of vibrations caused by sap cavitation in *Ricinus* xylem. *Planta*, **69**, 43–52.

Milburn, J.A. and McLaughlin, M.E. (1974) Studies of cavitation in isolated vascular bundles and whole leaves of *Plantago major* L. *New Phytol.*, **73**, 861–871.

Oertli, J.J. (1971) The stability of water under tension in the xylem. *Z. Pflanzenphysiol.*, **65**, 195–209.

Oertli, J.J. (1993) Effect of cavitation on the status of water in plants. In: *Water Transport in Plants Under Climate Stress* (eds M. Borghetti, J. Grace and A. Raschi). Cambridge University Press, Cambridge, pp. 27–40.

Peña, J. and Grace, J. (1986) Water relations and ultrasound emissions before, during and after a period of water stress. *New Phytol.*, **103**, 515–524.

Petty, J.A. (1978) Fluid flow through the vessels of birch wood. *J. Exp. Bot.*, **29**, 1463–1469.

Petty, J.A. and Palin, M.A. (1983) Permeability to water of the fibre cell wall material of two hardwoods. *J. Exp. Bot.*, **34**, 688–693.

Pickard, N.F. (1981). The ascent of sap in plants. *Progr. Biophys. Mol. Biol.*, **37**, 181–229.

Ratcliffe, R.G. (1991) Nuclear magnetic resonance in plant science research. *Bot. J. Scot.*, **46**, 107–120.

Renner, O. (1915) Theoretisches und Experimentelles zur Kohesionetheories der Wasserbewegung. *Jahrbuecher für Wissenschaftliche Botanisch*, **56**, 617–667.

Ritman, K.T. and Milburn, J.A. (1988) Acoustic emissions from plants: ultrasonic and audible compared. *J. Exp. Bot.*, **39**, 1237–1248.

Roberts, J. (1976) An examination of the quantity of water stored in mature *Pinus sylvestris* L. trees. *J. Exp. Bot.*, **27**, 473–479.

Robson, D.J. and Petty, J.A. (1993) A proposed mechanism of freezing and thawing in conifer xylem. In: *Water Transport in Plants Under Climatic Stress* (eds M. Borghetti, J. Grace and A. Raschi). Cambridge University Press, Cambridge, pp. 75–85.

Salleo, S., Lo Gullo, M.A. and Siracusano, L. (1984). Distribution of vessel ends in stems of some diffuse- and ring-porous trees and the nodal regions as 'safety zones' of the water conducting system. *Ann. Bot.*, **54**, 543–552.

Sandford, A.P. and Grace, J. (1985) The measurement and interpretation of ultrasound from woody stems. *J. Exp. Bot.*, **36**, 298–311.

Skaar, C. (1972) *Water in Wood.* Syracuse University Press, New York.

Sobrado, M.A., Grace, J. and Jarvis, P.G. (1992) Relationship between water content and xylem embolism recovery in *Pinus sylvestris* L. *J. Exp. Bot.*, **43**, 831–836.

Sperry, J. (1993) Effect of cavitation on the status of water in plants. In: *Water Transport in Plants Under Climatic Stress* (eds M. Borghetti, J. Grace and A. Raschi). Cambridge University Press, Cambridge, pp. 86–98.

Sperry, J.S., Holbrook, N.M., Zimmermann, M.H. and Tyree, M.T. (1987) Spring filling of xylem vessels in wild grapevine. *Plant Physiol.*, **83**, 414–417.

Sperry, J.S., Donnelly, J.R. and Tyree, M.T. (1988). A method of measuring hydraulic conductivity and embolism in xylem. *Plant Cell Environ.*, **11**, 35–40.

Sperry, J.S., Perry, A.H. and Sullivan, J.E.M. (1991) Pit membrane degradation and air-embolism formation in ageing xylem vessels of *Populus tremuloides* Michx. *J. Exp. Bot.*, **42**, 1399–1406.

Sucoff, E.I. (1969) Freezing of conifer xylem and the cohesion-tension theory. *Physiol. Plant.*, **22**, 424–431.

Tyree, M.T. and Dixon, M.A. (1983) Cavitation events in *Thuja occidentalis* L.? Ultrasonic acoustic emissions from sapwood can be measured. *Plant Physiol.*, **72**, 1094–1099.

Tyree, M.T. and Ewers, F.W. (1991) The hydraulic architecture of trees and other woody plants. *New Phytol.*, **119**, 345–360.

Tyree, M.T. and Sperry, J.S. (1988). Do woody plants operate near the point of catastrophic xylem dysfunction caused by dynamic water stress? Answers from a model. *Plant Physiol.*, **88**, 574–580.

Tyree, M.T. and Sperry, J.S. (1989). Vulnerability of xylem to cavitation and embolism. *Ann. Rev. Plant Physiol. Mol. Biol.*, **40**, 19–38.

Tyree, M.T. and Yang, S. (1990) Water-storage capacity of *Thuja*, *Tsuga* and *Acer* stems measured by dehydration isotherms. The contribution of capillary water and cavitation. *Planta*, **182**, 420–426.

Tyree, M.T., Dixon, M.A., Tyree, E.L. and Thompson, R.G. (1984a) Ultrasonic acoustic emissions from the sapwood of cedar and hemlock: an examination of three hypotheses regarding cavitation. *Plant Physiol.*, **75**, 988–992.

Tyree, M.T., Dixon, M.A. and Thompson, R.G. (1984b) ultrasonic acoustic emissions from the sapwood of *Thuja occidentalis* measured inside a pressure bomb. *Plant Physiol.*, **74**, 1046–1049.

Wang, J., Ives, M.J. and Lechowicz, M.J. (1992) The relation of foliar phenology to xylem embolism in trees. *Funct. Ecol.*, **6**, 469–475.

Waring, R.H. and Running, S.W. (1978) Sapwood water storage: its contribution to transpiration and effect upon water conductance through the stems of old-growth Douglas Fir. *Plant Cell Environ.*, **1**, 131–140.

Waring, R.H., Whitehead, D. and Jarvis, P.G. (1979) The contribution of stored water to transpiration in Scots pine. *Plant Cell Environ.*, **2**, 309–317.

Whitehead, D. and Jarvis, P.G. (1981) Coniferous forests and plantations. In: *Water Deficits and Plant Growth. vi Woody Plant Communities* (ed. T. T. Kozlowski). Academic Press, London, pp. 49–152.

Wodzicki, T.J. and Brown, C.L. (1970) Role of xylem parenchyma in maintaining the water balance of trees. *Acta Soc. Bot. Pol.* **39**, 617–621.

Zimmermann, M.H. (1983). *Xylem Structure and the Ascent of Sap.* Springer, New York.

Soil water deficits and atmospheric humidity as environmental signals

E.-D. Schulze

8.1 Introduction

The observation of the stomatal response to humidity (Lange *et al.*, 1971; Raschke and Kühl, 1970) introduced a significant change in the understanding of how plants respond to their environment. The concept of feedforward regulation of plants to environmental signals (Farquhar, 1978) replaced the earlier contention that feedback mechanisms predominantly regulated plant water relations. The observation that plants stabilize water status before large quantities of water are lost led to the hypothesis that similar principles of regulation may exist at the plant/soil interface (Schulze, 1986). Thus, the concept of root–shoot signals was a logical consequence of understanding the feedforward response to humidity.

Following this change in the understanding of control mechanisms for regulating plant water relations (Kramer, 1988; Schulze *et al.*, 1988) this chapter considers the following points:

(1) the present understanding of the humidity response at the cellular level;
(2) the evidence for root–shoot signals and the significance of plant internal cycles in modulating such signals; and
(3) the importance of stomatal response to air humidity in controlling water loss by entire plant canopies.

8.2 The response of stomata to air humidity

The response of stomata to air humidity was originally described for the isolated epidermis of *Tradescantia*, in which adjacent stomata closed and opened in relation to the vapour pressure deficit (VPD) in the ambient air (Lange *et al.*, 1971). For intact leaves, the humidity response was distinguished from a water

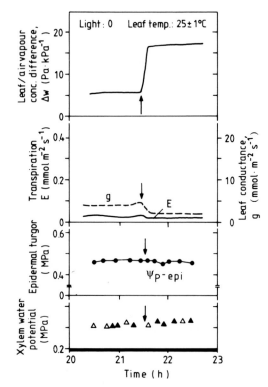

Figure 8.1. *Response of tran-spiration, leaf conductance, epider-mal turgor and xylem water potential during a change in vapour pressure deficit. The experiment was carried out in the dark. (From Frensch and Schulze, 1988.)*

stress response, in that transpiration may decrease at increasing VPD, and that water content may increase at reduced transpiration (Schulze *et al.*, 1972).

Lange *et al.* (1971) explained the response of stomata to humidity by peristo-matal transpiration (Maier-Maerker, 1983). The guard cells should lose water in relation to the ambient VPD by cuticular transpiration. This water loss was thought to be recharged mainly from the surrounding epidermis and/or from the subepidermal airspace. It was hypothesized that the resistance for these water flows would be such that a water deficit was increased in the guard cell as a function of peristomatal transpiration.

Following this interpretation of the original observation, many attempts have been made to demonstrate peristomatal water loss and its effect on guard cell turgor (Maier-Maerker, 1983). However, insight into the cellular mechanism of a humidity response became possible only after it was techni-cally possible to measure turgor *in situ* in different tissues (Zimmermann *et al.*, 1980). Using the pressure probe, Frensch and Schulze (1988) showed that peristomatal transpiration cannot explain the humidity response in *Tradescantia virginiana*. Epidermal turgor remained constant when humidity was changed in the dark at a time when the stomata were closed (Figure 8.1).

In contrast to these experiments in the dark, epidermal turgor was negatively correlated with transpiration in the light (Figure 8.2a), leaf conduc-tance increased with epidermal turgor (Figure 8.2b) and leaf conductance

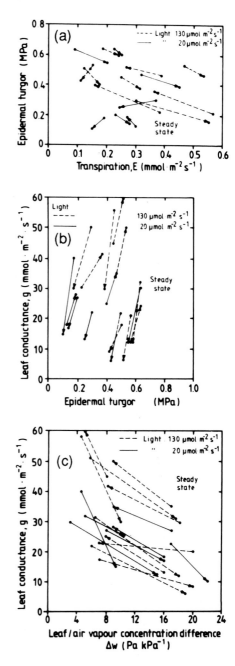

Figure 8.2. *Changes in epidermal turgor with transpiration (a), and of leaf conductance with epidermal turgor (b) and with VPD (c). Each line represents a separate experiment and connects the steady-state values before and after a step-change. Arrows indicate the direction of change. (From Frensch and Schulze, 1988.)*

decreased with increasing VPD (Figure 8.2c). Thus, in the light, the response of open stomata was closely correlated with epidermal turgor.

Closer inspection of the water relations parameters in different layers of the leaf tissue showed that turgor pressure was always lower in epidermal than in

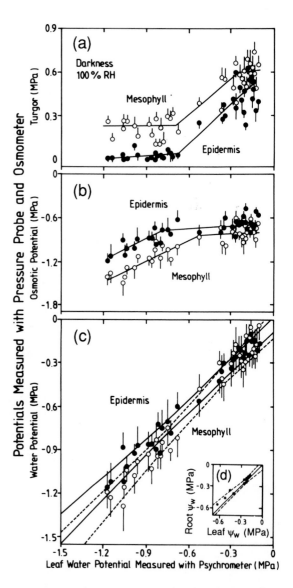

Figure 8.3. *Mesophyll and epidermal cell turgor (a), osmotic potential (b) and water potential (c) at various xylem water potentials of* Tradescantia *plants incubated at 100% RH in the dark for at least 1 day, but at different levels of soil drought (d). Each point represents the mean of 7–11 measurements on single cells; vertical bars indicate 95% confidence limits. Solid lines and dotted lines in (c) and (d) indicate the confidence band at* p < 0.01. *(From Nonami and Schulze, 1989.)*

mesophyll cells (Nonami and Schulze, 1989). This corresponded with higher osmotic potentials in the mesophyll than in the epidermis (Figure 8.3). Measured and calculated xylem water potentials were lower in mesophyll than in epidermal cells. We conclude that evaporation of water occurs mainly from mesophyll cells, and that peristomatal transpiration is less important than has been proposed previously. This does not exclude the possibility that there may be cases where peristomatal transpiration is related directly to the regulation of turgor in guard cells, because of specialized cuticular structures (Appleby and Davies, 1983).

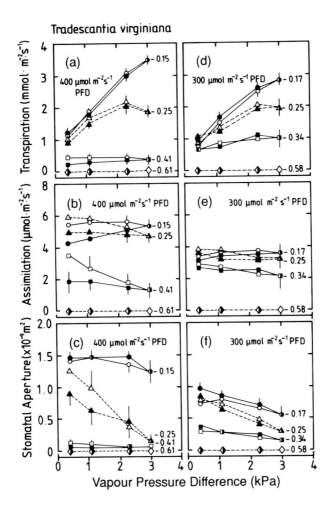

Figure 8.4. *Transpiration (a,d), assimilation (b,e) and stomatal aperture (c,f) of* Tradescantia *plants having various xylem water potentials in relation to vapour pressure difference between leaf and air at high light (left) and low light (right). (From Nonami et al., 1990.)*

The second criterion for a feedforward response of stomata to humidity would be a decrease in transpiration despite increasing VPD (Schulze *et al.*, 1972). If peristomatal transpiration is not regulating stomatal closure at low air humidity, then we need to determine whether our experimental plant species, *Tradescantia virginiana*, exhibits reduced transpiration at low air humidity. In well-watered plants the decrease in transpiration does not result from alterations in stomatal aperture (Figure 8.4). At high water stress, stomatal aperture is too low for transpiration to change with dry air. However, there is an intermediate level of plant water stress, where dry air induces a large range of stomatal opening (Figure 8.4). At this level of water stress, a decrease in transpiration with increasing VPD can be observed. The response

Tradescantia virginiana

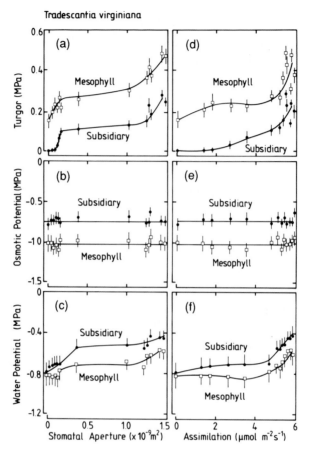

Figure 8.5. *Cell turgor (a,d), cell osmotic potential (b,e) and cell water potential (c,f) of* Trades-cantia *mesophyll and subsidiary cells in relation to stomatal aperture (left) and assimilation (right). (From Nonami et al., 1990.)*

is reversible, and we propose that all plant species will exhibit it (Sheriff, 1979) under appropriate conditions.

In *Tradescantia*, subsidiary cells exhibited lower turgor and water potential than mesophyll cells when stomata were wide open, and their turgor declined as the stomatal aperture decreased (Figure 8.5). This decrease was more pronounced in subsidiary cells than in mesophyll cells. When stomata closed, the water potential gradient between epidermis and mesophyll collapsed. An interpretation of this result was only possible after electron microscopy revealed that *Tradescantia* has an internal cuticle which extends from the outside into the stomatal cavity and covers the inner side of the guard cells, subsidiary cells and even part of the epidermal cells (Nonami *et al.*, 1990). The extensions of the internal cuticle terminate at the transitional border between epidermal and mesophyll cells. Apparently, during transpiration,

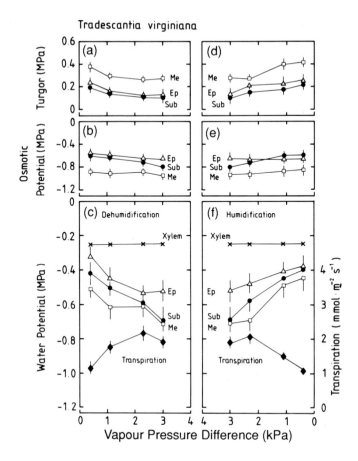

Figure 8.6. *Cell turgor (a,d), cell osmotic potential (b,e) and cell water potential (c,f) of mesophyll cells (Me), epidermal cells (Ep) and subsidiary cells (Sub) in relation to VPD when air humidity was decreased (left) and increased (right). (From Nonami et al., 1990.)*

water loss from the guard cells and the subsidiary cells is reduced significantly because of this cuticular cover, and evaporation takes place mainly at the cell walls of the mesophyll.

In order to establish how stomatal movement and cell water status are related to the leaf internal cuticle, we measured the humidity response of stomata and the cellular water relations in the range where a decrease of transpiration with increasing humidity was to be expected (Figure 8.6). Cell turgor was highest in mesophyll cells, but subsidiary cell turgor was similar to epidermal cell turgor. Turgor of all cells decreased when air humidity became lower and increased when air humidity became higher. In contrast to epidermal and mesophyll cells, osmotic potential changed only in subsidiary cells. The water potential was lowest in mesophyll cells under all conditions, indicating that water was lost mainly by evaporation from the mesophyll cells. When the VPD was increased, the subsidiary cell water potentials decreased further and became closer to that of mesophyll cells, despite the fact that the epidermal water potential recovered slightly. At this point transpiration decreased despite increasing VPD. We suggest that the observed decrease of transpiration at increasing VPD can be attributed to a withdrawal of water from the epidermis at high rates of transpiration from the mesophyll. This decreases the water supply to the guard cells from subsidiary cells, and turgor decreases more in guard cells than in mesophyll cells. The leaf internal cuticle appears to play a special role in channelling the internal water flow during transpiration.

We may conclude that the stomatal response to air humidity is not caused by peristomatal transpiration, at least in *Tradescantia*. Thus, following the definition of Farquhar (1978), the mechanism of the humidity response is not one of feedforward regulation, but is feedback control of cellular water relations due to changes of leaf internal water fluxes in the stomatal complex at low air humidity. Such a feedback regulation behaves under steady-state conditions as if it was a feedforward response Although I have explained the sequence of events of the stomatal humidity response in terms of cellular water relations, it is quite clear that the underlying mechanism of guard cell movement is physiologically regulated (Lösch and Schenk, 1978), and that abscisic acid is also involved in this response (Brinckmann *et al.*, 1990).

8.3 Stomatal response to soil water deficits

The classical view of the response of stomata to water stress was that stomatal aperture was regulated according to plant water status (see review by Schulze and Hall, 1982). At the cellular level of the stomatal apparatus it has been demonstrated that such feedback control does occur during responses to VPD. At the whole-plant level, however, there is little evidence for this feedback response (Schulze and Hall, 1982). A number of plant species, especially legumes, close stomata in dry soil even though the plant internal water status has not changed (Bates and Hall, 1981). The contention that plants respond to a signal from the root was supported by experiments using split roots (Blackman and Davies, 1984) or split pots (Coutts, 1981). In these experiments part of the root experienced drought, while the other part remained in wet soil, but stomata closed even though the plant water status was not affected (Davies and Zhang, 1991; see Chapter 9). This response of stomata may be regarded as a feedforward response, in which a signal from roots experiencing dry soil is transmitted to the leaf so that water loss is reduced before the plant experiences internal water stress.

The experiment which proved the existence of a root signal was carried out by Gollan *et al.* (1986). Plants were grown in a pressurized pot, which equilibrated any change in leaf water potential, so that the leaf was always fully turgid. It was shown (Figure 8.7) that fully turgid leaves responded in the same way as leaves which dried out naturally. The stomata of turgid and non-turgid leaves closed when the soil lost about 60% of the water available to the plant.

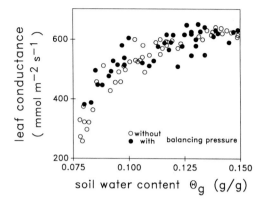

Figure 8.7. *Relationship between soil water content and leaf conductance, soil water suction and the balancing pressure in* Triticum aestivum. *Solid symbols, leaf conductance of plants that were maintained at high leaf water potential by applying the balancing pressure; open symbols, leaf conductance of control plants to which no balancing pressure was applied. (From Gollan et al., 1986, with permission from CSIRO Editorial Publications.)*

From very early on it had been proposed that the stress hormone abscisic acid (ABA) was produced by root tips and transported to the leaf via the xylem stream (Raschke, 1979, 1982). It appears that the root tip is the actual stress sensor, and there is evidence that the root tip experiences a loss in turgor earlier than the root because it is partially disconnected from the main xylem flow (Zhang and Davies, 1987). Thus, the sensing mechanism would be similar to that described in the leaf: a local feedback response initiates a metabolic reaction which results in a steady-state response of the whole plant, acting as if it was a feedforward response.

Indeed, changes in ABA concentration have been observed in many laboratory studies, but there have been only a few field experiments (but see also Chapter 9). For example, in *Prunus dulcis* maximum stomatal conductance decreased when ABA increased beyond a certain threshold level in the xylem sap (Figure 8.8). These data originate from a lysimeter experiment in which plants were grown for up to 4 years in plastic pots of 3 m diameter and 1, 2 or 3 m depth. The ABA response was independent of pot size. However, an interesting phenomenon emerged when comparing different age classes (Figure 8.9). Plants growing in pots which had been filled with homogenized loess responded with stomatal closure at a more negative plant water potential than plants which had gone through a repeated dry-out cycle. Drying causes local re-orientation of particles which leads to the formation of soil crumbs (Semmel *et al.*, 1990). The water transport from these denser particles is decreased, and this leads to a more sensitive response of plants in structured soils.

It is still unclear whether the concentration of ABA or total mass flow of ABA controls stomatal aperture. Zhang *et al.* (1987) demonstrated that it is apparently the total amount of ABA which is important, not the concentration. In addition, the protonation of ABA determines whether this plant hormone will migrate through the cell membrane of mesophyll cells (Schurr *et al.*, 1992). If ABA is undissociated, it will pass through the membrane, down the pH gradient, and enter into the chloroplast, where it will be metabolized. Only charged ABA should accumulate via the

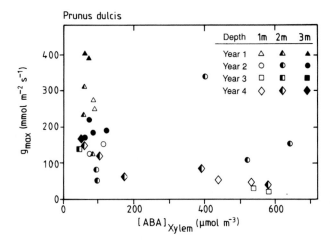

Figure 8.8. *Relationship between daily average xylem sap ABA concentration and maximum leaf conductance of* Prunus dulcis *at Avdat in 1987. (From Wartinger et al., 1990.)*

transpiration stream at the stomatal apparatus and affect the potassium channels. Should this then enter into the guard cells it would explain the very high concentrations of ABA in these organs (Brinckmann *et al.*, 1990).

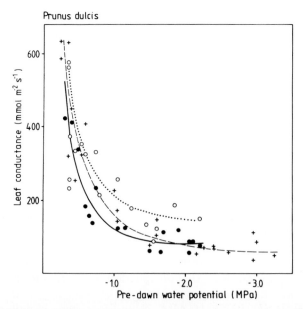

Figure 8.9. *The change of maximum leaf conductance with pre-dawn water potential in homogenized loess (circles), after one dry-out cycle (crosses), and after two dry-out cycles (dots). (From Wartinger, 1991.)*

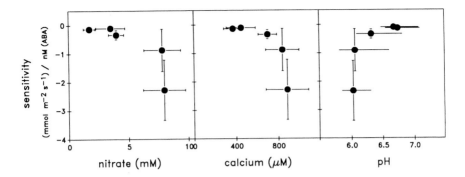

Figure 8.10. *Relation between the sensitivity of stomatal conductance for ABA from the xylem sap during drought and the concentration of calcium, nitrate and the pH of the xylem sap. Standard deviations are given for the concentration values and the sensitivity. (From Schurr et al., 1992, with permission from Blackwell Scientific Publications Ltd.)*

Evidence that the pH of the xylem sap will modulate the response of stomata to ABA was presented by Gollan *et al.* (1992) and Schurr *et al.* (1992). In sunflower, a decrease in nitrate and phosphate uptake with soil drought exceeded the change in cation uptake, such that the pH of xylem sap increased. In this case, stomatal sensitivity to ABA was dependent on the charge balance in the xylem sap (Figure 8.10).

ABA is not only transported in the xylem from the root to the shoot and redistributed in the leaf; a reverse flow of ABA also occurs in the phloem. Thus, there is a 'futile cycle' of ABA in the plant between root and shoot, so it must be considered whether the leaf or the root is the origin of increased xylem ABA. Schurr (1991) estimated the magnitude of this cycle and came to the conclusion that in well-watered plants the import into the leaf via the xylem equalled the rate of export via the phloem (Table 8.1). Under water stress, the import from the xylem increased dramatically while the export remained constant or decreased, because increasing water stress in the soil affected the unloading of the phloem. The sugar concentration increased and the recirculation of ABA decreased. This would immediately enhance the build-up of ABA in the leaf and might even be a

Table 8.1. *ABA transport in the phloem and xylem*

	Phloem	Xylem
Concentration (nM)		
21% soil water content	12.17	45.10
9% soil water content	38.20	1450.00
Transport per leaf area (pmol m^{-2} s^{-1})		
21% soil water content	4.40	4.10
9% soil water content	4.89	52.20

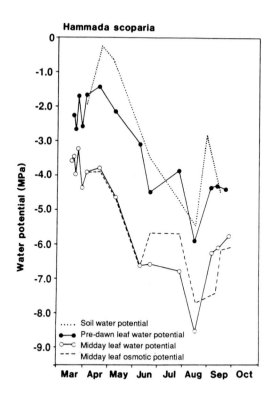

Figure 8.11. *Seasonal change in soil water potential, pre-dawn water potential, midday water potential, and midday osmotic potential in* Hammada scoparia. *(After Kappen et al., 1975.)*

more rapid signal than the transport of new ABA from the root to the shoot.

It has not been resolved whether ABA is the only stress signal for stomatal responses, or if increased ABA in the xylem and stomatal closure are coincidental events, with control regulated by other signals. It was pointed out by Schulze (1991) that a response to a signal which is produced in the root and carried to the shoot may be quite slow, especially in trees. In a 100 m high *Sequoia* tree it may take weeks before the water deficit signal reaches the shoot. Therefore, we have to consider additional types of signals. Munns (1990) showed that stomatal closure of wheat occurred in plants which were fed with xylem sap from which ABA had been excluded. Thus, other substances in the xylem sap may have a similar effect to ABA. Cessation of phloem circulation due to soil drought would also be an immediate turgor signal to the shoot and would cause a similar, but faster, reaction than ABA transport from the root to the shoot. However, it appears (Figure 8.11), that whole-plant function is regulated such that the internal circulation is maintained. Under natural conditions in *Hammada scoparia* the diurnal variation in water potential was constant while the absolute water potential decreased from −4 to −8 MPa (Kappen *et al.*, 1975). Stomatal closure, loss of assimilatory organs and osmotic adjustment maintain the internal circulation even at extreme water stress (see Chapter 15). Our present concept of leaf and root responses to water stress still needs to take this into account in order to understand the set-point of any control over stomatal functioning.

8.4 Canopy responses to air humidity

Although a large body of data is available about leaf responses to air humidity and soil drought, very little information exists about the extent to which stomata affect the water loss of whole canopies. This will depend on the ratio of aerodynamic to stomatal resistance. The coupling between atmosphere and vegetation will increase with the roughness of the canopy (McNaughton and Jarvis, 1983; see Chapter 18). Information on whole-canopy performance will be important for climate models and predictions of global climate change, as discussed in detail in Chapters 18 and 19.

One example of canopy performance has been collected for a primary *Nothofagus* forest in New Zealand (Köstner *et al.*, 1992). In this case the coupling factor of Jarvis was not constant. The canopy responded like a grassland in the morning and like a coniferous forest in the afternoon (Figure 8.12). The reason for this change in coupling was a response of stomata to VPD in the morning and to plant internal water stress in the afternoon.

These data are presented as half-hourly averages which do not allow much insight into the dynamic response of the regulatory mechanism. At a higher time resolution, xylem sap flow is by no means constant, but highly variable. It oscillates in such a way that several trees apparently lose water in phase (Figure 8.13). Spectral analysis of these oscillations reveals that these fluctuations in stem water flux are similar to those eddies measured above the canopy (Figure 8.14) (Hollinger *et al.*, 1993). Not only are tree transpiration and atmospheric fluxes characterized by fluctuations of similar length, but these fluctuations are also coherent (correlate over a range of frequencies, Figure 8.14b). Sap flow at the top of the tree is coherent not only with sap-flux variations at the tree base, but also with fluctuations of atmospheric water vapour density. The minimum at 10^{-2} Hz suggests that the maximum size of the coherent eddies may be about 100 m. Thus, we can visualize the eddies by measuring the xylem flux in the tree.

The intermittent nature of the turbulence means that foliage experiences a non-steady-state evaporation environment. If stomata do indeed respond to localized feedback from the water relations of the substomatal cavity, then transpiration fluctuations may act as the controlling mechanism that adjusts stomatal aperture to changes in atmospheric humidity. The changes in water flux could be the means by which the leaf 'senses' humidity. The short-term variation in VPD in the canopy, as imposed by large-scale eddies which enter the canopy by advection, will cause short-term changes in transpiration and local changes in water status in the stomatal apparatus, and hence stomatal closure. Since the closing response is much faster than the opening response, the stomata will always be somewhat more closed in the canopy than we would expect from steady-state experiments. During the daily course this type of

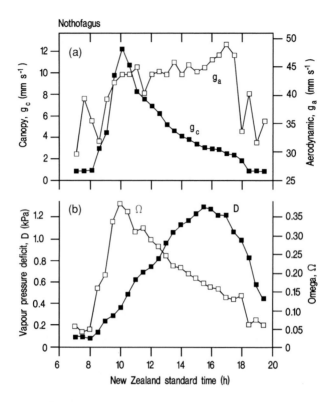

Figure 8.12. *(a) Half-hourly courses of canopy conductance (g_c) and total aerodynamic conductance (g_a) of emergent* Nothofagus fusca *trees, derived from sap flow, humidity and wind speed measurements. (b) The atmospheric coupling parameter (open symbols) and vapour pressure deficit (D, closed symbols) above the canopy. (From Köstner et al., 1992.)*

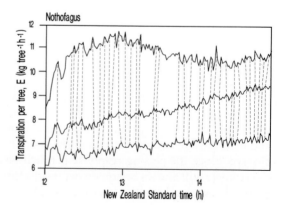

Figure 8.13. *Oscillations of xylem sap flow in adjacent trees.*

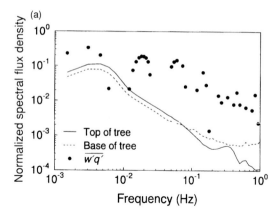

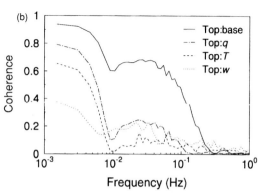

Figure 8.14. (a) Co-spectra of water vapour flux density above the canopy, as measured by the eddy correlation method, and power spectra of xylem sap flux at 1.5 m and 15 m for a 32 m tall tree. (b) Coherence between sap flux measured at 15 m (top) and at 1.5 m (base), as well as coherence between sap flux at 15 m and vapour density (q), air temperature (T) and vertical wind speed (w) as measured above the canopy. (From Hollinger et al., 1992.)

oscillation will cause stomata to close to such an extent that the total water loss from the canopy will be kept constant or even decrease at increasing vapour pressure deficit.

8.5 Conclusions

From the whole-plant perspective, stomatal response and plant water relations demonstrate different control mechanisms. Stomata of *T. virginiana* close at low air humidity due to a cellular feedback to water loss from the stomatal pore. The result of this feedback is a steady-state response which appears like a feedforward response. Stomata also respond to a drying signal from the root. There is increasing evidence that this signal is mediated by ABA. However, the effect of ABA seems to be modulated by nutrition, and additional mechanisms for root–shoot signals cannot be ruled out. A plant internal cycle maintains whole-plant integrity and it may enhance the signal communication between root and shoot. In fact, the maintenance of the plant internal circulation may be the actual controlling mechanism. Finally, short-term fluctuations between demand and supply of water during transpiration may, in fact, be the actual trigger that adjusts stomata in a variable environment.

References

Appleby, R.F. and Davies, W.J. (1983) The structure and orientation of guard cells in plants showing stomatal responses to changing vapor pressure difference. *Ann. Bot.*, **52**, 459–468.

Bates, L.M. and Hall, A.E. (1981) Stomatal closure with soil water depletion not associated with changes in bulk leaf water status. *Oecologia*, **50**, 62–65.

Blackman, P.G. and Davies, W.J. (1984) Age-related changes in stomatal response to cytokinins and abscisic acid. *Ann. Bot.*, **54**, 121–125.

Brinckmann, E., Hartung, W. and Watringer, M. (1990) Abscisic acid of individual leaf cells. *Physiol. Plant.*, **80**, 51–54.

Coutts, M.P. (1981) Leaf water potential and control of water loss in droughted sitka spruce seedlings. *J. Exp. Bot.*, **131**, 1193–1201.

Davies, W.J. and Zhang, J. (1991) Root signals and the regulation of growth and development of plants in drying soil. *Ann. Rev. Plant Physiol. Plant Mol. Biol.*, **42**, 55–76.

Farquhar, G.D. (1978) Feedforward responses of stomata to humidity. *Aust. J. Plant Physiol.*, **5**, 787–800.

Frensch, J. and Schulze, E.-D. (1988) The effect of humidity and light on cellular water relations and diffusive conductance of leaves of *Tradescantia virginiana* L. *Planta*, **173**, 554–562.

Gollan, T., Passioura, J.B. and Munns, R. (1986) Soil water status affects the stomatal conductance of fully turgid wheat and sunflower leaves. *Aust. J. Plant Physiol.*, **13**, 459–464.

Gollan, T., Schurr, U. and Schulze, E.-D. (1992) Stomatal response to drying soil in relation to changes in the xylem composition of *Helianthus annuus*. I. The concentration of cations, anions, amino acids in, and pH of the xylem sap. *Plant Cell Environ.*, **15**, 551–559.

Hollinger, D.Y., Kelliher, F.M. Schulze, E.-D. and Köstner, B.M. (1993) Coupling of tree transpiration to atmospheric turbulence. *Nature*, submitted.

Kappen, L., Oertli, J.L., Lange, O.L., Schulze, E.-D., Evenari, M. and Buschbom, U. (1975) Seasonal and diurnal courses of water relations of the arido active plant *Hammada scoparia* in the Negev desert. *Oecologia*, **21**, 174–192.

Köstner, B.M.M., Schulze, E.-D., Kelliher, F.M., Hollinger, D.Y., Beyers, J.N., Hunt, J.E., McSeveny, T.M., Meserth, R. and Weri, P.L. (1992) Transpiration and canopy conductance in a pristine broadleaf forest of *Nothofagus*: An anlysis of xylem sap flow and eddy correlation measurements. *Oecologia*, **91**, 350–359.

Kramer, P.J. (1988) Changing concepts regarding plant water relations. *Plant Cell Environ.*, **11**, 573–576.

Lange, O.L., Lösch, R., Schulze, E.-D. and Kappen, L. (1971) Responses of stomata to changes in humidity. *Planta*, **100**, 76–86.

Lösch, R. and Schenk, B. (1978) Humidity response of stomata and the potassium content of guard cells. *J. Exp. Bot.*, **29**, 781–787.

McNaughton, K.G. and Jarvis, P.G. (1983) Predicting effects of vegetation changes on transpiration and evaporation. In: *Water Deficits and Plant Growth*, Vol. 7 (ed. T.T. Karlowski). Academic Press, New York, pp. 1–47.

Maier-Maerker, U. (1983) The role of peristomatal transpiration in the mechanism of stomatal movement. *Plant Cell Environ.*, **6**, 369–380.

Munns, R. (1990) Chemical signals moving from roots to shoots: The case against ABA. In: *Importance of Root to Shoot Communication in the Responses to Environmental Stress* (eds W.J. Davies and B. Jeffcoat), British Society for Plant Growth Regulators Monograph, Vol. 21, pp. 175–183.

Nonami, H. and Schulze, E.-D. (1989) Cell water potential, osmotic potential, and turgor in the epidermis and mesophyll of transpiring leaves. *Planta*, **177**, 35–46.

Nonami, H., Schulze, E.-D. and Ziegler, H. (1990) Mechanisms of stomatal movement in response to air humidity, irradiance and xylem water potential. *Planta*, **183**, 57–64.

Raschke, K. (1979) Movements of stomata. *Encycl. Plant Physiol. N.S.*, **7**, 383–441.

Raschke, K. (1982) Involvement of abscisic acid in the regulation of gas exchange: evidence and inconsistencies. In: *Plant Growth Substances* (ed. P.F. Wareing). Academic Press, London, pp. 581–590.

Raschke, K. and Kühl, U. (1970) Stomatal responses to changes in atmospheric humidity and water supply: experiments with leaf sections of *Zea mays* in CO_2-free air. *Planta*, **87**, 36–48.

Schulze, E.-D. (1986) Carbon dioxide and water vapor exchange in response to drought in the soil. *Ann. Rev. Plant Physiol.*, **37**, 247–274.

Schulze, E.-D. (1991) Water and nutrient interactions with plant water stress. In: *Responses of Plants to Multiple Stresses* (eds H.A. Mooney, W.E. Winner and E.J. Pell). Academic Press, San Diego, pp. 89–103.

Schulze, E.-D. and Hall, A.E. (1982) Stomatal responses, water loss, and CO_2 assimilation rates of plants in contrasting environments. *Encycl. Plant Physiol. N.S.*, **12B**, 181–230.

Schulze, E.-D., Lange, O.L., Buschbom, U., Kappen, L. and Evenari, M. (1972) Stomatal responses to changes in humidity in plants growing in the desert. *Planta*, **108**, 259–270.

Schulze, E.-D., Steudle, E., Gollan, T. and Schurr, U. (1988) Response to Dr P.J. Kramer's article, 'Changing concepts regarding plant water relations'. **11**, (7), 656–658.

Schurr, U. (1991) Die Wirkung von Bodenaustrocknung auf den Xylem- und Phloemtransport von *Ricinus communis* und deren Bedeutung für die Interaktion zwischen Wurzel and Sproß. Doctoral thesis, Bayreuth, Germany.

Schurr, U., Gollan, T. and Schulze, E.-D. (1992) Stomatal response to drying soil in relation to changes in the xylem sap composition of *Helianthus annuus*. II. Stomatal sensitivity to abscisic acid imported from the xylem sap. *Plant Cell Environ.*, **15**, 561–567.

Semmel, H., Horn, R., Hell, U., Dexter, A.R. and Schulze, E.-D. (1990) The dynamics of soil aggregate formation and the effect on soil physical properties. *Soil Technol.*, **3**, 113–129.

Sheriff, D.W. (1979) Stomatal aperture and the sensing of the environment by guard cells. *Plant Cell Environ.*, **2**, 15–22.

Tardieu, F., Zhang, J., Katerji, N., Bethenod, O., Palmer, S. and Davies, E.J. (1992) Xylem ABA controls the stomatal conductance of field-grown maize subjected to compaction or soil drying. *Plant Cell Environ.* **15**, 193–198.

Wartinger, A. (1991) Der Einfluß von Austrocknungszyklen auf Blattleitfähigkeit, CO_2-Assimilation, Wachstum und Wassernutzung von *Prunus dulcis* (Miller) D.A. Webb. Dissertation, Bayreuth 1991.

Wartinger, A., Heilmeier, H., Hartung, W. and Schulze, E.-D. (1990) Daily and seasonal course of leaf conductance and abscisic acid in the xylem sap of almond trees (*Prunus dulcis* (Miller) D.A. Webb) under desert conditions. *New Phytol.*, **116**, 581–587.

Zhang, J. and Davies, W.J. (1987) Increased synthesis of ABA in partially dehydrated root tips and ABA transport from roots to leaves. *J. Exp. Bot.*, **38,** 2015–2023.

Zhang, J., Schurr, U. and Davies, W.J. (1987) Control of stomatal behaviour by abscisic acid which apparently originates in roots. *J. Exp. Bot.*, **38**, 1174–1181.

Zimmermann, U., Hüsken, D. and Schulze, E.-D. (1980) Direct turgor pressure measurements in individual leaves of *Tradescantia virginiana*. *Planta*, **149**, 445–453.

9

Root–shoot communication and whole-plant regulation of water flux

F. Tardieu and W.J. Davies

9.1 Introduction

Regulation of stomatal conductance (g_s) enables the plant to modify gas exchange with the atmosphere reversibly, with a time scale of only a few minutes. This short-term regulation may optimize both the dry matter production and the protection from xylem embolism or other detrimental effects of very low leaf water potential (Jones and Sutherland, 1991). Despite these considerations, the modelling of stomatal behaviour under field conditions has received relatively little attention. The main reason for this is that stomatal control, particularly in droughted plants, was thought to be straightforward and to depend mainly on leaf water status and evaporative demand (e.g. Slatyer, 1967). This was despite the fact that it had been known for many years (Berger–Landfelt, 1936; Stocker, 1956) that several 'isohydric' species, such as cereals, will show reduced conductances even when plant water status is unaffected by drought. 'Isohydric' behaviour has since been described in more detail for cowpeas, maize and tomato (Bates and Hall, 1981; Dwyer and Stewart, 1984; Katerji *et al.*, 1988; Tardieu *et al.*, 1992a). Other 'anisohydric' species, such as sugar beet, sunflower and sorghum, can show clear changes in leaf water status (Ψ_l) during drying cycles, so that changes in Ψ_l correlate with changes in g_s.

In the past decade, a theory has been developed to explain 'isohydric' behaviour (e.g. Gollan *et al.*, 1986; Jones, 1980; Turner *et al.*, 1985; see also Chapter 8). It is suggested that g_s can be controlled by a chemical message, originating in dehydrating roots and conveyed by the water flux towards the stomatal complex. Considerable evidence now supports this theory, which has been presented in several review articles (e.g. Davies and Zhang, 1991), and about which there is some controversy (Kramer, 1988; Passioura,

1988). Our aim in this chapter is not to present again the elements in favour of, or against, the theory of a chemical control of stomatal conductance, but to try to review elements for a more quantitative and physical approach to this control, which may enable us to model the stomatal behaviour in 'isohydric' plants. Following Kramer's (1988) suggestions, we have given particular attention to studies carried out under field conditions and to possible discrepancies which may exist between mechanisms observed in controlled environments and in the field.

9.2 Control of stomatal conductance: a combined effect of chemical messages and leaf water status?

There is now a consensus that chemical signalling may control stomata under some circumstances. The main remaining issues are the generality and the importance of such control, especially under field conditions. Which chemical compounds are involved in the message? How consistent are the responses of stomatal conductance to the message, under a variety of environmental conditions? Can we get a better prediction of stomatal behaviour using models which involve chemical signalling rather than using the classical hydraulic models?

9.2.1 *Which chemical compounds are involved in the message? Evidence for stomatal control by the concentration of ABA in the xylem sap*

Although some controversy remains about the role of ABA as the only antitranspirant in the xylem sap, at least for some species (Munns and King, 1988; Trejo and Davies, 1991), the concentration of ABA in the xylem sap is apparently a major factor accounting for stomatal closure in water-stressed plants. Feeding detached wheat leaves with xylem sap collected from unwatered plants produces the same quantitative effect as feeding them with artificial saps with similar ABA concentrations (Zhang and Davies, 1991; Figure 9.1a); this relationship remains unchanged, with increased g_s, if most of the ABA of the xylem sap is removed by passing the sap through an immunoaffinity column. In this experiment, there is no unexplained antitranspirant activity and, therefore, no need to suggest a substantial involvement of any other chemical substance. Comparison of the effects of endogenous and applied ABA in whole-plant experiments (Zhang and Davies, 1989, 1990; Figure 9.1b) also supports the hypothesis that ABA is the main compound with antitranspirant action in the xylem sap. Such comparisons have been carried out for wheat, maize, sunflower and lupin (Henson *et al.*, 1989; Zhang and Davies, 1989, 1990, 1991).

Convincing relationships between stomatal conductance and xylem ABA concentration have been observed under field conditions by Wartinger *et al.* (1990) in *Prunus dulcis*, and by Tardieu *et al.* (1992a) in maize, while relationships between the leaf water status and g_s for the same leaves were extremely loose. In an experiment combining soil drying and soil compaction, we

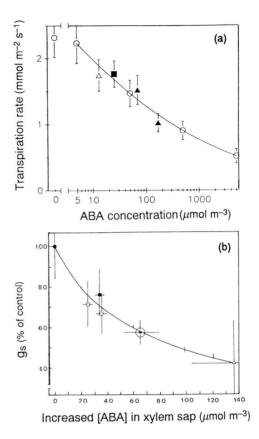

Figure 9.1. *Response of the transpiration rate or stomatal conductance to abscisic acid under laboratory conditions. (a) The transpiration rate of detached wheat leaves in response either to artificial ABA (open circles) or to endogenous ABA in the sap of plants grown with varying levels of soil water depletion (closed symbols). The open triangle represents the response to the sap of a droughted plant after removal of ABA by passing the sap through an immunoaffinity column. (b) Relative leaf conductance of maize plants as a function of increased ABA in the xylem sap of plants grown with varying levels of soil water depletion. The circled point in the middle is the result of ABA feeding of a well-watered plant.*

have established (Tardieu *et al.*, 1992a) that a common relationship between stomatal conductance and xylem ABA concentration applied, for variations in xylem ABA concentration linked to the decline with time of the soil water reserve, to simultaneous differences between plants grown on compacted,

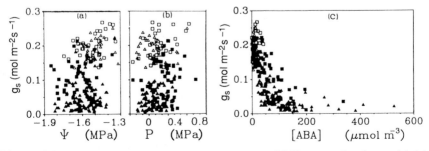

Figure 9.2. *Relationship between stomatal conductance (g_s) of field-grown maize plants and (a) leaf water potential, (b) leaf turgor (P), and (c) the ABA concentration of the xylem sap ([ABA]). Each point represents a coupled potential conductance or ABA conductance measurement for one leaf. ■, Non-irrigated, non-compacted; ▲, non-irrigated, compacted; □, irrigated, non-compacted; △, irrigated, compacted.*

non-compacted and irrigated soil, as well as to plant-to-plant variability (Figure 9.2). This relationship was observed as long as Ψ_l remained in a narrow range of values, and was consistent, over this range of Ψ_l, when artificial ABA was fed directly into the vascular tissue of field-grown maize plants (see Figure 9.3). The experimental relationship between xylem ABA concentration and g_s is therefore likely to be causal, and there is little need to invoke, for maize, antitranspirant compounds other than ABA in the sap. In particular, it is important to note that the effect of soil compaction on g_s could be accounted for by the common relationship between xylem ABA concentration and g_s, without the necessity for other root messages (Tardieu et al., 1992b). It seems likely that much of the extra xylem ABA that we detect in plants grown in drying or compacted soil originates in the roots and not in the shoots, since shoot water relations of plants grown on irrigated, non-irrigated and compacted soil were not detectably different.

9.2.2 Interaction between leaf water status and chemical message

The above-mentioned experiments suggest that, *for given experimental conditions*, the sensitivity of g_s to xylem ABA concentration is independent of the cause of variations in ABA concentration. However, this sensitivity is probably not a fixed characteristic in each species or cultivar, as suggested by the comparison between response curves obtained in the laboratory (Figure 9.1) and the field (Figure 9.2). Observed reductions in g_s due to ABA feeding or to drought are not greater than 60% in laboratory experiments, even when the xylem ABA concentration is increased by more than 1000 μmol m^{-3} (Zhang and Davies, 1990, 1991). Lower sensitivity (less than 20%) is observed when Ψ_l is maintained at high values by pressurizing the root system in a chamber (Schurr and Gollan, 1990). In contrast, the reduction in g_s can be greater than 90% for an increase in xylem ABA concentration of less than 200 μmol m^{-3} for plants in the field (Tardieu et al., 1992a; Wartinger et al., 1990).

Some variation in sensitivity between experimental conditions within a cultivar is suggested by the observations of Henson et al. (1989): g_s was only halved by feeding detached leaves of lupin with an artificial sap having an ABA concentration as high as 10 mmol m^{-3} [(+)-ABA], while it decreased by 80% when whole plants of the same lupin cultivar were grown in a drying soil in a greenhouse (unfortunately, the xylem ABA concentration was not measured in this experiment). Experiments in the field also suggest that stomatal sensitivity to xylem ABA concentration cannot be considered as a constant within a cultivar: in field-grown maize, sensitivity increased appreciably with the time of day, being relatively low during the morning (200 μmol m^{-3} ABA reduced g_s by 50%) and increasing later in the day (a reduction in g_s of 90% occurring with the same ABA concentration). In the experiment of Wartinger et al., apparent stomatal sensitivity to ABA differed markedly between trees (personal communication).

We have suggested (Tardieu and Davies, 1992) that ABA concentration and

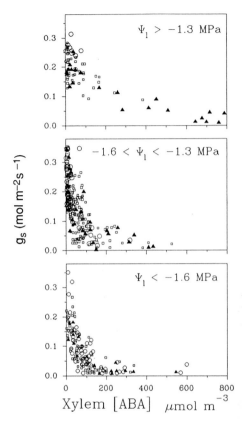

Figure 9.3. *Stomatal conductance* (g_s) *of field-grown maize plants as a function of the concentration of ABA in the xylem sap, for three ranges of leaf water potential* (Ψ_1). $\square$, *Leaves studied in 1990, natural variations in xylem ABA concentration, variations in Ψ_1 correspond to change with time;* $\blacktriangle$, *leaves of fed plants;* $\bigcirc$, *leaves studied in 1991, natural variations in xylem ABA concentration. Each point represents coupled values corresponding to one leaf.*

leaf water potential may interact in their effects on stomata. This interaction would provide an explanation for the commonly observed reduction in g_s shown by droughted plants during the afternoon hours (low Ψ_1) but not in the morning (high Ψ_1), in spite of the fact that the ABA signal is relatively constant, or even decreases, throughout the day (Tardieu *et al.*, 1992b). Stomata may open during the early hours of the day because of their relative insensitivity to this signal at this time. As the day progresses, evaporative demand will cause a decline in Ψ_1 and thus increase stomatal sensitivity to the ABA signal. An experiment where field-grown maize plants were fed with ABA (Figure 9.3) allowed this hypothesis to be tested, since feeding plants with ABA causes a decrease in g_s, but also an increase in Ψ_1. Leaves with Ψ_1 higher than -1.3 MPa were less sensitive to the xylem ABA concentration than those at lower Ψ_1, regardless of the cause of the high Ψ_1: ABA feeding with high evaporative demand, or low evaporative demand. Other explanations of the observed change in stomatal sensitivity during the day have been considered as unlikely. According to our data, changes in the pH or in the calcium composition of the sap could not account for observed changes in sensitivity. The possibility of a direct effect of vapour pressure deficit (VPD) being a major controlling factor of stomatal sensitivity is also unlikely, since joint variations in Ψ_l and in sensitivity were obtained in the feeding experiment without a change in VPD. Another candidate for higher sensitivity in the afternoon was the increased delivery of ABA from the xylem stream to the stomata: if the xylem ABA concentration is constant, increased transpiration rates at higher saturation deficits later in the day result in a greater ABA flux into the leaf. Again, this possibility fails to explain our observations, since an effect of Ψ_l was observed independently of the variations in evaporative demand in the feeding experiment. Neither is it supported by our calculations of ABA flux made on this dataset. An inter-action between the ABA signal and the leaf water potential is therefore the most likely explanation for the changing sensitivity during the day.

The causes of this interaction are not fully understood. They may involve a direct effect of Ψ_l on stomatal response, as suggested by an experiment (Tardieu and Davies, 1992) where detached epidermes of *Commelina communis* were incubated with a variety of ABA and polyethylene glycol concentrations. In the absence of ABA, stomatal aperture was not affected significantly by the water potential of the solution in the range between -0.3 and -1.5 MPa, but its sensitivity to ABA increased dramatically in this range of potential. This suggests a direct effect of the epidermal water status on stomatal sensitivity due, for instance, to an increased sensitivity of the ABA receptors in the guard cell, or to a non-linear relationship between the guard cell turgor and the stomatal aperture, which would result in a greater effect of ABA if the turgor is already affected by the water potential. Another possible explanation of the interaction in field conditions is an alteration in the apoplastic ABA con-centration due to a redistribution of ABA within the leaf, such that ABA moves to the guard cells from the symplast of mesophyll cells when leaves are droughted (Hartung and Slovik, 1991; Hartung *et al.*, 1988). A change of the transpiration-stream pathway between the xylem and the stomatal com-plex (Canny, 1990) may also affect the partitioning of root-sourced ABA between symplast and apoplast when Ψ_l decreases appreciably (see also Chapter 8).

9.2.3 *Modelling stomatal behaviour at the whole-plant level*

The arguments presented above lead us to consider that stomatal conductance could be modelled from a combination of xylem ABA concentration and current Ψ_l, for saturating PPFD:

$$g_s = g_{s\,min} + \alpha \exp\{[ABA]\,\beta \exp{(\delta\Psi_l)}\} \qquad 9.1$$

where $g_{s\,min}$ and $(\alpha + g_{s\,min})$ are the minimum and maximum g_s, respectively, and β and δ are fitted parameters. Experimental data for establishing this model and calculating the parameters (Tardieu *et al.*, 1993) are those presented in Figure 9.3 (data of 1990 only), and a response curve obtained with detached leaves in an experiment similar to that of Munns and King (1988) where Ψ_l

remained consistently at $-0.2\,\text{MPa}$. The model has been tested (Tardieu *et al.*, 1993) by using independent data, obtained in the laboratory (Figure 9.1b) and under field conditions. It predicted with good accuracy the behaviour of g_s with changing xylem ABA concentration, due to endogenous variations or to feeding, in the field as well as in the laboratory.

A new field of research will be to investigate to what extent this model could apply to species other than maize and, if it does, to calculate its parameters for several species or cultivars. This could be useful in research programmes comparing cultivars with contrasting stomatal behaviour or ability to synthesize ABA.

9.3 Variations in xylem ABA concentration in plants subjected to a water constraint: can they be accounted for by soil water status only?

In many annual plants, such as sunflower or maize, the concentration of ABA in the xylem sap usually ranges between 1 and $10\,\mu\text{mol m}^{-3}$ in well-watered plants and can increase by two orders of magnitude when plants are subjected to soil drying. Key questions for modelling are: which of the variables that change with soil drying will affect the xylem ABA concentration, and are we able to predict variations in xylem ABA concentration with changing environmental conditions?

9.3.1 *Evidence for an effect of the water flux on the ABA message*

Soil water status has been invoked by several authors as a trigger for the synthesis of root messages (Davies and Zhang, 1991; Gollan *et al.*, 1986). In simplified approaches used for modelling, Jones (1980) and Ludlow *et al.* (1990) have hypothesized a relationship between the root message and the soil water reserve. These views are supported by the results of Zhang and Davies (1989) and Henson *et al.* (1989) who reported good correlations between the soil water potential (Ψ_s) and the concentration of ABA in the xylem sap or in the roots. A relationship has also been found in field conditions between pre-dawn Ψ_l and xylem ABA concentration (Tardieu *et al.*, 1992b; Wartinger *et al.*, 1990) or root ABA concentration (Figure 9.4). These arguments therefore support the view that the message should be linked to the soil water status. However, two reasons have led us to consider that this view is not compatible with our knowledge of ABA synthesis and of soil–plant water relations.

Dependence of the message synthesis on the water flux. ABA is synthesized by dehydrating roots (Cornish and Zeevaart, 1985; Walton *et al.*, 1976), in non-growing tissues as well as in apices, and in the cortex as well as in the stele (Hartung and Davies, 1991). The synthesis is roughly proportional to the root water status in all parts of a mature root system of maize (unpublished data). The 'root message' in drying soil should therefore be linked to the

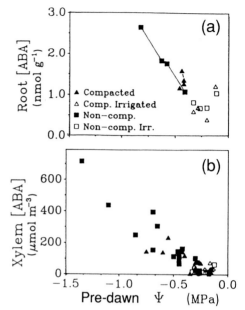

Figure 9.4. *ABA concentration in the roots and xylem sap collected before dawn, plotted against the pre-dawn leaf water potential (field-grown maize plants). Symbols as in Figure 9.2, plus:* ▽, *compacted soil, 1989;* ○, *non-compacted soil, 1989.*

water status of the root system. *This water status must differ from that in the bulk soil whenever water flows from soil to roots.* During the day, a gradient of soil water potential appears around each root, with gradients often being appreciable when the soil dries (see, for example, Dunham and Nye, 1973; Lafolie *et al.*, 1991). Another fall in water potential occurs at the soil–root interface (see, for example, Passioura, 1988; Tinker, 1976). The magnitude of the gradients of water potential in the water pathway depends on the water flux through the soil–plant–atmosphere continuum and on the resistance to water flow in the soil and at the soil–root interface. The root water status therefore depends on *both* the bulk soil water potential and the water flux. As a consequence, ABA synthesis by roots, depending on the root water status, cannot be independent of the water flux through the soil–plant continuum.

Dilution of the message. If the message is an ABA concentration in the xylem sap and not an ABA flux into the leaves (Gowing, 1991; see also Section 9.2 above), it should be related to the reciprocal of the water flux. During the time-course of the day, root water potential decreases from its pre-dawn value to an unknown intermediate value between pre-dawn and current leaf water potentials. ABA synthesis in the roots should therefore increase during the day. The ABA concentration in the xylem sap does not tend to increase in such a way (Tardieu *et al.*, 1992b). The observed tendency was towards a decline in the early morning followed by relative stability, also observed in *Prunus dulcis* by Wartinger *et al.* (1990). Dilution of root-source ABA by the water flow is, therefore, likely to have occurred. The message cannot be considered as depending solely on the soil water status.

9.3.2 *Modelling the xylem ABA concentration evolution*

The argument outlined above suggests that the root message should be related to the root water potential and inversely related to the water flux. In order to simplify our system, we shall consider a unique rate of ABA synthesis for the whole root system and shall not take into account the translocation of ABA from shoots to roots via the phloem, which is relatively less important in stressed than in non-stressed plants (Wolf *et al.*, 1990). We shall also consider a common root water potential in the root system. Appreciable synthesis of ABA is expected to take place only when the soil dries, so the resistance to water flow in the soil takes high values. Under these conditions, the relative importance of the resistance within the root system decreases, so a simplifying hypothesis is to consider one water potential for the whole root system (see the discussion in Tardieu *et al.*, 1992c). The concentration in ABA in the xylem sap would therefore be given by the equation:

$$[ABA] = J_{ABA}/(J_w + b) = a\Psi_r/(J_w + b) \qquad 9.2$$

where [ABA] is the concentration of ABA in the xylem sap, J_{ABA} is the flux of ABA considered to be linearly related to Ψ_r and diluted by the water flux. The basal level of [ABA] when Ψ_r is close to 0 is extremely low for maize (Tardieu *et al.*, 1992b), sunflower (Schurr and Gollan, 1990; Zhang and Davies, 1989) and *Prunus dulcis* (Wartinger *et al.*, 1990), but may be higher for cherry trees (Gowing, 1991), *Quercus* spp. and *Phaseolus* (Trejo and Davies, 1991).

9.3.3 *Can the 'synthesis–dilution' model account for experimental situations?*

Stability of the message during the day. The observed relative stability of xylem ABA concentration during the day can be expected from Equation 9.2. During this period:

$$-J_w = (\Psi_l - \Psi_r)/R_p \qquad 9.3$$

where R_p is the resistance to water flow in the plant, which can frequently be considered as independent of the water flux (Passioura, 1984; Simonneau, 1992). It follows that the term Ψ_r/J_w, controlling the synthesis of ABA, would be the sum of R_p and Ψ_l/J_w. This sum will undergo considerably smaller variations than Ψ_r in the time-course of the day.

Experiments with soil compaction. For a given soil water reserve or pre-dawn Ψ_l, daytime values of xylem ABA concentration were much higher in plants grown in compacted than in non-compacted soil, but pre-dawn xylem ABA concentrations did not differ (Figures 9.4b and 9.5a). This suggests that the message did not respond directly to root mechanical impedance, as suggested by Masle and Passioura (1987), since in such a case we would expect to observe changes in xylem ABA concentration during the night also. In addition, daytime increases in xylem ABA concentration occurred after silk-

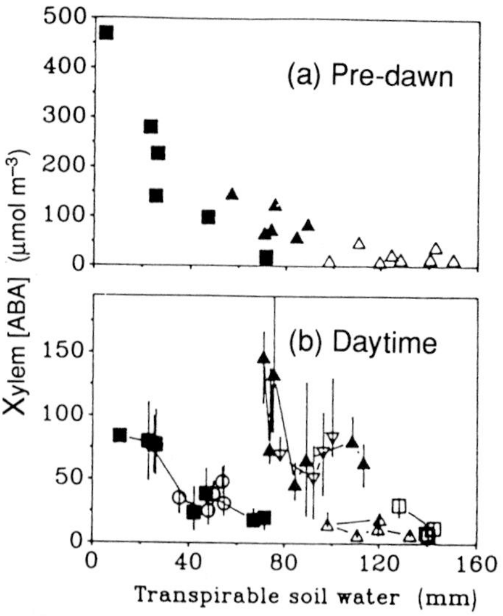

Figure 9.5. ABA concentration in the xylem sap of field-grown maize plants plotted against the transpirable soil water, at pre-dawn and daytime. Values in (b) correspond to the median, bars correspond to quartiles. Symbols as in Figure 9.4.

ing, when root growth had almost stopped, so mechanical impedance should be low regardless of the soil water status. An alternative explanation could depend on water relations. Soil compaction generates root clumping, thereby increasing the resistance to water flux in the soil and reducing the water uptake even with a high soil water reserve (Tardieu, 1988; Tardieu *et al.*, 1992c). As a consequence, the difference between soil and root water potentials, and therefore the synthesis of message for a given soil water status, should increase. The reduction in water flux, linked to lower water uptake, could also contribute to high levels of the message since it reduces its dilution. These effects can be simulated by modelling (see Section 9.4 and Figure 9.6).

'Split-root' experiments. A review of the results of 'split-root' experiments provides somewhat contradictory information. In such experiments, part of the root system can supply the plant with a high water flux while another part is subjected to soil drying. Different primary roots with their branches can be split between two different compartments (Zhang *et al.*, 1987) or different fractions of the same roots can be 'split' vertically (Neales *et al.*, 1989; Zhang and Davies, 1989). The latter situation is comparable to the general case in field conditions: the first soil layers are frequently close to 'wilting point' while deep roots are still in moist soil. While laboratory experiments show that subjecting part of the root system to low water potential can increase xylem ABA concentration by up to one order of magnitude, this is not observed in field situations as long as part of the root system is located in deep, wet soil layers (Tardieu and Katerji, 1991). Xylem ABA concentration does not increase appreciably until the soil water reserve is almost depleted and the pre-dawn leaf water potential decreases to low values (Tardieu *et al.*,

1992b; Wartinger *et al.*, 1990). This behaviour is consistent with the common observation (Katerji *et al.*, 1988; Tardieu and Katerji, 1991) that stomatal conductance is often linked, in field conditions, to the soil water availability (as measured, for instance, by the pre-dawn leaf water potential) rather than linked to the water potential of the driest soil layers. It can be argued that stomatal control by the water status of the driest layer would not be reasonable under field conditions, since drying of upper layers is frequent and would lead to a drastic effect of short droughts on photosynthesis. Furthermore, plants would adapt their conductance in the same way in a shallow soil with a low water reserve and in a deep soil which still has high water availability.

Why could 'split-root' situations cause increased xylem ABA concentrations in laboratory conditions and not in field conditions? Part of the answer could be in the ability of shallow roots, subjected to frequent stress, to synthesize ABA, or in the absence of appreciable water flow from these roots to the shoot, limiting ABA flux to the shoot. Taking into account the dilution of ABA by the water flux may also help in interpreting these differences. The same increase in ABA synthesis by roots will have a considerably greater effect on xylem ABA concentration if the water flux is slow, as in most laboratory experiments, than if flux is fast, as in many field conditions. Dehydration of part of the root system, causing an extra synthesis of ABA, could thereby have an appreciable effect on xylem ABA concentration and g_s in laboratory conditions and a negligible effect in field conditions.

Even if the approach leading to Equation 9.2 is an oversimplification of reality, we think that taking into account both the synthesis of the message and its dilution in the water flow is more correct than the approach taken in most published work, where only the soil water status and the synthesis of the message are considered. Our approach leads to simple explanations of the daily time-course of the xylem ABA concentration or of its change with time under contrasting soil conditions (see Figure 9.6). As in the paragraph about stomatal response, we are led to combine chemical and hydraulic approaches for modelling root-to-shoot communication.

9.4 A preliminary model integrating hydraulic and chemical approaches of root-to-shoot communication

The system described in the first two sections of this chapter has multiple feedbacks: stomatal conductance, which controls the water flux through the plant, would be controlled by a water-flux-dependent message, with a sensitivity that depends on the leaf water potential, whose value differs from the soil water potential by a gradient that depends on the water flux. We have investigated whether this system is conceivable: is there any solution to such a system? If so, is this solution unique? Can its variation with evaporative demand and soil water potential be considered as reasonable?

In order to answer these questions, we have combined Equations 9.1 and 9.2 with the classical equations of water transfer, in order to describe a complete system (Tardieu, 1993): (1) Van den Honert equations, which give relationships between soil water potential, root water potential, leaf water potential and water flux, and (2) the Penman–Monteith equation relating the water flux to stomatal conductance.

The most important result of these simulations is probably that the experimentally observed multiple interactions and feedbacks have a unique solution in the wide range of conditions investigated.

A second result is that the system not only provides a sensible control of stomatal conductance, but also provides control of the leaf water potential and of the concentration of ABA in the xylem sap. Figure 9.6 presents two examples of the daily pattern of the output, for an initial soil water potential of -0.3 MPa. The first simulates a 'favourable' root system with a low resistance to the water flow in the soil–root pathway (R_{sp}); the second simulates a root system grown in compacted soil, where R_{sp} is multiplied by 20 (see Section 9.3.3). Such a difference in R_{sp} generates a reduction in g_s with the characteristic non-symmetrical daily pattern which is generally observed in the field (Figure 9.6a). This daily pattern of g_s is also observed in 'favourable' root systems for lower soil water potential, and is due in our system to the increase in stomatal sensitivity to ABA concentration with daytime decrease in Ψ_l. The system also provides a regulation of the daily time-course of xylem ABA concentration and leaf water potential. Xylem ABA concentration is stable except in the early morning, as it probably is in field-grown maize (Tardieu *et al.*, 1992b). Interestingly, an increase in R_{sp} provokes an increase in xylem ABA concentration. This is also observed for longer-term simulations (Figure 9.7), as it is in experimental data (see Section 9.3.3). In spite of changes in g_s, water flow and xylem ABA concentration, the leaf water potential undergoes relatively small variations. This behaviour is, again, similar to that observed under field conditions for maize (Tardieu *et al.*, 1992b).

9.5 Conclusions

The classical controversy over the mechanism of stomatal control, whether chemical or hydraulic, may reflect two faces of a single phenomenon. Convincing relationships can be found which support either of these types of control, but they fail to apply to all conditions, so it is necessary to argue for a more complex control. Water relations are probably involved, to some extent, in the generation and in the transport of root messages, and in stomatal sensitivity to the message, so we can expect stomatal conductance to be apparently more related to hydraulic variables in some cases, or to chemical variables in other cases, depending on species ('isohydric' vs. 'anisohydric' species) or environmental conditions. Common regulation by hydraulic and chemical messages may help us to understand this complexity and appears to account for experimental observations. We are aware that some processes

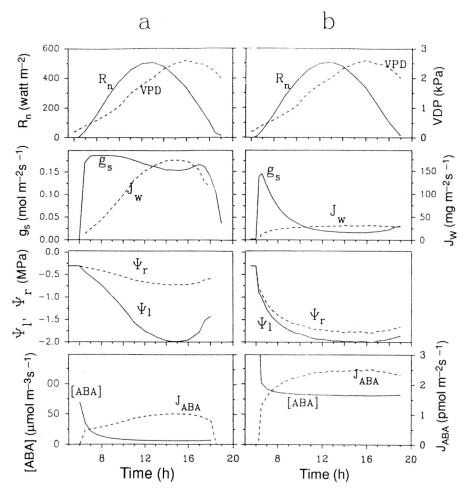

Figure 9.6. *Simulations of the daily pattern of stomatal conductance (g_s), water flux in the soil–plant–atmosphere continuum (J_w), leaf and root water potentials (Ψ_l and Ψ_r), ABA concentration in the xylem sap ([ABA]) and ABA flux to the leaf (J_ABA). Soil water potential (Ψ) is −0.3 MPa, meteorological conditions (net radiation, R_n, and air vapour pressure deficit, VPD) are those measured on 26 July 1990. Soil characteristics: a silty clay loam. (a) Calculations with values of soil–root resistance to water transfer (R_sp) simulating a root system in favourable conditions. (b) Calculations with a value of R_sp multiplied by 20, simulating unfavourable characteristics of the root system for water uptake (such as root clumping).*

described here are oversimplifications of reality. Nevertheless, the framework of a common feedback regulation could remain correct and provide the basis for further models.

It is appropriate that the controlling systems are modified by signals that vary with a similar time scale to the process being controlled. A gradually increasing xylem ABA concentration in drying soil, with a relatively stable signal on a daily basis, may provide reliable information on the functioning of the root system (soil water status and resistances to water flux) and act as

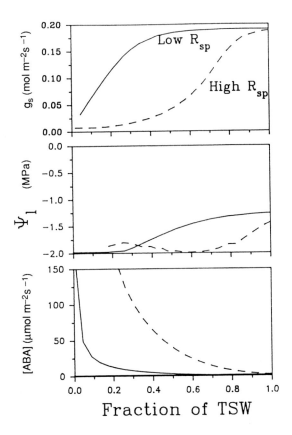

Figure 9.7. *Long-term evolution of stomatal conductance (g_s), leaf water potential (Ψ_l), and ABA concentration in the xylem sap ([ABA]) as a function of the level of the soil water reserve, expressed as a fraction of the transpirable soil water (TSW).*

a suitable developmental regulator. The variation in sensitivity of the shoot response provides the capacity for a dynamic response to any rapid change in the atmospheric environment. The generation of, and response to, an ABA signal can therefore play a central part in the day-to-day regulation of stomatal behaviour. This will be the case particularly in drying soil, but, even when the plant is well supplied with water, stomatal response to changing air vapour pressure deficit might be mediated by changing sensitivity to a baseline ABA supply through the xylem.

References

Bates, L.M. and Hall, A.E. (1981) Stomatal closure with soil depletion not associated with changes in bulk leaf water status. *Oecologia*, **50**, 62–65.

Berger-Landfelt, U. (1936) Der Wasserhaushalt der Alpenpflanzen. *Bibl. Bot.*, **1936**, H115.

Canny, M.J. (1990) What becomes of the transpiration stream? *New Phytol.*, **114**, 341–368.

Cornish, K. and Zeevaart, J.A.D. (1985) Abscisic acid accumulation by roots of *Xanthium strumarium* L. and *Lycopersicon esculentum* Mill. in relation to water stress. *Plant Physiol.*, **79**, 653–658.

Davies, W.J. and Zhang, J. (1991) Root signals and the regulation of growth and development of plants in drying soil. *Ann. Rev. Plant Physiol. Plant Mol. Biol.*, **42**, 55–76.

Dunham, R.J. and Nye, P.H. (1973) The influence of water content on the uptake of ions by roots. I Soil water content gradient near a plane of onion roots. *J. Appl. Ecol.*, **10**, 585–598.

Dwyer, L.M. and Stewart, D.W. (1984) Indicators of water stress in corn. *Can. J. Plant Sci.*, **64**, 537–546.

Gollan, T., Passioura, J.B. and Munns, R. (1986) Soil water status affects the stomatal conductance of fully turgid wheat and sunflower leaves, *Aust. J. Plant Physiol.*, **13**, 459–464.

Gowing, D.J.G. (1991) The sensing of drying soil by roots and the involvement of ABA as a signal. PhD thesis, University of Lancaster.

Hartung, W. and Davies, W.J. (1991) Drought-induced changes in physiology and ABA. In: *Abscisic acid, Physiology and Biochemistry* (eds W.J. Davies and H.G. Jones). BIOS Scientific Publishers, Oxford, pp. 63–79.

Hartung, W. and Slovik, S. (1991) Physicochemical properties of plant growth regulators and plant tissues determine their distribution and redistribution: stomatal regulation by abscisic acid in leaves. *New Phytol.*, **119**, 361–382.

Hartung, W., Radin, J.W. and Hendrix, D.L. (1988) Abscisic acid movement into the apoplastic solution of water stressed cotton leaves. Role of apoplastic pH. *Plant Physiol.*, **86**, 908–913.

Henson, I.E., Jensen, C.R. and Turner, N.C. (1989) Leaf gas exchange and water relations of lupin and wheat. III Abscisic acid and drought-induced stomatal closure. *Aust. J. Plant Physiol.*, **16**, 429–442.

Jones, H.G. (1980) Interaction and integration of adaptative responses to water stress: the implications of an unpredictable environment. In: *Adaptation of Plants to Water and High Temperature Stress* (eds N.C. Turner and P.J. Kramer). Wiley, New York, pp. 353–365.

Jones, H.G. and Sutherland, R.A. (1991) Stomatal control of xylem embolism. *Plant Cell Environ.*, **14**, 607–612.

Katerji, N., Itier, B. and Ferreira, I. (1988) Etude de quelques critères indicateurs de l'état hydrique d'une culture de tomate en région semi aride. *Agronomie*, **8**, 425–433.

Kramer, P.J. (1988) Changing concepts regarding plant water relations. *Plant Cell Environ.*, **11**, 565–568.

Lafolie, F., Bruckler, L. and Tardieu, F. (1991) Modelling the root water potential and soil–root water transport in the two-dimensional case. 1 Model presentation. *Soil Sci. Soc. Am. J.*, **55**, 1203–1212.

Ludlow, M.M., Sommer, K.J. and Muchow, R.C. (1990) Agricultural implications of root signals. In: *Importance of Root to Shoot Communication in the Response to Environmental Stress* (eds W.J. Davies and B. Jeffcoat). BSPGR monograph 21, Bristol, p. 251.

Masle, J. and Passioura, J.B. (1987) The effect of soil strength on the growth of young wheat plants. *Aust. J. Plant Physiol.*, **14**, 643–656.

Munns, R. and King, R.W. (1988) Abscisic acid is not the only stomatal inhibitor in the transpiration stream of wheat plants. *Plant Physiol.*, **88**, 703–708.

Neales, T.F., Masia, A., Zhang, J. and Davies, W.J. (1989) The effect of partially drying part of the root system of *Helianthus annuus* on the abscisic acid content of roots, xylem sap and leaves. *J. Exp. Bot.*, **40**, 1113–1120.

Passioura, J.B. (1984) Hydraulic resistance of plants. I. Constant or variable? *Aust. J. Plant Physiol.*, **11**, 333–339.

Passioura, J.B. (1988) Water transport in and to roots. *Ann. Rev. Plant Physiol. Plant Mol. Biol.*, **39**, 245–265.

Schurr, U. and Gollan, T. (1990) Composition of xylem sap of plants experiencing root water stress – a descriptive study. In *Importance of Root to Shoot Communication in the Response to Environmental Stress* (eds W.J. Davies and B. Jeffcoat). BSPGR monograph 21, Bristol, pp. 201–214.

Simonneau, Th. (1992) PhD thesis, Institut National Agronomique, Paris, Grignon.

Slatyer, R.O. (1967) *Plant Water Relationships*, Academic Press, London.

Stocker, O. (1956) Die Abhängigkeit der Transpiration von den Umweltfaktoren. In: *Encyclopedia of Plant Physiology* (ed. W. Ruhland) Vol. III. Springer-Verlag, Berlin, pp. 436–488.

Tardieu, F. (1988) Analysis of the spatial variability of maize root density. III. Effect of a wheel compaction on water extraction. *Plant Soil*, **109**, 257–262.

Tardieu, F. (1993) Will progresses in understanding soil–root relations and root signalling substantially alter water flux models? *Phil. Trans. R. Soc. London*, 338.

Tardieu, F. and Davies, W.J. (1992) Stomatal response to ABA is a function of current plant water status. *Plant Physiol.*, **98**, 540–545.

Tardieu, F. and Katerji, N. (1991) Plant response to the soil water reserve: consequences of the root system environment. *Irrigation Sci.*, **12**, 145–152.

Tardieu, F., Zhang, J., Katerji, N., Bethenod, O., Palmer, S. and Davies, W.J. (1992a) Xylem ABA controls the stomatal conductance of field-grown maize subjected to soil compaction or soil drying. *Plant, Cell Environ.*, **15**, 193–197.

Tardieu, F., Zhang, J. and Davies, W.J. (1992b) What information is conveyed by an ABA signal from maize roots in drying field soil? *Plant, Cell Environ.*, **15**, 185–191.

Tardieu, F., Bruckler, L. and Lafolie, F. (1992c) Root clumping may affect the root water potential and the resistance to soil–root water transport. *Plant Soil*, **140**, 291–301.

Tardieu, F., Zhang, J. and Gowing, D.J.G. (1993) Stomatal control by both [ABA]in the xylem sap and leaf water status: test of a model and of alternative hypotheses for droughted or ABA-fed field-grown maize. *Plant Cell Environ.*, **16**, 413.

Tinker, P.B. (1976) Transport of water to plant roots in soil. *Phil. Trans. R. Soc. Lond. B*, **273** 445–461.

Trejo, C. and Davies, W.J. (1991) Drought induced closure of *Phaseolus vulgaris* stomata precede leaf water deficit and any increase in xylem ABA concentration. *J. Exp. Bot.*, **42**, 1507–1515.

Turner, N.C., Schulze, E.D. and Gollan, T. (1985) The response of stomata and leaf gas exchange to vapour pressure deficits and soil water content. II In the mesophytic herbaceous species *Helianthus annuus*. *Oecologia*, **65**, 348–355.

Walton, D.C., Harrison, M.A. and Cote, P. (1976) The effects of water stress on abscisic acid levels and metabolism in roots of *Phaseolus vulgaris* and other plants. *Planta*, **131**, 141–144.

Wartinger, A., Heilmeier, H., Hartung, W. and Schultze, E.D. (1990) Daily and seasonal courses of leaf conductance and abscisic acid in the xylem sap of almond trees (*Prunus dulcis* M.) under desert conditions. *New Phytol.*, **116**, 581–587.

Wolf, O., Jeschke, W.D. and Hartung, W. (1990) Long distance transport of abscisic acid in salt stressed *Lupinus albus* plants. *J. Exp. Bot.*, **41**, 593–600.

Zhang, J. and Davies, W.J. (1989) Sequential response of whole plant water relations to prolonged soil drying and the involvement of xylem sap ABA in the regulation of stomatal behaviour of sunflower plants. *New Phytol.*, **113**, 167–174.

Zhang, J. and Davies, W.J. (1990) Changes in the concentration of ABA in xylem sap as a function of changing soil water status can account for changes in leaf conductance and growth. *Plant Cell Environ.*, **13**, 277–285.

Zhang, J. and Davies, W.J. (1991) Antitranspirant activity in xylem sap of maize plants. *J. Exp. Bot.*, **42**, 317–321.

Zhang, J., Schurr, U. and Davies, W.J. (1987) Control of stomatal behaviour by abscisic acid which apparently originates in the roots. *J. Exp. Bot.*, **192**, 1174–1181.

Rectifier-like behaviour of root–soil systems: new insights from desert succulents

Park S. Nobel and Gretchen B. North

10.1 Introduction

Although a primary function of roots is water uptake (Chapter 2), a problem for plants in general, and for desert succulents in particular, is how to limit water movement from roots to the soil when the roots and shoot have a higher water potential than the soil. Bidirectional water movement can be both temporal and spatial, as for some desert species whose deep roots in moist soil take up water that is subsequently lost to drier soil near the soil surface at night, only to be taken up again during the next daytime (Caldwell and Richards, 1989; Mooney et al., 1980; Richards and Caldwell, 1987). For shallow-rooted desert succulents–indeed, for most perennial plants with roots in temporarily dry soil–the reduction of water loss by roots during drought is mandatory. Water conservation without permanent decreases in root hydraulic conductivity would be beneficial, particularly for desert plants that respond quickly to episodic rainfall.

Root systems gain or lose water depending on the water potentials and hydraulic conductances of the three components of the root–soil pathway: the root, an air gap that can occur between the root and the soil, and the soil. Which of these conductances governs water exchange depends primarily on the soil moisture content. For wet soil, root hydraulic conductivity is generally the primary limiter of water uptake (Blizzard and Boyer, 1980; Nobel and Cui, 1992a; Passioura, 1988). As soil dries, roots of many species shrink radially (Faiz and Weatherley, 1978; Rowse and Goodman, 1981; Taylor and Willatt, 1983), and the conductance of the resulting air gap between roots and the soil can become limiting (Nobel, 1992; Nobel and Cui, 1992a). As the soil dries further, its hydraulic conductance decreases greatly and the soil becomes the primary limitation on water movement in the root–soil pathway. Hydraulic conductances of roots, the root–soil air gap and soil have been investigated for the monocotyledonous leaf succulent *Agave deserti* and two dicotyledonous stem succulents, the barrel cactus *Ferocactus acanthodes*, and the prickly-pear cactus *Opuntia ficus-indica*. The middle of the root zone of the first two species is approximately 0.10 m below the soil surface in their native Sonoran Desert, and roots of *O. ficus-indica*, which is cultivated world-wide, are typically about twice as deep (Nobel, 1988). Roots of all three species can be in soils with water potentials (Ψ_{soil}) lower than -10 MPa for several months, yet their shoot water potentials rarely fall below -0.7 MPa (Nobel, 1988).

To account for the small amount of water lost from the shoot of *A. deserti* during a 6-month drought, the hydraulic conductance of the root–soil system must decrease by over five orders of magnitude (Schulte and Nobel, 1989), and similar decreases are expected for the other two species. Decreases in hydraulic conductance are reversible in *A. deserti*, *F. acanthodes* and the sympatric species *Opuntia basilaris*, as indicated by water uptake and stomatal opening within 1–2 days after rainfall (Nobel, 1976, 1977; Szarek *et al.*, 1973). A decrease in hydraulic conductance that limits plant water loss during drought followed by a rapid increase in conductance when soil moisture is restored has been called rectification (Nobel and Sanderson, 1984; Shone and Clarkson, 1988). Changes in the components of the root–soil pathway that underlie such rectification for *A. deserti*, *F. acanthodes* and *O. ficus-indica* are analysed in this chapter.

10.2 Root hydraulic conductivity — responses to drying

In the steady state, the volume of water flowing across each component of the root–soil pathway is constant. The volumetric water flux density into a root (J_V; $m^3 m^{-2} s^{-1} = m s^{-1}$) is related to root hydraulic conductivity (L_P; $m s^{-1} MPa^{-1}$) as follows (Nobel, 1991):

$$J_V = L_P(\Psi_{surface} - \Psi_{xylem}) \qquad 10.1$$

where $\Psi_{surface}$ is the water potential at the root surface and Ψ_{xylem} is the water potential of the root xylem. L_P has been determined for individual roots in solution and in soil by measuring J_V induced by applying a partial vacuum to the distal end of an excised root (Nobel *et al.*, 1990). Plants were grown in the glasshouse, in large containers of sandy desert soil, and drought was imposed by withholding water.

For wet soil, L_P for roots of *A. deserti*, *F. acanthodes* and *O. ficus-indica* is within the range of $1 \times 10^{-8} - 5 \times 10^{-7} m s^{-1} MPa^{-1}$ reported for other species (Fiscus, 1977; North and Nobel, 1991, 1992; Salim and Pitman, 1984; Steudle *et al.*, 1987). Despite the morphological and anatomical specialization of shoots of these desert succulents, their roots are relatively unspecialized, with no structural characteristics that might lead to unusually high or low L_P, at least in the absence of drought. For young main roots

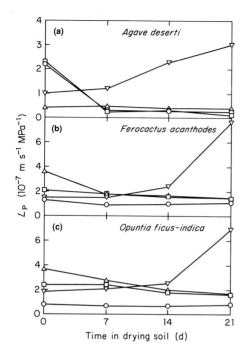

Figure 10.1. *Effects of drought on root hydraulic conductivity* (L_P) *for* Agave deserti (*a*), Ferocactus acanthodes (*b*), *and* Opuntia ficus-indica (*c*) *for young main roots 4–8 weeks old* (○), *older main roots 12 months old* (△), *lateral roots about 4 weeks old* (□) *and lateral root–young main root junctions* (▽). Ψ_{soil} *was −0.5, −1.5, −3.6 and −7.7 MPa at 0, 7, 14 and 21 days of drought, respectively. Data are means for roots from four plants; standard errors averaged 9% of the means.*

up to 2 months in age and for lateral roots occurring as branches on main roots, L_P is similar for the three species, averaging $2.0 \times 10^{-7}\,\mathrm{m\,s^{-1}\,MPa^{-1}}$ (Figure 10.1). During 21 days of drying, L_P for young main roots of *A. deserti* decreases tenfold, in part because dehydration and collapse of cortical cells produce lacunae that interrupt radial water uptake (North and Nobel, 1991) and embolism reduces axial water flow in xylem vessels. In response to drying, sheaths composed of soil particles, mucilage and root hairs develop around young main roots of the two cactus species, helping to prevent the desiccation of underlying root tissues and the consequent reduction in L_P (North and Nobel, 1992). Soil sheaths can form around roots of other species, including dune grasses, which are also subjected to extremely dry soil conditions (Sprent, 1975; Wullstein and Pratt, 1981).

In contrast to the case for young main roots, L_P for older main roots of *A. deserti* is unaffected by drying (Figure 10.1a), yet it decreases by more than 50% for the older main roots of *F. acanthodes* and *O. ficus-indica* (Figures 10.1b, c). For the data considered, the older roots of *A. deserti* had been subjected to drought previously, whereas those of the two cactus species had been continuously well-watered. For all three species, increases in suberization of the hypodermis, endodermis and/or periderm, and the drying of suberized layers, apparently decrease L_P (North and Nobel, 1991, 1992).

Unlike L_P for lateral roots and main roots, L_P for segments containing a junction between a lateral root and a main root increases two- to fivefold during drought (Figure 10.1). The largest increase for all three species occurred at 21 days of soil drying, when about 50% or more of the lateral roots had abscised. This increase in L_p apparently reflects cellular changes related to abscission, including the shrinkage and separation of unsuberized parenchyma cells at the junction between suberized layers of the lateral root and the main root. Lateral root abscission helps to prevent local water loss caused by the increased L_p at lateral root–main root junctions.

10.2.1. *Root axial hydraulic conductance*

Root hydraulic conductivity (L_p) can be partitioned into both axial and radial components. The volumetric flow of water per unit time in the root xylem, $Q_V (m^3 s^{-1})$, is related to the axial hydraulic conductance per unit pressure gradient, $K_h (m^4 s^{-1} MPa^{-1})$:

$$K_h = \frac{Q_V}{\Delta P / \ell} \qquad 10.2$$

where the pressure drop, ΔP, is applied across the length, ℓ, of the root segment (Gibson *et al.*, 1984).

Young main roots and young lateral roots of *A. deserti*, *F. acanthodes* and *O. ficus-indica* have a lower K_h than older roots, averaging $2.5\times 10^{-12} m^4 s^{-1} MPa^{-1}$ (North and Nobel, 1991, 1992). The lower K_h is due to the smaller diameter of the protoxylem and the early metaxylem vessels and to the immaturity of the larger late metaxylem vessels, as is also the case for *Zea mays* (St Aubin *et al.*, 1988) and *Hordeum vulgare* (Sanderson *et al.*, 1986). K_h for older main roots of *F. acanthodes* and *O. ficus-indica* $(9.6 \times 10^{-11} m^4 s^{-1} MPa^{-1})$ is about fiftyfold greater than that for young roots, but about eightyfold lower than the K_h for older main roots of *A. deserti* $(K_h = 8.0 \times 10^{-9} m^4 s^{-1} MPa^{-1}$; North and Nobel, 1991, 1992). Despite much secondary xylem in older roots of the two cactus species, the smaller diameter of their late metaxylem vessels $(26 \mu m)$ compared with those of *A. deserti* $(67 \mu m)$ causes K_h to limit their root hydraulic conductivity, even under wet conditions (North and Nobel, 1992).

The extent of air embolism in the xylem can be indicated by the ratio of axial conductance measured on segments before (K_h^{init}) and after (K_h^{max}) pressurization in solution to remove the emboli (Sperry, 1986; see also Chapter 7). Under wet conditions, K_h^{init}/K_h^{max} ranges from 0.82 to 1.01 for root segments of the three species and from 0.72 to 0.92 for stem segments (Figure 10.2). Drying in soil for 21 days causes no further embolism in stems for the three species. Main roots of *A. deserti* are similar to stems in that K_h^{init}/K_h^{max} is 0.76 at 21 days of drought, indicating only a moderate amount of embolism (Figure 10.2a). For lateral roots of the three species, K_h^{init}/K_h^{max} averages 0.24 at 21 days of soil drying; main roots of the two cactus species show a similar reduction in conductivity caused by embolism (Figure 10.2b, c).

Throughout soil drying, junctions between lateral roots and main roots and between main roots and stems are more embolized than are the roots or the

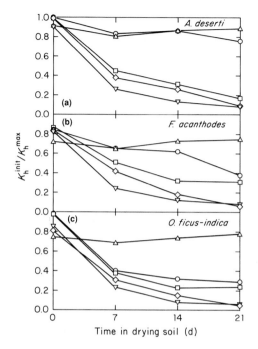

Figure 10.2. *Effects of drought on root axial conductance per unit pressure gradient measured before* (K_h^{init}) *and after* (K_h^{max}) *pressurization to remove emboli for A. deserti (a), F. acanthodes (b) and O. ficus-indica (c) for young main roots* ($\circ$), *lateral roots* ($\square$), *lateral root–main root junctions* ($\triangledown$), *root–stem junctions* ($\diamond$) *and stems* ($\triangle$). *Root ages and* Ψ_{soil} *are as for Figure 10.1. Data are means for roots from four plants; standard errors averaged 12% of the means.*

stems. At 21 days of drought, both types of junction for all three species are more than 90% embolized (Figure 10.2). Xylem at the junctions differs from that elsewhere in the root systems; specifically, abundant small tracheary elements occur, with large pits in the cell walls for *A. deserti* and reticulate secondary thickenings in the cell walls for the two cactus species. Such cells are more susceptible to embolism than are other tracheary elements, perhaps because of the increased likelihood of air-seeding through pores in their extensive, unlignified primary cell walls (Tyree and Sperry, 1989). When embolized during drought, junctions restrict axial water flow from the stem to main roots, from main roots to lateral roots, and thus ultimately from the plant to the dry soil. In addition, embolism is confined to junctions by the absence (for *A. deserti*) or the infrequency (for *F. acanthodes* and *O. ficus-indica*) of vessels continuing through junctions and by the small size of the junction tracheary elements.

10.2.2 *Root radial hydraulic conductivity*

Besides K_h, the other component of L_P is the radial hydraulic conductivity (L_R; $m\,s^{-1}\,MPa^{-1}$), which equals the volume flux density of water at the root surface divided by the drop in water potential from the root surface to the root xylem. Using measured values of K_h and L_P together with the length (ℓ) and the radius (r_{root}) of root segments, root radial conductivity averaged over the entire root segment is calculated as follows (Landsberg and

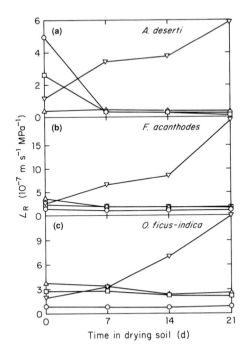

Figure 10.3. *Effects of drought on root radial conductivity* (L_R) *for* A. deserti (*a*), F. acanthodes (*b*) *and* O. ficus-indica (*c*). *Panels, symbols and* Ψ_{soil} *are as for Figure 10.1.* L_R *was calculated from root dimensions and mean values of* L_P *and* K_h *for roots from four plants.*

Fowkes, 1978):

$$L_R = L_P \alpha \ell / \tanh (\alpha \ell) \qquad 10.3$$

where α equals $(2\pi r_{\text{root}} L_R / K_h)^{1/2}$. Equation 10.3 is solved iteratively, with L_R initially set equal to L_P and then gradually increased until convergence occurs.

As the disparity between L_R and L_P increases, K_h becomes more limiting to root hydraulic conductivity. Under wet conditions, low K_h due to immaturity of the xylem means that the relatively high L_R for young main roots of *A. deserti* determines L_P (Figure 10.3a). The difference between L_R and L_P for young main roots of *A. deserti* diminishes as the soil dries (Figures 10.1a vs. 10.3a), until at 21 days of drought L_R is essentially the same as L_P, as is the case for older roots of *A. deserti* under both wet and dry conditions. Development of cortical lacunae and increases in the number of suberized layers in the exodermis and outside the endodermis for roots of *A. deserti* cause L_R to decrease, both during drought and with root ageing (North and Nobel, 1991).

For lateral roots and main roots of *F. acanthodes* and *O. ficus-indica*, the difference between L_P and L_R increases slightly during drought (Figures 10.1 and 10.3), reflecting decreases in K_h due to embolism. Decreases in L_R due to dehydration of the periderm (Vogt *et al.*, 1983) are less important than are accompanying decreases in K_h for older main roots of the two cactus species, because older roots consist chiefly of secondary xylem, with only periderm to the outside (North and Nobel, 1992).

In contrast to the decreases in L_R for lateral roots and main roots of the three species, L_R for junctions between lateral roots and main roots increases two- to fivefold during 21 days of drought (Figure 10.3). These increases are even greater than drought-induced increases in L_P for lateral root–main root junctions (Figure 10.1) and would permit substantial loss of water from the root xylem to drier soil were it not for the simultaneous decreases in K_h due to embolism.

10.3 Hydraulic conductance of the root–soil air gap

Roots of *A. deserti*, *F. acanthodes* and *O. ficus-indica* shrink in response to drought (Figure 10.4). The extent of shrinkage increases gradually as the soil water potential decreases from −0.01 MPa to −0.3 MPa, the latter being similar to the root xylem water potential, Ψ_{xylem}, under hydrated conditions (Nobel and Lee, 1991). Below −0.3 MPa, young roots of all three species shrink substantially (Figure 10.4), as water moves out of the cortical cells. In soil with a water potential of −10 MPa, reductions in root diameter range from 43% for 3-week-old roots to 6% for 12-month-old roots, with a mean of about 20% shrinkage for 2-month-old roots (Figure 10.4) of the three species (Nobel, 1992; Nobel and Cui 1992a). Thus, a 0.2 mm annular air gap can develop around a young root 2 mm in diameter as it shrinks away from the soil during a drought that reduces the soil water potential to −10 MPa, as can occur over a 30-day period for loamy sand (Nobel, 1988; Young and Nobel, 1986).

In the steady state and under isothermal conditions, the water vapour conductance of a root–soil air gap (L_{gap}; m s^{-1} MPa^{-1}) is related to J_V, the volumetric flux density of water, as follows (Nobel and Cui, 1992a):

$$J_V = L_{gap}(\Psi_{gap} - \Psi_{surface}) \qquad 10.4$$

$$= \frac{L'}{r_{root} \ln(1 + \Delta x_{gap}/r_{root})}(\Psi_{gap} - \Psi_{surface})$$

$$\cong \frac{L'}{\Delta x_{gap}}(\Psi_{gap} - \Psi_{surface})$$

where Ψ_{gap} is the soil water potential at the soil surface at the outer edge of the

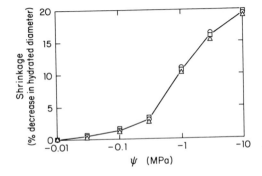

Figure 10.4. *Shrinkage of young main roots of* A. deserti *(○),* F. acanthodes *(△) and* O. ficus-indica *(□) in response to decreases in water potential (Ψ). Data are means for four main roots approximately 2 months old and 1.7–2.6 mm in diameter; standard errors averaged 8% of the means. (Adapted from Nobel, 1992; Nobel and Cui, 1992a; and unpublished observations.)*

air gap; Ψ_{surface} is the water potential at the root surface; L' equals $\bar{V}_{\text{w}}^2 D_{\text{wv}} P_{\text{wv}}^* / (RT)^2$, where $\bar{V}_{\text{w}}$ is the partial molal volume of water, D_{wv} is the diffusion coefficient of water vapour in air, P_{wv}^* is the saturation partial pressure of water vapour, R is the gas constant and T is the absolute temperature; and Δx_{gap} is the distance across the gap. At 25°C, L' is $4.18 \times 10^{-12} \, \text{m}^2 \, \text{s}^{-1} \, \text{MPa}^{-1}$.

Equation 10.4 represents an analytical solution for L_{gap} under isothermal conditions for a root concentrically located within a gap. Because of local cooling due to water evaporation at the root surface, the measured L_{gap} is 20–80% lower than that predicted using Equation 10.4 (Nobel and Cui, 1992b). Conversely, L_{gap} is larger for roots eccentrically located in the gaps, more than doubling for a root touching the soil at one location on the perimeter of the gap. Despite such over- and underpredictions, Equation 10.4 indicates the inverse relationship between L_{gap} and Δx_{gap}, the latter increasing due to root shrinkage during drought. To incorporate both the lowering of root surface temperature as water evaporates, which decreases water loss, and the typical eccentric location of roots in the air gaps, which increases water loss, L_{gap} for the concentric isothermal condition (Equation 10.4) is henceforth multiplied by 0.6 (P.S. Nobel and M. Cui, unpublished observations).

Under wet conditions, when roots are turgid and fully in contact with the soil, L_{gap} (Figure 10.5) is infinite, and L_{P} is the main limiter of root–soil water exchange. Beginning at about 7 days of drought ($\Psi_{\text{soil}} \cong -0.1 \, \text{MPa}$), L_{gap} is the primary limiter of water uptake by the roots. At 11 days of drought, Ψ_{soil} decreases below the minimum water potential of the shoot and of the main roots during the early phase of drought, $-0.3 \, \text{MPa}$ (Nobel, 1988; Nobel and Lee, 1991); L_{gap} then limits the loss of water vapour from the roots (Figure 10.5). Indeed, for the next 13 days of drought, L_{gap} helps isolate the relatively hydrated roots from the drier bulk soil, assisted in the case of young roots of *F. acanthodes* and *O. ficus-indica* by the presence of soil sheaths.

10.4 Soil hydraulic conductance

Assuming cylindrical symmetry, the volumetric flux density of water in the soil toward a root (J_{V}) can be represented by Darcy's law in cylindrical coordinates (Nobel, 1991):

$$J_{\text{V}} = \frac{1}{r_{\text{root}}} L_{\text{soil}} \frac{(\Psi_{\text{distant}} - \Psi_{\text{gap}})}{\ln(r_{\text{distant}}/r_{\text{gap}})} \qquad 10.5$$

where L_{soil} is the soil hydraulic conductivity coefficient ($\text{m}^2 \, \text{s}^{-1} \, \text{MPa}^{-1}$), Ψ_{distant} is the soil water potential away from the root, r_{distant} is the radial distance from which water is taken up by a root and r_{gap} is the radial distance from the centre of the root to the outer side of the root–soil air gap ($r_{\text{root}} + \Delta x_{\text{gap}}$). The effective soil hydraulic conductance ($L_{\text{soil}}^{\text{eff}}$) is (Nobel and

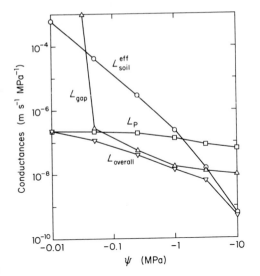

Figure 10.5. *Effects of water potential on the hydraulic conductances of the root* (L_P), *the root–soil air gap* (L_{gap}) *and the soil* (L_{soil}^{eff}) *as a loamy sand dries over a 30-day period. The mean L_P for young main roots of the three species (Figure 10.1) is replotted on a logarithmic scale. Equation 10.4 is used to calculate L_{gap} based on the mean shrinkage observed for young main roots of the three species (Figure 10.4), multiplied by 0.6 to take into account the eccentricity of roots in air gaps and non-isothermal conditions caused by water evaporation. Equation 10.6 is used to calculate L_{soil}^{eff} based on L_{soil} at a depth of 0.10 m and setting r_{gap} equal to 2 mm and r_{soil} equal to 30 mm. The three conductances in series are summed as reciprocals to obtain the reciprocal of the conductance of the overall pathway, $L_{overall}$ (Equation 10.7).*

Cui, 1992a):

$$L_{soil}^{eff} = \frac{J_V}{\Psi_{distant} - \Psi_{gap}} \qquad 10.6$$

$$= \frac{L_{soil}}{r_{root} \ln(r_{distant}/r_{gap})}.$$

Commonly, $r_{distant}$ is set equal to the lesser of 30 mm and half of the inter-root spacing (Alm and Nobel, 1991; Caldwell, 1976). L_{soil} markedly depends on Ψ_{soil}, decreasing over 10^6-fold as Ψ_{soil} decreases from 0.01 MPa to -10 MPa for a sandy loam (Hillel, 1982; Young and Nobel, 1986), the type of soil in which desert succulents typically occur (Nobel, 1988).

Because of its direct dependence on L_{soil} (Equation 10.6), mean L_{soil}^{eff} for root systems of *A. deserti*, *F. acanthodes* and *O. ficus-indica* in soil from a Sonoran Desert site decreases over 10^6-fold as Ψ_{soil} decreases from -0.01 MPa to -10 MPa over a 30-day period (Figure 10.5). The steep decline in L_{soil}^{eff} causes it to become the predominant limiter of water movement in the root–soil pathway after about 24 days of drought, when Ψ_{soil} is approximately -4 MPa. L_{soil}^{eff} is also limiting for water movement of *Gossypium hirsutum* in loamy sand during drought (Taylor and Klepper, 1975).

10.5 Overall hydraulic conductance

The overall hydraulic conductance of a root–soil system, $L_{overall}$

$(m\,s^{-1}\,MPa^{-1})$, can be represented as:

$$J_V = L_{overall}(\Psi_{distant} - \Psi_{xylem}) \qquad 10.7$$

where $L_{overall}$ is the reciprocal of the sum of the reciprocals of the hydraulic conductances for each of the three root–soil components (Nobel, 1992).

For roots of *A. deserti*, *F. acanthodes* and *O. ficus-indica* in a drying soil for 30 days, $L_{overall}$ decreases over 10^3-fold compared with a wet soil (Figure 10.5). Under wet conditions, the principal determinant of $L_{overall}$ is L_P, here represented as an average for young main roots of the three species (Figure 10.5). Although L_P decreases less than do the other two conductances as the soil dries, its two- to threefold drop during the first week of drying (Figure 10.1), when L_P is the limiting conductance for water movements, is critical in reducing root water loss. Furthermore, the decrease in L_P for individual roots during drought is accompanied by abscission of lateral roots, which under wet conditions represent about 30% of the total root length for *A. deserti* and 35% for *F. acanthodes* (Hunt and Nobel, 1987). At only 7 days of soil drying, L_{gap} becomes the lowest conductance and thus has the greatest influence on $L_{overall}$. Finally, after just over 3 weeks of drought, $L_{overall}$ becomes essentially equal to L_{soil}^{eff}, sharply restricting further water movement in the root–soil pathway (Figure 10.5).

10.6 Root–soil conductances: responses to rewetting

In addition to decreases in L_P, L_{gap} and L_{soil}^{eff} in response to drought, rectification for root systems involves a rapid increase in conductance when soil moisture is restored. For *A. deserti*, *F. acanthodes* and *O. ficus-indica*, all three components of the root–soil pathway behave as rectifiers (Lopez and Nobel, 1991; Nobel and Cui, 1992a; Nobel and Sanderson, 1984; North and Nobel, 1991, 1992).

For Sonoran Desert soil, L_{soil}^{eff} rapidly returns to its pre-drought level upon rewetting (Young and Nobel, 1986) and is no longer limiting to $L_{overall}$. For the first 5 days after Ψ_{soil} is raised to -0.01 MPa by rewetting, L_{gap} is the primary limiter of water movement (Figure 10.6). L_{gap} increases steadily as roots become rehydrated and hence expand to fill the air gap (Nobel, 1992). After 5 days of rewetting, L_P again becomes the limiting factor for the young main roots represented here (Figure 10.6). Such roots re-attain pre-drought L_P within 6 days of rewetting (Figures 10.1, 10.6). For young main roots of *A. deserti*, the increase in L_P after rewetting is due to the rapid reduction in size of cortical lacunae following cellular rehydration, and perhaps also to rehydration of unsuberized cells in the exodermis and the endodermis (North and Nobel, 1991). For the two cactus species, recovery of young main roots is facilitated by the presence of soil sheaths, which help prevent large decreases in L_P during drought. In addition, K_h^{init}/K_h^{max} for both young and older main roots of all three species returns to pre-drought values within a week after rewetting, indicating complete reversal of embolism (North and Nobel, 1991, 1992).

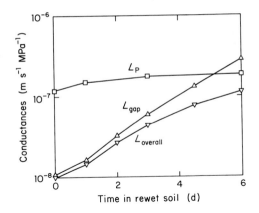

Figure 10.6. *Effects of rewetting time after 30 days of drought on hydraulic conductances.* L_P *is based on mean values for young main roots of A. deserti, F. acanthodes and O. ficus-indica (North and Nobel, 1991, 1992).* L_{gap} *is calculated using Equation 10.4 times 0.6 and data for the reversal of root shrinkage (Nobel and Cui, 1992a).* $L_{overall}$ *is calculated using Equation 10.7, assuming that* L_{soil}^{eff} *equals* $5 \times 10^{-4}\ m\ s^{-1}$, *as is appropriate for a* Ψ_{soil} *of* $-0.01\ MPa$.

In contrast to the case for young main roots, rewetting after 21 days of drought causes L_P for older main roots of *F. acanthodes* and *O. ficus-indica* to increase to only 60% of pre-drought levels, despite the recovery in K_h (North and Nobel, 1992). Thus, the drought-induced decreases in L_R for these roots are apparently not fully reversible. The low L_P yet high K_h for previously droughted older main roots of *A. deserti* also suggests that soil drying produces irreversible changes in the radial pathway. Moreover, L_P for lateral roots of *A. deserti* after rewetting is only 25% of pre-drought L_P, and cortical lacunae persist (North and Nobel, 1991). Evidence from roots and other organs of other species implicates the dehydration of suberized cell walls as the cause of non-reversible decreases in radial conductivity (Clarkson *et al.*, 1987; Vogt *et al.*, 1983).

10.7 Conclusions

The processes that decrease L_P, L_{gap} and L_{soil}^{eff} for root–soil systems of *A. deserti*, *F. acanthodes* and *O. ficus-indica* during drought probably occur for most species. Although L_{soil}^{eff} for a clay may decrease more slowly than that for a sandy loam, the low conductance of dry soils is the major limiter of water loss from roots during extended drought. Gaps between roots and soil may occur more readily in porous soils (Passioura, 1988); nevertheless, root shrinkage in response to drought lessens root contact in many types of soil (Tinker, 1976). Certain anatomical and morphological changes underlying decreases in L_P may be more species-dependent than are changes in L_{soil}^{eff} and L_{gap}, but reductions in K_h due to embolism and in L_R due to dehydration of cortical cells and suberized layers can occur for roots of most species. Changes that reduce overall hydraulic conductance during drought are readily reversible by rewetting, except those involving suberization and lateral root abscission. Thus, the root–soil system is exquisitely tuned to act as a rectifier — reversible changes in hydraulic conductance occurring in the root, at the root–soil air gap and in the soil — all of which facilitate water uptake under wet conditions but prevent water loss from a plant to the soil during drought.

Acknowledgements

Financial support for new observations and for the preparation of this chapter were provided by National Science Foundation grant DCB 90-02333 and the Environmental Sciences Division, Office of Health and Environmental Research, Department of Energy contract DE-FC03-87-ER60615.

References

Alm, D.M. and Nobel, P.S. (1991) Root system water uptake and respiration for *Agave deserti*: observations and predictions using a model based on individual roots. *Ann. Bot.*, **67**, 59–65.

Blizzard, W.E. and Boyer, J.S. (1980) Comparative resistance of the soil and the plant to water transport. *Plant Physiol.*, **66**, 809–814.

Caldwell, M.M. (1976) Root extension and water absorption. In: *Water and Plant Life. Problems and Modern Approaches. Ecological Studies*, Volume 19 (eds O.L. Lange, L. Kappen and E.-D. Schulze). Springer, Berlin, pp. 63–85.

Caldwell, M.M. and Richards, J.H. (1989) Hydraulic lift: water efflux from upper roots improves effectiveness of water uptake by deep roots. *Oecologia*, **79**, 1–5.

Clarkson, D.T., Robards, A.W., Stephens, J.E. and Stark, M. (1987) Suberin lamellae in the hypodermis of maize (*Zea mays*) roots; development and factors affecting the permeability of hypodermal layers. *Plant Cell Environ.*, **10,** 83–93.

Faiz, S.M.A. and Weatherley, P.E. (1978) Further investigations into the location and magnitude of the hydraulic resistances in the soil : plant system. *New Phytol.*, **81**, 19–28.

Fiscus, E.L. (1977) Determination of hydraulic and osmotic properties of soybean root systems. *Plant Physiol.*, **59**, 1013–1020.

Gibson, A.C., Calkin, H.W. and Nobel, P.S. (1984) Xylem anatomy, water flow, and hydraulic conductance in the fern *Cyrtomium falcatum*. *Am. J. Bot.*, **71**, 564–574.

Hillel, D. (1982) *Introduction to Soil Physics*. Academic Press, New York.

Hunt, E.R. Jr and Nobel, P.S. (1987) Allometric root/shoot relationships and predicted water uptake for desert succulents. *Ann. Bot.*, **59**, 571–577.

Landsberg, J.J. and Fowkes, N.D. (1978) Water movement through plant roots. *Ann. Bot.*, **42**, 493–508.

Lopez, F.B. and Nobel, P.S. (1991) Root hydraulic conductivity of two cactus species in relation to root age, temperature, and soil water status. *J. Exp. Bot.*, **42**, 143–149.

Mooney, H.A., Gulmon, S.L., Rundel, P.W. and Ehleringer, J. (1980) Further observations on the water relations of *Prosopis tamarugo* of the northern Atacama Desert. *Oecologia*, **44**, 177–180.

Nobel, P.S. (1976) Water relations and photosynthesis of a desert CAM plant, *Agave deserti*. *Plant Physiol.*, **58**, 576–582.

Nobel, P.S. (1977) Water relations and photosynthesis of a barrel cactus, *Ferocactus acanthodes*, in the Colorado Desert. *Oecologia*, **27**, 117–133.

Nobel, P.S. (1988) *Environmental Biology of Agaves and Cacti*. Cambridge University Press, New York.

Nobel, P.S. (1991) *Physicochemical and Environmental Plant Physiology*. Academic Press, San Diego.

Nobel, P.S. (1992) Root–soil responses to water pulses in dry environments. In: *Exploitation of Environmental Heterogeneity by Plants: Ecophysiological Processes Above and Below Ground* (eds M.M. Caldwell and R.W. Pearcy). Academic Press, San Diego.

Nobel, P.S. and Cui, M. (1992a) Hydraulic conductances of the soil, the root–soil air gap, and the root: changes for desert succulents in drying soil. *J. Exp. Bot.*, **43**, 319–326.

Nobel, P.S. and Cui, M. (1992b) Prediction and measurement of gap water vapor conductance for roots located concentrically and eccentrically in air gaps. *Plant Soil*, **145**, 157–166.

Nobel, P.S. and Lee, C.H. (1991) Variations in root water potentials: influence of environmental factors for two succulent species. *Ann. Bot.*, **67**, 549–554.

Nobel, P.S. and Sanderson, J. (1984) Rectifier-like activities of roots of two desert succulents. *J. Exp. Bot.*, **35**, 727–737.

Nobel, P.S., Schulte, P.J. and North, G.B. (1990) Water influx characteristics and hydraulic conductivity for roots of *Agave deserti* Engelm. *J. Exp. Bot.*, **41**, 409–415.

North, G.B. and Nobel, P.S. (1991) Changes in hydraulic conductivity and anatomy caused by drying and rewetting roots of *Agave deserti* (Agavaceae). *Am. J. Bot.*, **78**, 906–915.

North, G.B. and Nobel, P.S. (1992) Drought-induced changes in hydraulic conductivity and structure in roots of *Ferocactus acanthodes* and *Opuntia ficus-indica*. *New Phytol.*, **120**, 9–19.

Passioura, J.B. (1988) Water transport in and to roots. *Ann. Rev. Plant Physiol. Plant Mol. Biol.*, **39**, 245–265.

Richards, J.H. and Caldwell, M.M. (1987) Hydraulic lift: substantial nocturnal water transport between soil layers by *Artemisia tridentata* roots. *Oecologia*, **73**, 486–489.

Rowse, H.R. and Goodman, D. (1981) Axial resistance to water movement in broad bean (*Vicia faba*) roots. *J. Exp. Bot.*, **32**, 591–598.

St Aubin, G., Canny, M.J. and McCully, M.E. (1986) Living vessel elements in the late metaxylem of sheathed maize roots. *Ann. Bot.*, **58**, 577–588.

Salim, M. and Pitman, M.G. (1984) Pressure-induced water and solute flow through plant roots. *J. Exp. Bot.*, **35**, 869–881.

Sanderson, J., Whitbread, F.C. and Clarkson, D.T. (1988) Persistent xylem cross-walls reduce the axial hydraulic conductivity in the apical 20 cm of barley seminal root axes: implications for the driving force for water movement. *Plant Cell Environ.*, **11**, 247–256.

Schulte, P.J. and Nobel, P.S. (1989) Responses of a CAM plant to drought and rainfall: capacitance and osmotic pressure influences on water movement. *J. Exp. Bot.*, **40**, 61–70.

Shone, M.G.T. and Clarkson, D.T. (1988) Rectification of radial water flow in the hypodermis of nodal roots of *Zea mays*. *Plant Soil*, **111**, 223–229.

Sperry, J.S. (1986) Relationship of xylem embolism to xylem pressure potential, stomatal closure and shoot morphology in the palm *Rhapis excelsa*. *Plant Physiol.*, **80**, 110–116.

Sprent, J.I. (1975) Adherence of sand particles to soybean roots under water stress. *New Phytol.*, **74**, 461–463.

Steudle, E., Oren, R. and Schulze, E.-D. (1987) Water transport of maize roots. Measurement of hydraulic conductivity, solute permeability, and of reflection coefficients of excised roots using the root pressure probe. *Plant Physiol.*, **84**, 1220–1232.

Szarek, S.R., Johnson, H.B. and Ting, I.P. (1973) Drought adaptation in *Opuntia basilaris*. Significance of recycling carbon through Crassulacean acid metabolism. *Plant Physiol.*, **23**, 539–541.

Taylor, H.M. and Klepper, B. (1975) Water uptake by cotton root systems: an examination of the assumptions in the single root model. *Soil Sci.*, **120**, 57–67.

Taylor, H.M. and Willatt, T. (1983) Shrinkage of soybean roots. *Agron. J.*, **75**, 818–820.

Tinker, P.B. (1976) Roots and water: transport of water to plant roots in soil. *Phil. Trans. R. Soc. Lond. B*, **273**, 445–461.

Tyree, M.T. and Sperry, J.S. (1989) The vulnerability of xylem to cavitation and embolism. *Ann. Rev. Plant Physiol. Plant Mol. Biol.*, **40**, 19–38.

Vogt, E., Schönherr, J. and Schmidt, H.W. (1983) Water permeability of periderm membranes isolated enzymatically from potato tubers (*Solanum tuberosum* L.). *Planta*, **158**, 294–301.

Wullstein, L.H. and Pratt, S.A. (1981) Scanning electron microscopy of rhizosheaths of *Oryzopsis hymenoides*. *Am. J. Bot.*, **68**, 408–419.

Young, D.R. and Nobel, P.S. (1986) Predictions of soil–water potentials in the north-western Sonoran Desert. *J. Ecol.*, **74**, 143–154.

Tissue hydraulic properties and the water relations of desert shrubs

Paul J. Schulte

11.1 Introduction

The role of plant hydraulic architecture and tissue properties such as elasticity and capacitance in the adaptation of plants to arid environments is not well understood. Desert plants display a wide array of drought adaptations (Smith and Nobel, 1986), and many of these adaptations are associated with a great diversity in the hydraulic properties of the plant. For example, succulent species such as cacti have large volumes of stored water and highly elastic cells, while desert shrubs may have relatively little stored water and inelastic cells (Nobel and Jordan, 1983). A number of studies utilizing either direct-measurement or simulation-modelling approaches have demonstrated the importance of the hydraulic properties of succulent plants in both the acquisition of water from the soil, and in the internal redistribution of water during periods of drought (Ruess *et al.*, 1988; Schulte and Nobel, 1989; Schulte *et al.*, 1989; Smith *et al.*, 1987). Similar approaches applied to other succulent and non-succulent species will be important for understanding the role of various tissue properties and structural features in the growth and survival of plants in arid environments.

11.2 Modelling approach

As our ability to study the water transport characteristics of plant tissues has grown, we face increasing difficulty in coupling an understanding of the parts of a plant with the behaviour of the whole. Herein lies the value of modelling. One possible approach to studying water flow through plant tissues uses models that describe water flow in terms of water potential as a driving force, incorpor-

Table 11.1. *Equivalence of units for electrical circuits and water relations. Translations assume that the basic unit quantity of water is volume (m^3), analogous to the coulomb as the basic unit quantity for electrical charge. Equally valid units could be derived assuming a basic unit quantity for water of mass (g)*

Quantity	Electrical units	Water model units
Flow		
basic unit	coulomb s^{-1}	$m^3 s^{-1}$
derived unit	ampere (A)	
Driving force		
basic unit	$kg\, m^2\, s^{-3}\, A^{-1}$	$kg\, m^{-1}\, s^{-2}$
derived unit	volt (V)	pascal (Pa)
Resistance		
basic unit	$kg\, m^2\, s^{-3}\, A^{-2}$	$Pa\, s\, m^{-3}$
derived unit	ohm	
Capacitance		
basic unit	coulomb V^{-1}	$m^3\, Pa^{-1}$
derived unit	farad	

ating the various hydraulic characteristics of the plant tissues as resistances and capacitances. Such models may be developed as electrical circuit analogues. Assuming that the language of plant–water relations can be translated into electrical circuit terms (Table 11.1), these models are relatively easy to solve using electrical circuit simulation programs. Several commercially available software packages are derived from the circuit simulator developed at the University of California, Berkeley, called SPICE (Nagel, 1975). The models described in this chapter were solved with the PSPICE program (Tuinenga, 1988; Microsim Corp., Irvine, CA, USA) running on an IBM PS/2 Model 70 486 microcomputer. The objective of these modelling studies will be to understand how the resistances and capacitances of tissues along the transpiration stream affect plant–water relations. Ultimately, it will also be important to consider the roles of these hydraulic properties not as constants, but as variables over short or long time periods.

11.3 Model development and testing

A potted shrub of *Chilopsis linearis* (desertwillow; Bignoniaceae) was chosen for development of the model. The above-ground portion of the plant was mapped (Figure 11.1) and all branch segments were numbered. To develop a complete model of this plant, the hydraulic resistance of the xylem and the surface area of all leaves need to be measured for each branch segment. In addition, the hydraulic parameters of the leaves, such as xylem resistance, tissue capacitance and resistance to flow through the parenchyma tissues, must be determined or estimated. The root system should be considered at a corre-

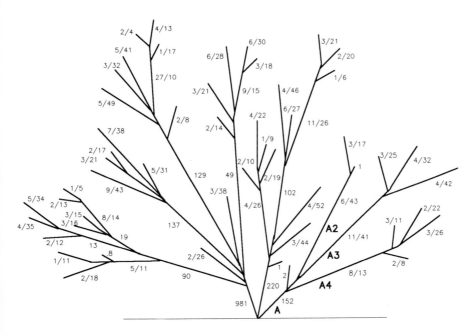

Figure 11.1. *Hydraulic map of the* Chilopsis linearis *shoot system. The branch relative lengths are approximate, but angles from vertical have been increased somewhat for illustration purposes. Total shrub height was 0.98 m. Single quantities next to branches, or the numerator if a fraction, indicate hydraulic conductance* ($10^{-9} m^3 s^{-1} MPa^{-1}$). *The number of leaves (if any) on each segment is shown as the denominator of a fraction. The portion of the plant incorporated into the model is indicated by bold branch codes A and A2, A3, and A4.*

sponding level of detail; however, for this initial model the root system was simplified to a single component describing the resistance to flow encountered by water passing from the soil into the root.

The branch structure of this modelled plant contains five orders of branching (Figure 11.1). It was judged not necessary to consider all of this structure for addressing the objectives discussed earlier. Therefore, a model will be presented that considers only one major branch (Figure 11.1) and two orders of branching. The equivalent electrical circuit (Figure 11.2) contains a voltage source representing soil water potential (set to −0.5 MPa as a typical value) and various resistances for water flow along the transpiration stream. Water stored within the leaf is represented as a capacitor that exchanges water with the transpiration stream. Leaf capacitance is lumped into a single parameter located at the midpoint, hydraulically, between the xylem and the point of water loss from the leaf. A current source is used for transpiration, and the flow rate is set to match a measured transpiration course. Transpiration over a 24 h period was estimated from measurements made with a steady-state diffusion porometer (LiCor LI-1600).

Resistances along the xylem can be either measured or estimated, as noted in

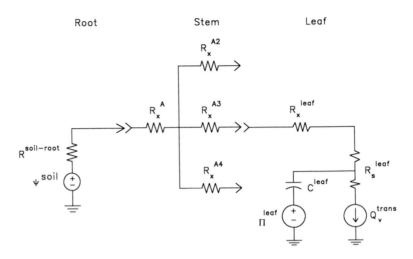

Figure 11.2. *Electrical circuit model for the* Chilopsis linearis *shrub. The circuit was constructed in modules for the roots, stems and leaves. Voltage sources represent soil water potential (Ψ^{soil}) and leaf osmotic pressure (Π), and a current source represents transpiration (Q_v^{trans}). Subscripts refer to xylem (x) or storage tissue (s). Leaf water storage (capacitance; C^{leaf}) is coupled to the transpiration stream midway between the leaf xylem and the site of evaporation within the leaf.*

Chapters 6, 7 and 10. The Hagen–Poiseuille equation considers non-turbulent flow of incompressible fluids for a simple ideal conduit:

$$Q_v = \frac{\pi d^4}{128\eta}\frac{dP}{dx}, \quad \text{or} \quad K_h = \frac{\pi d^4}{128\eta} \qquad 11.1$$

where Q_v is the volumetric flow of water ($m^3 s^{-1}$), K_h is the unit length hydraulic conductance ($m^4 s^{-1} MPa^{-1}$), d is the conduit diameter (m), P is the pressure (MPa) across a xylem segment of length x (m), and η is the viscosity of water ($1.00 \times 10^{-9} MPa\,s$ at 20°C; also called absolute or dynamic viscosity). One approach involves measuring the diameters of all vessels in a transverse section of the xylem and calculating conductance (K_h divided by stem segment length) from Equation 11.1. For non-circular conduits, a version of Equation 11.1 for elliptical cross-sections may be used (see Gibson *et al.*, 1985).

Another approach to estimating xylem resistance (inverse conductance) involves actual measurement using a pressure-flow system (see also Chapters 7 and 10). Water is forced through the xylem at known pressure and the flow velocity is measured (Calkin *et al.*, 1985; Schulte *et al.*, 1987).

$$K_h = \frac{Q_v}{dP/dx}. \qquad 11.2$$

This method was applied to segments of stems collected from *Chilopsis linearis* branches. For the development of the simplified model, the xylem conductance of all stem segments in the aboveground shoot was not measured. Instead, a relationship between conductance and stem diameter was developed from measurements of four stems. A second-order polynomial fitted to these data was used to estimate xylem conductances for all stems in the plant, based upon diameter measurements of stem segments measured midway along the segment. The resultant hydraulic map, although approximate because all segments were not directly measured, suggests that conductances decline considerably as branch order increases (Figure 11.1).

The methods used to estimate xylem resistances of stems were also applied to midveins of leaves. In this case, estimates of conductance by both direct measurement and the Hagen–Poiseuille method yielded fairly similar results; measured conductance averaged 90% of that predicted. The vessels in the leaf midveins appear to conduct similarly to ideal conduits.

Leaf storage capacitance estimates were obtained from pressure–volume curves. This method is described elsewhere (Richter *et al.*, 1980; Tyree and Richter, 1981), but essentially provides estimates of leaf water potential and its components (turgor pressure and osmotic pressure) as a function of leaf relative water content. Capacitance is defined by the change in water volume per unit change in water potential (Molz and Ferrier, 1982), and this parameter can be calculated directly from the pressure–volume curve. Pressure–volume data were analysed as described by Schulte and Hinckley (1985).

Leaf storage resistance was estimated from the time constant for water exchange using a method described by Nobel and Jordan (1983) and Morse (1990). Individual leaves are rehydrated as for the pressure–volume method and inserted into the pressure chamber. An initial balance pressure (pressure at which water returns to the cut petiole surface) is measured and the pressure is then increased by 0.5 MPa. Water is forced from the cut petiole and collected. After a given period of time, the pressure is reduced and the balance point is redetermined. The process is repeated, and the balance pressure gradually approaches the overpressure. For a storage component treated as a simple resistance–capacitance network, the balance pressure should approach the overpressure as an exponential function, and the time constant is a product of resistance and capacitance. The data set is fitted with an exponential function:

$$P = (P^{over} - P^{bal}) * (1 - e^{-t/\tau}) + P^{bal} \qquad 11.3$$

where

$$\tau = RC \qquad 11.4$$

and the estimated time constant (τ) is used to calculate the storage resistance, given prior values for capacitance.

Root resistance can be estimated by placing the entire root system in a solution-filled pressure chamber (see Mees and Weatherly, 1957; Oosterhuis and Wullschleger, 1987). Pressure is applied and the volumetric flow of water out of the cut stem surface is measured. As noted earlier, a more complete treatment of the root system would require measurements of individual roots and separate consideration of the axial (xylem) versus radial components of resistance.

Table 11.2. *Electrical circuit component values for the* Chilopsis linearis *model*

Component	Value
Soil water potential	-0.5 MPa
Soil–root resistance	6.25×10^7 MPa s m^{-3}
Xylem resistance	
stem A	6.27×10^5 MPa s m^{-3}
stem A2	5.99×10^7 MPa s m^{-3}
stem A3	2.70×10^7 MPa s m^{-3}
stem A4	3.60×10^7 MPa s m^{-3}
Leaf midvein resistance	
stem A2	1.49×10^7 MPa s m^{-3}
stem A3	6.72×10^6 MPa s m^{-3}
stem A4	1.18×10^7 MPa s m^{-3}
Leaf storage resistance	
stem A2	2.54×10^8 MPa s m^{-3}
stem A3	1.14×10^8 MPa s m^{-3}
stem A4	2.00×10^8 MPa s m^{-3}
Leaf capacitance	
stem A2	1.77×10^{-7} m^3 MPa^{-1}
stem A3	3.93×10^{-7} m^3 MPa^{-1}
stem A4	2.25×10^{-7} m^3 MPa^{-1}
Leaf osmotic pressure	
fixed	-1.41 MPa
variable	$(0.199 * \text{water potential}) - 1.41$
Transpiration rate (variable)	
maximum, A2	3.08×10^{-9} m^3 s^{-1}
maximum, A3	6.86×10^{-9} m^3 s^{-1}
maximum, A4	3.96×10^{-9} m^3 s^{-1}

Leaf osmotic pressure is estimated for the bulk leaf by the pressure–volume curve method as described earlier. This parameter can be estimated as a function of leaf water content or as a function of leaf water potential. The model may treat osmotic pressure as a constant, or as a function of leaf water potential. The final parameters for the model are summarized in Table 11.2.

A partial test of the model was conducted by measuring leaf water potential for several field plants on the same day the transpiration course was measured. The model produced estimates of leaf water potential that are slightly lower than the measured values (Figure 11.3), but the correspondence was reasonably good, particularly considering the incomplete treatment of the root system.

11.4 Results from modelling studies

11.4.1 *Potential gradients along the plant*

Estimates of the water potential at points along the transpiration stream were used to determine where the water potential gradients were greatest and hence which plant components were of greatest significance in determining the water

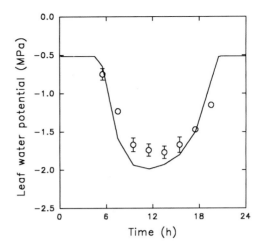

Figure 11.3. *Estimated water potential for leaves over a 24 h period as predicted by the model (solid line) and as measured with the pressure chamber (symbols, ±SE).*

potential ultimately achieved in the leaf. The water potential decline was relatively large at the root and within the leaf after the transpiration stream left the xylem, and relatively small along the xylem of the stem and the leaf midvein (Figure 11.4). This result is based on xylem resistances for a well-watered plant early in the growing season; changes in xylem resistance during the season due to cavitation of xylem conduits have not yet been considered in this model.

11.4.2 *Water flows within the plant*

The flow of water into the root system over the 24 h course closely mirrored cumulative transpiration for the leaves on all three branches (Figure 11.5). The contribution of water from storage within the leaf does not appear to be significant (Figure 11.5). At the time of peak water flow from storage, stored

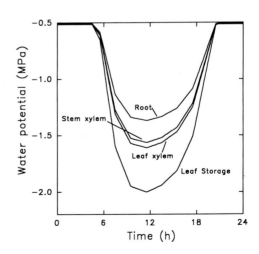

Figure 11.4. *Predicted water potentials at various points along the transpiration stream over the course of a day.*

183

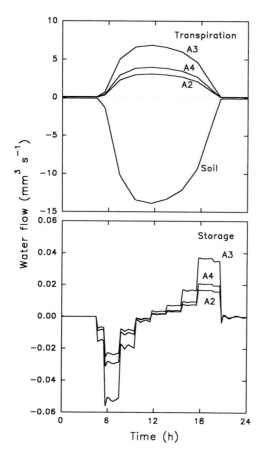

Figure 11.5. *Water flow rates through the plant. Transpiration is shown separately for each second-order stem (A2, A3, A4), but soil water uptake is for the entire stem (upper panel). Water flows from storage (lower panel) utilize a different scale because of the large difference between storage flow and transpiration stream flow.*

water was accounting for only 4.7% of transpiration. Flow from storage integrated over the period of discharge (00.00–11.30 h) was $1.19 \times 10^{-6}\,\text{m}^3$ compared to $2.31 \times 10^{-4}\,\text{m}^3$ transpired, or only 0.5% from storage. The abrupt changes apparent in flow rate from storage followed by periods of constant flow (Figure 11.5) result from the discontinuous slope of the transpiration curve. Transpiration was estimated every 2 h, and was assumed to change linearly between measurements. Thus the transpiration curve used in the model has abrupt slope changes at the times of measurement. It is also likely that these abrupt changes in slope resulted in the overshoot observed just after a change in storage flow rate, particularly at 06.00 h. One solution to this problem would be to provide a smooth transpiration curve by fitting a cubic spline to the data before applying them to the model.

11.4.3 *Significance of capacitance*

The contribution of water stored in leaf cells to transpiration is minor, indicating that leaf capacitance in *Chilopsis linearis* is not significant in buffering the leaf against water potential changes, contrary to the results one might expect for a succulent plant (Schulte and Nobel, 1989; Schulte et al., 1989; Smith et al., 1987). A similar conclusion can be reached by considering the short time constant measured for the leaves of this species. The time constant suggests that leaves would equilibrate to changes in transpiration rate within a few minutes. In comparison, time constants for succulent leaves of *Agave deserti* are in the order of 1 h (calculated from Smith et al., 1987), and from 1 h to as much as 13 h for various storage tissues in succulent stems of *Ferocactus acanthodes* (calculated from Schulte et al., 1989).

The minor significance of stored water in leaves as a contributor to transpiration in *Chilopsis linearis* can also be demonstrated independently of the model. The typical desertwillow leaf lost about 0.0355 g of water ($3.55 \times 10^{-8} \, m^3$) going from a water potential near zero to oven-dry (data collected as part of pressure–volume curve development). The same leaf would lose about 1.94 g of water ($1.94 \times 10^{-6} \, m^3$) to transpiration during 1 day (from integrating measured transpiration). Storage could therefore only provide about 1.8% of the transpired water.

11.4.4 *Accounting for variation in tissue osmotic pressure*

The need to consider changes in leaf cell osmotic pressure as water potential and water content change can be demonstrated by comparing the results of a simulation where osmotic pressure is fixed, with a simulation where osmotic pressure is allowed to vary as a function of leaf water potential. This relationship between leaf osmotic pressure and water potential was determined from the pressure–volume curves discussed earlier.

The primary effect of leaf osmotic pressure on the water relations of the leaf tissue involves the maintenance of cell turgor. When osmotic pressure was allowed to decline as the leaf water potential fell during the day, turgor pressure in the cells did not drop as low as when the osmotic pressure was fixed at the saturated tissue value (Figure 11.6). Of course, the choice of fixing osmotic pressure at the value for a saturated leaf as opposed to a leaf at turgor loss was

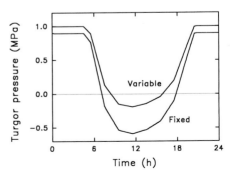

Figure 11.6. *Effect of the modelling method employed for leaf osmotic pressure. Osmotic pressure was either fixed at saturated leaf osmotic pressure (−1.41 MPa), or allowed to vary as a function of leaf water potential.*

an arbitrary one. Had the comparison between fixed and variable osmotic pressure used a fixed potential for a leaf at turgor loss (-1.76 MPa), the leaves in the case of fixed osmotic pressure would have had night-time turgor pressures higher than those in the case of variable osmotic pressure. This comparison is intended, therefore, to demonstrate that the changes in cell osmotic pressure that we assume to occur as cells lose or gain water have a significant effect on cell turgor and should be considered in any model of living tissues.

11.4.5 *Significant resistances along the transpiration stream*

In Section 11.4.1 above, it was suggested that transpiration stream resistances at the soil–root interface and across living cells in the leaf were dominant in the total pathway through the plant. One common technique in modelling considers the sensitivity of the estimated variables (leaf water potential in this case) to changes in these model components. Doubling or halving the values of the leaf storage resistance, a parameter describing the transpiration path between the leaf xylem and the sites of evaporation near the leaf surface, produced changes of -0.4 and 0.2 MPa, respectively, in leaf water potential (Figure 11.7). Doubling or halving the value of the soil–root resistance produced changes of -0.9 and 0.4 MPa, respectively, in leaf water potential (Figure 11.7). Such changes in water potential also produced quantitatively similar changes in turgor pressure (data not shown). Thus these parameters are quite important in determining leaf water potential and need to be determined carefully, particularly in more detailed studies of the root system.

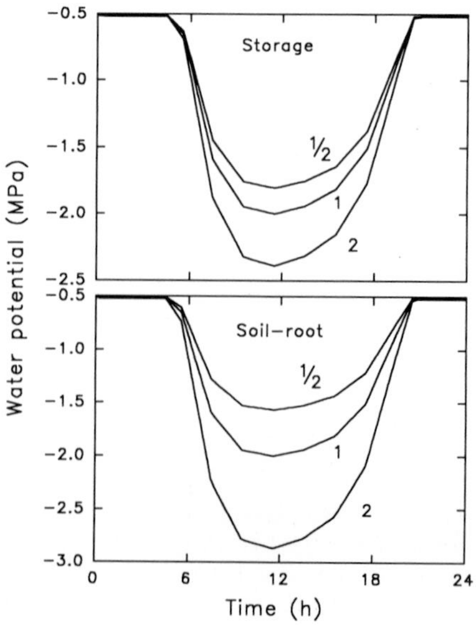

Figure 11.7. *Sensitivity analysis of the effects on leaf water potential of either doubling or halving leaf storage resistance (upper panel) or the soil–root resistance (lower panel).*

11.4.6 *The role of leaf tissue elasticity*

The elastic modulus of plant cells has, by definition, a profound effect on the relationship between the water content of cells and their water potential, particularly the turgor pressure component of water potential. This property of cells has been studied from two directions. The advent of the cell pressure probe has allowed estimates to be made of the elastic modulus for individual cells (Hüsken *et al.*, 1978; Steudle *et al.*, 1977; Zimmermann and Hüsken, 1979). Such studies have also shown that wall elasticity is not constant, but depends on cell volume or cell turgor pressure. In addition, studies of the responses of leaves or whole plants to drought have suggested a number of water relations parameters that may change with drought, including the osmotic pressure of cells and the elasticity of their walls. After noting a change in the elastic modulus, speculations are often made as to its significance. Curiously, the elastic modulus has been described as increasing for some species (Davies and Lakso, 1979; Elston *et al.*, 1976) but decreasing for others (Melkonian *et al.*, 1982; Nunes *et al.*, 1989) in response to drought. If these measurements are correct, then changes in elasticity, in combination with other tissue properties, may be a component in drought adaptation. A decrease in the elastic modulus for xeric white oak has been associated with the drought resistance of this species (Parker *et al.*, 1982). Salleo (1983) suggests that both increases and decreases in tissue elasticity may be associated with different 'strategies' in drought response.

If the elastic modulus decreases (cells become more elastic), perhaps cells can provide more water from storage. Capacitance has been calculated from:

$$C = \frac{V}{\epsilon + \Pi} \qquad\qquad 11.5$$

and a decrease in ϵ (elastic modulus) would increase C. Thus the tissue could more completely buffer changes in water potential resulting from changes in transpiration. Alternately, if the elastic modulus increases, it has been suggested that when leaves lose a little water, as transpiration begins or as the soil dries, the water potential of their cells will decline further, thus enhancing the ability of the plant to acquire water from the soil (Bowman and Roberts, 1985).

One of the goals of this modelling project was to consider the possible effects of changes in leaf tissue elasticity on plant–water relations. For the shrub studied here, leaf water storage represents less than 1% of the water transpired in one day; water flow is completely dominated by transpiration and stored water cannot buffer changes in leaf water potential to any significant extent. Therefore if leaf capacitance is doubled or halved to simulate a change in elasticity, the flow of water from storage changes slightly, but the diurnal course of water potential for the leaf is essentially unchanged (change $\leq 1\%$; data not shown).

Still, it is interesting to consider what effects elasticity changes would have

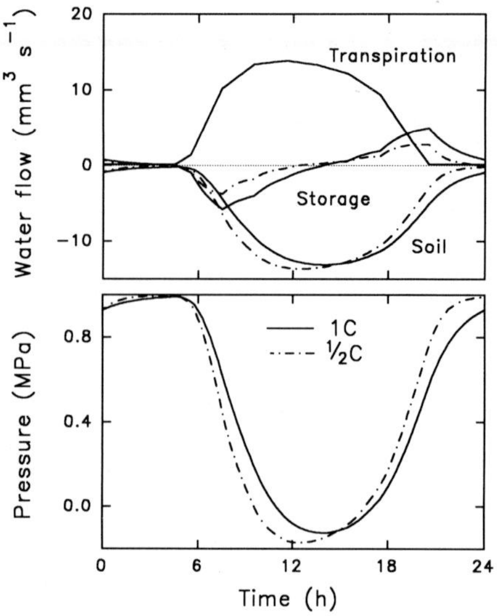

Figure 11.8. *Water flows (upper panel) and leaf turgor pressure (lower panel) for a hypothetical plant with significant water storage. For comparison with the standard case (solid lines), leaf capacitance was halved to simulate an approximate doubling in leaf cell elasticity (dash-dot lines).*

if storage were significant. In order to visualize such effects, the capacitance of leaf tissue was increased arbitrarily by two orders of magnitude. For this artificial plant, flow from storage now makes an observable contribution to storage (Figure 11.8). Water flows out of storage in the morning as transpiration increases and flows back into storage in the afternoon and at night. We can now simulate an increasing elasticity by halving capacitance (Equation 11.5). The contribution of storage to transpiration is reduced and the soil uptake curve peaks closer to the transpiration peak (reduced phase lag). The leaf tissue is somewhat less buffered, and turgor pressures drop lower during midday (Figure 11.8). Consider a similar halving of capacitance, but occurring over a period of 4 days (Figure 11.9). Storage flows decrease with time, as was also suggested by the one-day simulation. During early morning when transpiration is increasing, the leaf water potential drops faster for the leaves

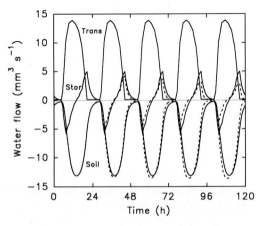

Figure 11.9. *Water flows for a plant experiencing a gradual decrease in cell elasticity (increase in elastic modulus) over a period of 4 days (dashed lines) starting at the end of the first day (t = 24 h). Compare with the constant elasticity condition (solid lines).*

with reduced capacitance (increased elastic modulus), and water uptake from the soil is enhanced. However, the opposite happens in the afternoon, when water potential rises faster for the low-capacitance plant and afternoon water uptake is reduced. If these flows are integrated over the entire day, uptake from the soil is identical with or without elasticity changes (see also Schulte, 1992).

The effects of changes in cell elasticity may be significant for succulent species in a manner distinct from water uptake from the soil. Increases in elasticity for plants with significant buffering capacity may further increase buffering. Water flow between tissues within succulent desert plants may be affected by changes in elasticity under conditions where water movement is driven by internal gradients of water potential that are not generated by transpiration. The transport of water between storage and chlorenchyma tissues of the desert succulents *Opuntia ficus-indica* and *A. deserti* depends on changes in tissue hydraulic properties and osmotic pressure during periods of extended drought when stomata remain closed throughout 24 h periods (Goldstein *et al.*, 1991; Schulte and Nobel, 1989). Tritiated water experiments have also demonstrated substantial redistribution of water between tissues in crassulacean acid metabolism (CAM) plants (Tissue *et al.*, 1991).

11.5 Conclusions and future considerations

The transpiration pathway from the soil through the leaves for this non-succulent desert shrub appears to be dominated by resistances encountered at two locations: (1) between the soil and the root xylem, and (2) within the leaf between the xylem and sites of evaporation. Simulations of small changes in these parameters indicate that their measurement, as well as proper model design, will be very important for modelling studies of plant–water relations. Further work is needed to determine how root resistance changes as the soil dries (Nobel and Cui, 1992). In addition, questions remain over changes in root resistance with water flow (Passioura, 1982; Pospíšilová, 1972; Sheriff, 1984) and perhaps during the day because of soil drying immediately adjacent to the root (Wallace and Biscoe, 1983). Such changes may slow recovery at night because of higher resistance to water uptake.

The apparent insignificance of resistances in the stem xylem make this tissue appear 'overbuilt'. However, these estimates are for a well-watered plant early in the growing season. The changes in stem xylem resistance that undoubtedly occur as water stress develops later in the growing season will be an important future consideration (Salleo and Lo Gullo, 1989; Sperry and Tyree, 1990; Sperry *et al.*, 1988).

The turgor pressure of leaf cells is highly dependent on their osmotic pressure. Changes in osmotic pressure with water potential and hence water content of cells is therefore a characteristic of cell–water relations that should be considered in modelling, at least if predictions of cell turgor are desired. In addition to such short-term changes in osmotic pressure, long-term changes in osmotic pressure of leaf cells during drought may be important for turgor maintenance (Kuang *et al.*, 1990).

Leaf capacitance appears to play a relatively minor role in the water relations of the shrub *Chilopsis linearis*. Modelling results, even for a plant with a far greater capacitance, suggest that increases in leaf elasticity cannot enhance water uptake during the 24 h period. The amount of water taken up by a plant over a time period longer than the time constant for storage exchange is determined by transpiration, and not by elasticity or water storage.

Finally, as models become more detailed, it will be useful to be able to estimate hydraulic parameters at physical scales smaller than whole organs. The pressure probe applied to individual cells will be an important tool for measuring storage resistances and cell wall elasticities, from which capacitance can be estimated.

References

Bowman, W.D. and Roberts, S.W. (1985) Seasonal changes in tissue elasticity in chaparral shrubs. *Physiol. Plant.*, **65**, 233–236.

Calkin, H.W., Gibson, A.C. and Nobel, P.S. (1985) Xylem water potentials and hydraulic conductances in eight species of ferns. *Can. J. Bot.*, **63**, 632–637.

Davies, F.S. and Lakso, A.N. (1979) Diurnal and seasonal-changes in leaf water potential components and elastic properties in response to water-stress in apple-trees. *Physiol. Plant.*, **46**, 109–114.

Elston, J., Karamanos, A.J., Kassam, A.H. and Wadsworth, R.M. (1976) The water relations of the field bean crop. *Phil. Trans. R. Soc. Lond. B*, **273**, 581–591.

Gibson, A.C., Calkin, H.W. and Nobel, P.S. (1985) Hydraulic conductance and xylem structure in tracheid-bearing plants. *Int. Assoc. Wood Anat. Bull.*, **6**, 293–302.

Goldstein, G., Ortega, J.K.E., Nerd, A. and Nobel, P.S. (1991) Diel patterns of water potential components for the Crassulacean acid metabolism plant *Opuntia ficus-indica* when well-watered or droughted. *Plant Physiol.*, **95**, 274-280.

Hüsken, D., Steudle, E. and Zimmermann, U. (1978) Pressure probe technique for measuring water relations of cells in higher plants. *Plant Physiol.*, **61**, 158–163.

Kuang, J.-B., Turner, N.C. and Henson, I.E. (1990) Influence of xylem water potential on leaf elongation and osmotic adjustment of wheat and lupin. *J. Exp. Bot.*, **41**, 217–221.

Mees, G.C. and Weatherly, P.E. (1957) The mechanism of water absorption by roots. 1. Preliminary studies on the effects of hydrostatic pressure gradients. *Proc. R. Soc. Lond. B*, **147**, 367–380.

Melkonian, J.J., Wolfe, J. and Steponkus, P.L. (1982) Determination of volumetric modulus of elasticity of wheat by pressure–volume relations and the effect of drought conditioning. *Crop Sci.*, **22**, 116–123.

Molz, F.J. and Ferrier, J.M. (1982) Mathematical treatment of water movement in plant cells and tissue: a review. *Plant Cell Environ.*, **5**, 191–206.

Morse, S.R. (1990) Water balance in *Hemizonia luzulifolia*: the role of extracellular polysaccharides. *Plant Cell Environ.*, **13**, 39–48.

Nagel, L. (1975) SPICE: A Computer Program to Simulate Semiconductor Circuits, ERL-M520. Electronics Research Laboratory, University of California, Berkeley.

Nobel, P.S. and Cui, M. (1992) Hydraulic conductances of the soil, the root–soil air gap, and the root: changes for desert succulents in drying soil. *J. Exp. Bot.*, **43**, 319–326.

Nobel, P.S. and Jordan, P.W. (1983) Transpiration stream of desert species: resistances and capacitances for a C_3, a C_4, and a CAM plant. *J. Exp. Bot.*, **34**, 1379–1391.

Nunes, M.A., Catarino, F. and Pinto, E. (1989) Strategies for acclimation to seasonal drought in *Ceratonia siliqua* leaves. *Physiol. Plant.*, **77**, 150–156.

Oosterhuis, D.M. and Wullschleger, S.D. (1987) Water flow through cotton roots in relation to xylem anatomy. *J. Exp. Bot.*, **38**, 1866–1874.

Parker, W.C., Pallardy, S.G., Hinckley, T.M. and Teskey, R.O. (1982) Seasonal changes in tissue water relations of three woody species of the *Quercus–Carya* forest type. *Ecology*, **63**, 1259–1267.

Passioura, J.B. (1982) Water in the soil–plant–atmosphere continuum. In: *Physiological Plant Physiology II. Encyclopedia of Plant Physiology*, New Series, Vol. 12B (eds O.L. Lange, P.S. Nobel, C.B. Osmond and H. Ziegler). Springer, Berlin, pp. 5–33.

Pospísilová , J. (1972) Variable resistance to water transport in leaf tissue of Kale. *Biol. Plant.*, **14**, 293–296.

Richter, H., Duhme, F., Glatzel, G., Hinckley, T.M. and Karlic, H. (1980) Some limitations and applications of the pressure–volume curve technique in ecophysiological research. In: *Plants and their Atmospheric Environment* (eds J. Grace, E.D. Ford and P.G. Jarvis). Blackwell Scientific Publications, Oxford, pp. 263–272.

Ruess, B.R., Ferrari, S. and Eller, B.M. (1988) Water economy and photosynthesis of the CAM plant *Senecio medley-woodii* during increasing drought. *Plant Cell Environ.*, **11**, 583–589.

Salleo, S. (1983) Water relations parameters of two Sicilian species of *Senecio* (groundsel) measured by the pressure bomb technique. *New Phytol.*, **95**, 179–188.

Salleo, S. and Lo Gullo, M.A. (1989) Different aspects of cavitation resistance in *Ceratonia siliqua*, a drought-avoiding Mediterranean tree. *Ann. Bot.*, **64**, 325–336.

Schulte, P.J. (1992) The units of currency for plant water status. *Plant Cell Environ.*, **15**, 7–10.

Schulte, P.J. and Hinckley, T.M. (1985) A comparison of pressure–volume curve data analysis techniques. *J. Exp. Bot.*, **36**, 1590–1602.

Schulte, P.J. and Nobel, P.S. (1989) Responses of a CAM plant to drought and rainfall: capacitance and osmotic pressure influences on water movement. *J. Exp. Bot.*, **40**, 61–70.

Schulte, P.J., Gibson, A.C. and Nobel, P.S. (1987) Xylem anatomy and hydraulic conductance of *Psilotum nudum*. *Am. J. Bot.*, **74**, 1438–1445.

Schulte, P.J., Smith, J.A.C. and Nobel, P.S. (1989) Water storage and osmotic pressure influences on the water relations of a dicotyledonous desert succulent. *Plant Cell Environ.*, **12**, 831–842.

Sheriff, D.W. (1984) Phases in water uptake during leaf rehydration, experiments and a heuristic model. *Ann. Bot.*, **53**, 865–873.

Smith, J.A.C., Schulte, P.J. and Nobel, P.S. (1987) Water flow and storage in *Agave deserti*: osmotic implications of crassulacean acid metabolism. *Plant Cell Environ.*, **10**, 639–648.

Smith, S.D. and Nobel, P.S. (1986) Deserts. In: *Photosynthesis in Contrasting Environments* (eds N.R. Baker and S.P. Long). Elsevier, Amsterdam, pp. 13–62.

Sperry, J.S. and Tyree, M.T. (1990) Water-stress-induced xylem embolism in three species of conifers. *Plant Cell Environ.*, **13**, 427–436.

Sperry, J.S., Donnelly, J.R. and Tyree, M.T. (1988) Seasonal occurrence of xylem embolism in sugar maple (*Acer saccharum*). *Am. J. Bot.*, **75**, 1212–1218.

Steudle, E., Zimmermann, U. and Lüttge, U. (1977) Effect of turgor pressure and cell size on the wall elasticity of plant cells. *Plant Physiol.*, **59**, 285–289.

Tissue, D.T., Yakir, D. and Nobel, P.S. (1991) Diel water movement between parenchyma and chlorenchyma of two desert CAM plants under dry and wet conditions. *Plant Cell Environ.*, **14**, 407–413.

Tuinenga, P.W. (1988) *SPICE: A Guide to Circuit Simulation and Analysis Using PSpice*. Prentice-Hall, Englewood Cliffs, NJ.

Tyree, M.T. and Richter, H. (1981) Alternative methods of analysing water potential isotherms: some cautions and clarifications. I. The impact of non-ideality and of some experimental errors. *J. Exp. Bot.*, **32**, 643–653.

Wallace, J.S. and Biscoe, P.V. (1983) Water relations of winter wheat. 4. Hydraulic resistance and capacitance in the soil-plant system. *J. Agric. Sci.*, **100**, 591–600.

Zimmermann, U. and Hüsken, D. (1979) Theoretical and experimental exclusion of errors in the determination of the elasticity and water transport parameters of plant cells by the pressure probe technique. *Plant Physiol.*, **64**, 18–24.

Drought tolerance and water-use efficiency

H.G. Jones

12.1 Introduction

Over the past 30 years major efforts have been made by plant physiologists and breeders to improve the drought tolerance of a wide range of agricultural and horticultural crops. There are many processes that affect the 'fitness' of a plant in water-limited situations, but those such as survival that may be appropriate in natural ecosystems are often not applicable to agricultural crops, where productivity is usually of the greatest interest (see Chapter 19). Even in agricultural situations there is no generally applicable definition of drought tolerance, as stability of yield may be a component in some situations, though for most crops drought-tolerant genotypes are those that have the highest absolute yield in a particular water-limited environment.

A useful classification of the various types of response that can favour drought tolerance is presented in Table 12.1. Unfortunately there is a conflict between most mechanisms involved in drought tolerance and the achievement of high yield, with most drought tolerance mechanisms tending to lower yield potential (Table 12.1). For example, drought avoidance by restriction of water loss through stomatal closure also decreases photosynthesis, while increasing water uptake by increased investment in root growth also has a cost in terms of the amount of carbohydrate available for harvestable yield. Similarly, it is likely, though generally not proven, that increased tolerance of low plant water potential is only achieved at the expense of potential yield.

One widely adopted approach to breeding for drought tolerance has been to concentrate on increasing what has come to be known as the water-use efficiency (WUE) of the crop (even though it is not strictly an efficiency). Depending on the circumstances, WUE can refer to the ratio between any

Table 12.1. *Drought-tolerance mechanisms (see Jones, 1992)*

Tolerance mechanism	Costs
1. Avoidance of plant water deficits	
(a) Drought escape – short growth cycle, dormant period	Short season
(b) Water conservation – small leaves, limited leaf area, stomatal closure, low cuticular conductance, low light absorption	Available water not used
(c) Maximal water uptake – good root system	Structural costs
2. Tolerance of plant water deficits	
(a) Turgor maintenance – osmotic adaptation, low elastic modulus	Metabolic costs
(b) Protective solutes, desiccation-tolerant enzymes, etc.	Metabolic costs
3. Efficiency mechanisms	
(a) Efficient use of available water	Low maximum rate
(b) Maximal harvest index	?

of yield, biomass or assimilation and either transpiration or evaporation (see also Chapter 16). It is timely to reconsider the most appropriate target for improving drought tolerance and whether it is sensible to continue to concentrate on WUE, so the main aim of this chapter is to assess the value of the concept of water-use efficiency as a component of drought tolerance in natural and agricultural situations, and to consider alternative approaches.

12.2 Water-use efficiency

The emphasis on WUE has been based on the assumption that, *other things being equal,* a plant with a high water-use efficiency should have a greater productivity under water-limited conditions than would a plant with low water-use efficiency. Unfortunately, as we shall see below, other things are not usually equal and a high WUE is normally associated with low absolute production. Extensive studies in the 1920s of a wide range of species grown in pots (Shantz and Piemeisel, 1927) showed that genotypic variation in WUE existed. It has since been noted, however, that the greatest differences were between plants having different photosynthetic pathways, rather than between plants having the same pathway. In practice, when integrated over the whole life cycle, differences between plants growing in one environment tend to be rather small. There are a number of reasons for this, but an important one is that the overall WUE is dominated by the value when stomata are open.

12.2.1 *Resistance analogues*

The simple resistance analysis of Bierhuizen and Slatyer (1965) has been used to study variation in WUE. In this analysis, the assimilation rate (A) is given by:

$$A = \frac{p_a - p_\Gamma}{P_a(r_a' + r_s' + r_m')}$$ 12.1

where p_a and p_Γ, respectively, are the atmospheric and 'internal' partial pressures of CO_2 (the latter is often assumed equal to the CO_2 compensation partial pressure), P_a is the atmospheric pressure, and r_a', r_s' and r_m' are the boundary layer, stomatal and mesophyll resistances, respectively, (with the prime referring to CO_2). Similarly for transpiration, or evaporation (E):

$$E = \frac{e_\ell - e_a}{P_a(r_a + r_s)}$$ 12.2

where e_ℓ and e_a are the saturation water vapour pressure at leaf temperature and the water vapour pressure of the air. If one assumes constant environmental conditions, and approximating $(e_\ell - e_a)$ by δe (the water vapour pressure deficit of the air), one can combine Equations 12.1 and 12.2 (see Jones, 1992) to give:

$$\frac{A}{E} \propto \frac{r_a + r_s}{r_a' + r_s' + r_m'}.$$ 12.3

Since the denominator contains an extra term as compared with the numerator, this equation indicates that WUE might be expected to increase as stomata close, and has, for example, been used to justify the extensive effort that has gone into the development of antitranspirants.

Unfortunately a number of important approximations have been made in this approach: (1) it assumes that the mesophyll limitation remains constant; (2) it assumes that $(e_\ell - e_a)$ is constant and equal to δe in one environment; (3) it is based on instantaneous gas exchange and does not integrate over time (and, especially importantly, it ignores night-time respiratory losses); and (4) it ignores soil evaporation. Some consequences of these assumptions are discussed further below.

12.2.2 *Carbon isotope discrimination: basic theory*

In recent years there has been a resurgence of interest in WUE because of the suggestion by Farquhar *et al.* (1982) that the degree of carbon isotope discrimination in different plants might be directly related to WUE, and should provide a rapid screening method. The basis of this proposal is the observation that photosynthesis tends to discriminate against the heavier isotopes of carbon, so that the ratio of $^{13}C/^{12}C$ in dry matter tends to be somewhat lower than the corresponding ratio for the CO_2 in air. The amount of this discrimination, usually expressed in parts per thousand (‰), is described by Δ, defined by:

$$\Delta = \frac{({}^{13}C/{}^{12}C)_{reactants}}{({}^{13}C/{}^{12}C)_{products}} - 1. \qquad 12.4$$

The value of Δ for plant material is commonly in the range 13–28‰ for C_3 plants and between -1 and 7‰ for C_4 plants. Farquhar *et al.* (1982) showed that for C_3 plants the value of Δ would be expected to be dependent on the value of the intercellular space partial pressure of CO_2 (p_i) approximately according to:

$$\Delta = 0.0044 + 0.0256 p_i/p_a. \qquad 12.5$$

This prediction has now been confirmed experimentally in a number of cases for C_3 plants (see Farquhar *et al.*, 1988), while Δ is largely insensitive to p_i/p_a in C_4 plants (Evans *et al.*, 1986). A set of typical results indicating the expected positive correlation between Δ and p_i/p_a is shown in Figure 12.1.

The relevance of carbon isotope discrimination to WUE arises because the value of p_i depends on the assimilation rate and the total gas-phase resistance to CO_2 according to Jones (1992):

$$A(r_a' + r_s') = (p_a - p_i)/(P_a) \qquad 12.6a$$

which can also be written in terms of the total gas-phase conductance to CO_2 (g') as:

$$A/g' = p_a(1 - p_i/p_a)/(P_a). \qquad 12.6b$$

It is clear from these equations that the intercellular partial pressure of CO_2 can be used as a measure of the ratio between assimilation and the total gas-phase conductance with A/g' increasing as p_i decreases. It follows that where two plants have the same conductance, the assimilation rate and hence the WUE

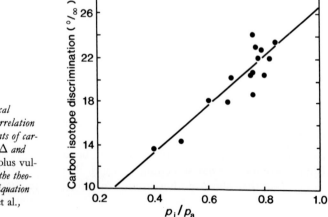

Figure 12.1. *A typical example of the positive correlation between on-line measurements of carbon isotope discrimination Δ and p_i/p_a for leaves of* Phaseolus vulgaris. *The line represents the theoretical relationship from Equation 12.5. (After Ehleringer* et al., *1991.)*

must be higher for the plant with the lower value of p_i, since from Equation 12.6:

$$[A/g']_1/[A/g']_2 = (1 - p_{i1}/p_a)/(1 - p_{i2}/p_a). \qquad 12.7$$

Substituting from Equation 12.5 and rearranging gives for plants with equal total gas-phase conductance:

$$WUE_1/WUE_2 = (0.03 - \Delta_1)/(0.03 - \Delta_2). \qquad 12.8$$

Alternatively one can combine Equations 12.2, 12.5 and 12.6 and approximate $(e_\ell - e_a)$ by δe to give:

$$A/E \simeq p_a(0.03 - \Delta)/(0.04096\delta e). \qquad 12.9$$

12.2.3 Indirect selection for water-use efficiency

In view of the above relationships, it is not surprising that there has been much recent interest in the use of Δ as a possible indirect screening method for the selection of plants differing in water-use efficiency as a means of breeding for drought tolerance (e.g. Condon et al., 1990; Ehleringer et al., 1991; Farquhar and Richards, 1984; Hall et al., 1992; Ismail and Hall, 1992; Martin and Thorstenson, 1988; Read et al., 1991; White et al., 1990; see also Chapters 16 and 17). Most of these studies have indeed provided evidence that the expected negative relationship between WUE and carbon isotope discrimination does exist (e.g. Figure 12.2), though the correlation is often not very strong.

Perhaps the greatest advantage of carbon isotope discrimination as a

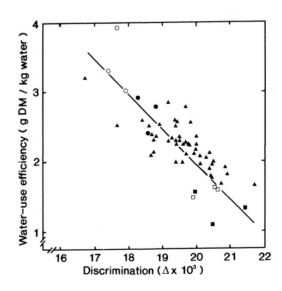

Figure 12.2. An example of the negative relationship between water-use efficiency and carbon isotope discrimination in dried peanut leaf tissue. Open symbols refer to water-stressed plants and closed symbols to well-watered plants. Different symbols refer to different cultivars. (Data from Hubick et al., 1988.)

potential indicator of WUE is that when it is based on the $^{13}C/^{12}C$ ratio of total dry matter it integrates both over time and over space. Such integration is extremely difficult using gas-exchange measurements which are essentially instantaneous. Unfortunately, as for the other methods for estimating WUE, it neither allows for respiratory loss, nor for evaporation from the soil. Furthermore it also assumes that δe is a good measure of $(e_\ell - e_a)$, although Farquhar *et al.* (1988) have suggested that it may be possible to estimate the vapour pressure difference $(e_\ell - e_a)$ weighted by stomatal conductance using information on the stable isotopic compositions of hydrogen and oxygen in organic material.

Some of these difficulties with the use of simple gas-exchange theory or carbon isotope discrimination as indicators of water-use efficiency are considered in more detail below.

12.2.4 *Energy balance and 'coupling'*

Unless comparisons are restricted to plants having equal total gas-phase conductances (as in Equation 12.8, see also Hall *et al.*, 1992), the simple resistance-analogue approach and carbon isotope discrimination both depend on the assumption that $(e_\ell - e_a)$ can be approximated by δe. Unfortunately this can lead to quite large errors because $(e_\ell - e_a)$ is dependent on the evaporation rate as well as on a range of environmental factors. The magnitude of the error depends on the degree to which the plant or crop is 'coupled' to the environment (see Jones, 1992; McNaughton and Jarvis, 1983; see also Chapter 18).

A plant is said to be well coupled when mass and energy exchange between the plant and the bulk atmosphere is efficient, so that leaf temperature closely follows air temperature. In such a situation $\delta e \simeq (e_\ell - e_a)$, and $E \propto g_\ell$ (where g_ℓ is the leaf conductance to water vapour), so that Equations 12.3 and 12.9, for example, are approximately valid. Where, however, coupling is not perfect, the effect of evaporation from plants upwind on the atmospheric humidity, and the fact that leaf temperature diverges from air temperature, both mean that E is not necessarily proportional to the stomatal conductance. In fact, in extreme conditions of very poor coupling, as occur with large boundary layer resistance (e.g. low wind speeds and large areas of short, smooth crops), E becomes independent of g_ℓ and proportional to incoming energy (see Chapter 18). Because assimilation is less sensitive to the degree of coupling than is evaporation, changes in stomatal conductance do not necessarily affect the instantaneous WUE in the way predicted by Equation 12.3.

Jones (1976) analysed the interactions between boundary layer resistance, radiation and stomatal resistance in their effects on WUE for single leaves and plants, and showed that these energy-balance considerations resulted in there often being an optimal stomatal conductance for maximal WUE (Figure 12.3). It was also pointed out that dark respiratory losses and non-

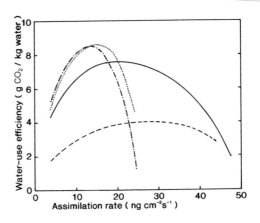

Figure 12.3. Theoretical relationships between water-use efficiency and assimilation rate (approximately proportional to stomatal conductance) for typical C_3 leaves in high light (———, -----) or low light (— · — · —, ...) and with high boundary layer resistance (------, ...) or low boundary layer resistance (———, — · — · —). (After Jones, 1976.)

stomatal evaporation (e.g. through the cuticle, or directly from the soil) both tended to increase the optimal stomatal conductance.

An important consequence of coupling that is often neglected in discussion of WUE is that of scale (Jarvis and McNaughton, 1986; see also Chapter 18). Breeders and physiologists often do their selection on single plants in controlled environments, or on isolated single plants in the field. In each case the plants are well coupled to the environment and changes in stomatal conductance would be expected to have a nearly proportional affect on E, with decreases in conductance tending to increase WUE. When, however, crops are grown on a large scale in the field, especially aerodynamically smooth short crops, coupling is poor, and one might expect selection for reduced stomatal conductance to lower WUE (because of the potentially greater sensitivity of A than E to changes in stomatal conductance in decoupled situations), or at least to give variable results (Farquhar et al., 1989). Where selection is for differences in assimilation capacity, differences may be maintained over a range of scales, including the field. There is a clear need for further testing of Δ as an indicator of WUE on a field scale, as is further discussed in Chapters 16 and 17.

12.2.5 *Relationship of WUE to drought tolerance*

The usual reason for attempting to improve WUE is as a means to improve drought tolerance. Unfortunately the tendency for WUE to increase as stomata close and productivity declines (e.g. under drought) means that selecting for high WUE may often be the same as selecting for low productivity, whereas the objective in breeding for drought tolerance is normally to achieve high yield.

Although effort was originally concentrated on the instantaneous value of WUE, it was quickly realized that maximizing WUE over the long term does not necessarily involve continuous maximization of the instantaneous ratio between assimilation and transpiration (Cowan and Farquhar, 1977;

Jones, 1976). Subsequently, the analyses have been extended to take account of uncertainties in the future environment and to investigate optimal patterns of stomatal behaviour in such circumstances (e.g. Cowan, 1982, 1986; Givnish, 1986; Jones, 1981).

12.3 Water use and drought tolerance in natural environments

In natural environments there is likely to be no advantage for an individual plant to conserve water if, as a consequence, competitors could then use the available soil water (see Chapter 17). Evolution therefore seems unlikely to have favoured water conservation behaviours or even high water-use efficiency. It is only where allelopathic exclusion of competing root systems from the root zone occurs that such 'conservative' or 'water-saving' responses might be favoured. Nevertheless, once the water has been captured, there could be an advantage in using that water more efficiently. The evolution of crassulacean acid metabolism may have been on this basis, as it is normally coupled with a water-storing or succulent life form. In an agricultural situation, where all plants may be of the same species, it is also possible for water conservation to be a suitable and achievable strategy.

It seems logical to assume that, in general, a more important factor in evolutionary success than maximizing water-use efficiency (the assimilation per unit water *used*) would be to maximize the amount of assimilation per unit water *available*. It is, however, rather difficult to envisage how a plant might achieve this. Not only does it imply a sensing of the rate of assimilation (possibly feasible by the sensing of carbohydrate accumulation?) and of the rate of evaporation (possibly feasible through a detection of leaf to soil water potential differences?), but it would also be necessary to have target values, which would depend on the environment and, more importantly, on the probability of future rainfall. This seems rather unlikely. On the other hand, it does seem possible for a simple mechanism to have evolved that might produce similar results based on maximizing assimilation. This must, of course, involve the avoidance of damage resulting from low plant water potentials, and would involve a trade-off between having open stomata for high assimilation rate and closing the stomata to prevent plant water potentials falling below the critical threshold at which damage occurs (that might lower assimilation). It would be particularly critical to avoid damage to the water-conducting system, particularly as it appears that xylem embolism tends to be irreversible (Tyree and Sperry, 1989; see Chapter 7).

12.3.1 *The role of stomata?*

If one hypothesizes that stomata operate to maximize assimilation while avoiding, or at least minimizing, damage to the conducting system (rather than operating in a fashion that maximizes water-use efficiency), one can envisage two contrasting approaches.

(1) The most conservative behaviour would be for stomatal closure to occur to prevent leaf or shoot water potential falling below the threshold at which xylem cavitations occur. In the absence of stomatal regulation, once cavitations occur, the consequent loss of functional xylem vessels would decrease the hydraulic conductivity (see Chapters 6 and 7). It follows that leaf water potential would need to decline to maintain the same transpiration rate (because, of course, transpiration is determined by the gas-phase rather than by the liquid-phase resistance). This in turn would cause further xylem cavitations, further decline in leaf water potential, and so on, in a catastrophic spiral of decline. Such catastrophic xylem failure (Tyree and Sperry, 1988) could be avoided by stomatal closure, and the fact that there is so little evidence for such catastrophic xylem failure actually occurring in natural environments suggests that this may be a major function of stomata.

(2) As an alternative function that might be optimized in a well-adapted plant, Jones and Sutherland (1991) speculated that a stomatal behaviour might exist that achieves higher productivity than (a) above by allowing some controlled xylem embolism. Jones and Sutherland (1991) proved that situations could exist where productivity could be maximized by allowing some xylem vessels to be embolized. They showed that the frequency of such situations increases as soil water potential declines, to the extent that once soil water potential falls to the threshold for damage, in all cases there should be some advantage in a proportion of xylem loss.

It is interesting that the very simple model used by Jones and Sutherland (1991) was able to explain a number of well-known features of the stomatal response, including stomatal regulation by soil water status and atmospheric humidity, rather than by leaf water potential. In another recent study, Friend (1991) used a simple model that adjusts stomatal aperture to maximize short-term carbon gain and that includes a treatment of the trade-off between increasing assimilation rate and declining leaf water potential (and hence decreasing photosynthetic capacity) as stomata open. This model was found to simulate observed stomatal behaviour reasonably well over a range of environments. Together these papers lend some credence to the idea that stomatal behaviour may have evolved to maximize carbon gain, while at the same time avoiding the occurrence of damaging water deficits.

Concrete evidence that stomata do operate in a fashion that maximizes dry-matter production by minimizing or regulating damage, rather than by maximizing production per unit water loss, has yet to be obtained. Although it is likely to be difficult to discriminate between these two options, this remains an important challenge for the future.

12.4 Conclusions

It is clear from the discussion above and from Chapter 17 that the concept of water-use efficiency is of only limited applicability to discussion of the fitness of plants for water-limited environments. This is particularly true for natural ecosystems where the dominant factors relating to survival or competitiveness include processes such as the effectiveness of mechanisms (such as stomatal closure, or differences in the exploration of soil by roots) that prevent the occurrence of damaging plant water deficits. Even in an agricultural or horticultural context, it is still efficiency of use of *available* soil water rather than water-use efficiency *per se* that appears to be important. Although Δ can be a useful indicator of WUE, it is not always reliable for this purpose, and even where it is, it follows from the above discussion that it is not necessarily a good indicator of success in water-limited environments. More often than not, a high WUE (or a low Δ) is related to low productivity and is therefore of only limited value to plant breeders. As described in Chapter 17, Δ is probably of more value for ecological studies in relation to its use as an indicator of p_i/p_a, or possibly as an indicator of stomatal conductance.

References

Bierhulzen, J.F. and Slatyer, R.O. (1965) Effect of atmospheric concentration of water vapour and CO_2 in determining transpiration–photosynthesis relationships of cotton leaves. *Agric. Meterol.*, **2**, 259–270.

Condon, A.G., Farquhar, G.D. and Richards, R.A. (1990) Genotypic variation in carbon-isotope discrimination and transpiration efficiency in wheat: leaf gas exchange and whole plant studies. *Aust. J. Plant Physiol.*, **17**, 9–22.

Cowan, I.R. (1982) Water-use and optimization of carbon assimilation. In: *Encyclopedia of Plant Physiology*, new series, Vol. 12B (eds O.L. Lange, P.S. Nobel, C.B. Osmond and H. Ziegler). Springer, Berlin, pp. 589–613.

Cowan, I.R. (1986) Economics of carbon fixation in higher plants. In: *On the Economy of Plant Form and Function* (ed. T.J. Givnish). Cambridge University Press, Cambridge, pp. 133–170.

Cowan, I.R. and Farquhar, G.D. (1977) Stomatal function in relation to leaf metabolism and environment. *Symp. Soc. Exp. Biol.*, **31**, 471–505.

Ehleringer, J.R., Klassen, S., Clayton, F., Sherrill, D., Fuller-Holbrook, Fu, Q. and Cooper, T. (1991) Carbon-isotope discrimination and transpiration efficiency in common bean. *Crop Sci.*, **31**, 1611–1615.

Evans, J.R., Sharkey, T.D., Berry, J.A. and Farquhar, G.D. (1986) Carbon-isotope discrimination measured concurrently with gas-exchange to investigate CO_2 diffusion in leaves of higher plants. *Aust. J. Plant Physiol.*, **13**, 281–292.

Farquhar, G.D. and Richards, R.A. (1984) Isotopic composition of plant carbon correlates with water-use efficiency of wheat genotypes. *Aust. J. Plant Physiol.*, **11**, 121–137.

Farquhar, G.D., O'Leary, M.H. and Berry, J.A. (1982) On the relationship between carbon-isotope discrimination and intercellular carbon dioxide concentration. *Aust. J. Plant Physiol.*, **9**, 121–137.

Farquhar, G.D., Hubick, K.T., Condon, A.G. and Richards, R.A. (1988) Carbon-isotope fractionation and plant water-use efficiency. In: *Stable Isotopes in Ecological Research* (eds P.W. Rundel, J.R. Ehleringer and K.A. Nagy). Springer, New York, pp. 21–40.

Farquhar, G.D., Ehleringer, J.R. and Hubick, K.T. (1989) Carbon-isotope discrimination and photosynthesis. *Ann. Rev. Plant Physiol., Plant Mol. Biol.*, **40**, 503–537.

Friend, A.D. (1991) Use of a model of photosynthesis and leaf microenvironment to predict optimal stomatal conductance and leaf nitrogen partitioning. *Plant, Cell Environ.*, **14**, 895–905.

Givnish, T.J. (1986) Optimal stomatal conductance, allocation of energy between leaves and roots, and the marginal cost of transpiration. In: *On the Economy of Plant Form and Function* (ed. T.J. Givnish). Cambridge University Press, Cambridge, pp. 171–213.

Hall, A.E., Mutters, R.G. and Farquhar, G.D. (1992) Genotypic and drought-induced differences in carbon-isotope discrimination and gas exchange of cowpea. *Crop Sci.*, **32**, 1–6.

Hubick, K.T., Shorter, R. and Farquhar, G.D. (1988) Heritability and genotype x environment interactions of carbon isotope discrimination and transpiration efficiency in peanut (*Arachis hypogaea* L.). *Aust. J. Plant Physiol.*, **15**, 799–813.

Ismail, A.M. and Hall, A.E. (1992) Correlation between water-use efficiency and carbon-isotope discrimination in diverse cowpea genotypes and isogenic lines. *Crop Sci.*, **32**, 7–12.

Jarvis, P.G. and McNaughton, K.G. (1986) Stomatal control of transpiration: scaling up from leaf to region. *Adv. Ecol. Res.*, **15,** 1–49.

Jones, H.G. (1976) Crop characteristics and the ratio between assimilation and transpiration. *J. Appl. Ecol.*, **13**, 605–622.

Jones, H.G. (1981) The use of stochastic modelling to study the influence of stomatal behaviour on yield–climate relationships. In: *Mathematics and Plant Physiology* (eds A. Rose and D. Charles-Edwards). Academic Press, New York, pp. 231–244.

Jones, H.G. (1992) *Plants and Microclimate: a Quantitative Approach to Environmental Plant Physiology*. Cambridge University Press, Cambridge.

Jones, H.G. and Sutherland, R.A. (1991) Stomatal control of xylem embolism. *Plant Cell Environ.*, **6**, 607–612.

Martin, B. and Thorstenson, Y.R. (1988) Stable carbon-isotope composition ($\delta^{13}C$), water-use efficiency, and biomass production of *Lycopersicon esculentum, Lycopersicon pennellii*, and the F_1 hybrid. *Plant Physiol.*, **88**, 213–217.

Read, J.J., Johnson, D.A., Asay, K.H. and Tieszen, L.L. (1991) Carbon-isotope discrimination, gas exchange, and water-use efficiency in crested wheatgrass clones. *Crop Sci.*, **31**, 1203–1208.

Shantz, H.L. and Piemeisel, L.N. (1927) The water requirements of plants at Akron, Colorado. *J. Agric. Res.*, **34**, 1093–1190.

Tyree, M.T. and Sperry, J.S. (1988) Do woody plants operate near the point of catastrophic xylem dysfunction caused by dynamic water stress? *Plant Physiol.*, **88**, 574–580.

Tyree, M.T. and Sperry, J.S. (1989) Vulnerability of xylem to cavitation and embolism. *Ann. Rev. Plant Physiol., Plant Mol. Biol.*, **40**, 19–38.

White, J.W., Castillo, J.A. and Ehleringer, J.R. (1990) Associations between productivity, root growth and carbon-isotope discrimination in *Phaseolus vulgaris* under water deficit. *Aust. J. Plant Physiol.*, **17**, 189–198.

13

Nitrogen assimilation and its role in plant water relations

J.A. Raven and J.I. Sprent

13.1 Introduction

Our intention in this chapter is to relate some aspects of nitrogen assimilation by photosynthetic organisms to their water relations. Despite our general predilections, we do not attempt to be encyclopaedic in our coverage; nevertheless, we encompass both poikilohydric and homoiohydric organisms, and both the thermodynamic and kinetic aspects of water relations.

13.2 Nitrogen in compatible solutes

Compatible solutes are employed in all major taxa except Archaeobacteria to generate osmolarity (decrease osmotic potential) to values outwith the range of values generated by non-toxic concentrations of inorganic ions. The toxic concentration of such ions is lower for metabolically complex compartments such as cytosol, plastid stroma and mitochondrial matrix than for metabolically less diverse compartments such as vacuoles. Nevertheless, compatible solutes occur in vacuoles as well as in cytosol, stroma and matrix (see Raven, 1985).

The three categories of compatible solutes are polyols (including, probably, certain non-reducing disaccharides) and two groups of zwitterionic compounds, the nitrogen-based betaines (and proline) and the sulphur-based sulphonium compounds. These compounds have a taxonomic distribution indicating multiple evolutionary origins, although they do have some taxonomic utility. They all clearly have energy and carbon costs of synthesis and, through the carbon costs, the implication of a water cost in increased transpiratory water loss per unit 'basic biomass' (i.e. components other than compatible solutes) synthesized. Only the nitrogenous zwitterions have a direct nitrogen cost in terms of presence of nitrogen in the compatible solute. Thus, while all compatible solutes increase the plant's demand for light energy and, in land plants, water per unit 'basic biomass' synthesized, only the betaines and proline increase the nitrogen demand. While some habitats for which compatible solutes are required are light-limiting (deep in the sea, for example) and many are water-limiting for plant growth, a number are nitrogen-limiting: examples of saline nitrogen-limiting habitats are many salt marshes (Stewart *et al.*, 1979) and the open ocean (cf. Smith, 1984). There does not seem to be a universal correlation between nitrogen-poor environments and the use of non-nitrogenous compatible solutes on the part of the organisms native to them. This has been shown for salt marshes (Stewart *et al.*, 1979) and can be deduced from references cited in Kirst (1990) in the case of marine phytoplankton.

These data suggest that some plants in nitrogen-poor environments use nitrogen-containing compatible solutes, and thus require more nitrogen (N) to synthesize a unit of 'basic biomass'. This in turn means that more N is needed to produce a unit of viable propagule (a proxy for selective fitness) in such organisms than in sympatric, competing organisms which use compatible solutes that do not contain nitrogen. How could this cost, in terms of an additional need for a limiting resource in plants using nitrogen-containing compatible solutes, be offset? One possibility is that the use of nitrogen-containing compatible solutes gives some advantage in facilitating metabolism relative to other compatible solutes. The evidence as to the universal applicability of this suggestion is scant or absent. Thus, glycine betaine is as good as sorbitol, but better than proline, in supporting CO_2 fixation activity in isolated chloroplasts of *Spinacia oleracea*, a plant that uses glycine betaine as its compatible solute (Larkum and Wyn Jones, 1979). Here the efficacy of the native, N-containing compatible solute is equalled by a non-N-containing compatible solute (sorbitol) and exceeds that of the N-containing proline.

Another possible function of the zwitterionic compatible solutes containing N and sulphur (S) is that of ultraviolet-B (UV-B; 280–320 nm) screening, i.e. decreasing (by absorption) the fraction of the UV-B flux incident on unit area of the cell surface which reaches such UV-B-sensitive components as the nuclear DNA near the centre of diatom cells (Karentz *et al.*, 1991; Raven, 1991). This possibility is topical in view of the increased UV-B flux due to depletion of ozone in the stratosphere which is not always countered by increased ozone levels in the troposphere. This possibility is negated for the well-characterized compatible solutes containing N or S (e.g. glycine, betaine, proline, dimethylsulphonium propionate) which lack significant UV-B absorption. Even the mycosporine-like amino acids have UV absorption spectra centred on the ultraviolet-A (UV-A) region (320–400 nm), but they do have significant UV-B absorption. What is not known is if they have any of the characteristics of compatible solutes. This possibility is worth exploring, especially in view of the possibility that there are no non-N-containing (e.g. S-containing) analogues of the mycosporine-like amino acids which could double as compatible solutes and UV-B screens (Raven, 1991).

13.3 Nitrogen in proteins found in all plants: the case of Rubisco

The advent of molecular biological techniques for altering the extent of expression of specific proteins permits the experimenter to manipulate the nitrogen allocation to a given protein. The example used here is Rubisco (ribulose 1,5-bisphosphate carboxylase-oxygenase), the core carboxylase involved in producing reduced carbon in all O_2-evolvers (regardless of whether they have C_3, C_4 or crassulacean acid metabolism). This single protein accounts for about 23% of the *total* protein nitrogen in the leaf of a C_3 sun plant (Evans and Seeman, 1989). This large fractional allocation can be rationalized in terms of the relatively large M_r and low specific reaction rate (mol CO_2 fixed per mol enzyme per second) of Rubisco relative to other enzymes of CO_2 fixation, and its oxygenase activity. In addition, the enzyme has a CO_2 affinity such that atmospheric-equilibrium CO_2 concentrations are subsaturating, a circumstance exacerbated by diffusive limitation of CO_2 supply such as occurs *in vivo* (Lorimer, 1981). The use of antisense RNA to the small subunit of Rubisco permitted Quick *et al.* (1991) and Stitt *et al.* (1991) to reduce the expression of Rubisco in leaves of the C_3 plant *Nicotiana tabacum*. These workers investigated the effects of the reduced Rubisco expression on photosynthetic performance and the photon and water costs (as mol photon absorbed or mol water lost per mol carbon fixed) and the nitrogen cost (mol N needed to assimilate one mol CO_2 per second). The predicted effects of reduced expression of Rubisco are a decrease in the nitrogen cost of CO_2 fixation, if there is indeed an 'excess' of Rubisco over what is needed to account for the observed CO_2 fixation rate. Further, there is a predicted increase in the water cost of photosynthesis as a result of a decreased fractional limitation of photosynthesis by CO_2 diffusion when the reduced Rubisco activity increases fractional limitation by biochemical processes, provided that the 'matching' of biochemical and diffusive conductances (Cowan and Farquhar, 1977) is disrupted by the reduced expression of Rubisco.

The predictions were fulfilled, in that the reduced expression of Rubisco decreased the nitrogen cost of photosynthesis and increased that of water (Quick *et al.*, 1991; Stitt *et al.*, 1991). Hudson *et al.* (1992) report partially overlapping data, arrived at independently, again using antisense RNA to the small subunit of Rubisco in a *Nicotiana* species, but including $^{13}C/^{12}C$ data relating to the increased water cost of photosynthesis. The data obtained in these studies show that the apparent 'excess' of Rubisco in unmodified C_3 plants permits a lower water (transpiration) cost of photosynthesis at the expense of an increased nitrogen cost relative to plants which have been engineered to have lower Rubisco levels. The 'excess' Rubisco also permits a rather lower photon cost of photosynthesis under light-limiting conditions in normal air (Quick et al., 1991; Stitt et al., 1991).

The approach used by Quick et al. (1991), Stitt et al. (1991) and Hudson et al (1992) shows that the apparent 'excess' of Rubisco has important interacting effects on photosynthetic costs in terms of photons, nitrogen and water. In essence, the high Rubisco levels permit lower water costs (mol H_2O lost per mol C fixed) of photosynthetic CO_2 fixation, at the expense of a higher N cost (mol N needed to permit the fixation of one mol C per second). While these data refer to individual leaves and isolated plants, and should be extrapolated with caution to the natural environment (see Jarvis and McNaughton, 1986), the work is of great significance, and further work along these lines will undoubtedly yield important data which will clarify the relationships between H_2O supply and loss and N acquisition and allocation in plants.

13.4 The water cost of growth as a function of nitrogen source for growth

13.4.1 *Theoretical costs*

Raven (1985) modelled the effect of nitrogen source for growth (NH_4^+, NO_3^- or N_2) on the water cost of organic N accumulation in biomass (mol H_2O lost in transpiration per unit (mol) N assimilated). The method used was to take the known transport and biochemical processes involved in N assimilation to estimate the additional CO_2 fixation (additional to that needed for the C skeletons or organic N, and all processes and syntheses not related to organic N synthesis) required to provide the respiratory substrate needed for transport and reduction processes in the roots and in shoot tissues in the dark which are specific to NO_3^- and N_2 as N sources and, for NO_3^- assimilation, the C skeletons required for acid–base regulation (OH^- removal). This additional net CO_2 fixation in the light (regardless of whether it is retained in the harvested plant material as organic C, or is lost in respiration) has a water (transpiration) loss depending on the ratio of (stomatal + boundary layer conductance) to (biochemical conductance) to CO_2 and the difference between the leaf intercellular space H_2O vapour concentration and the bulk air H_2O vapour concentration. Making reasonable assumptions as to these gas exchange parameters, Raven (1985, Table 5 therein) tabulates the additional transpiratory H_2O cost per mol N assimilated, and the total cost per mol N assimilated including the costs not related to different N sources, i.e. adding the H_2O cost of growth with the least costly N source, NH_4^+. The results of these computations are summarized in Table 13.1. Also shown in Table 13.1 is

Table 13.1. H_2O costs of growth predicted by Raven (1985) on the basis of (1) the additional H_2O lost per mol N assimilated from NO_3^- or N_2 relative to the least water-costly N source, NH_4^+; (2) the total H_2O cost per mol N assimilated; and (3) (not explicitly computed by Raven, 1985) the H_2O cost per mol C assimilated and retained until harvest. The assumed H_2O cost for growth with NH_4^+ as N source in a plant with a $C:N$ (molar) ratio of 20 is 1.4×10^4 mol H_2O lost per mol N assimilated. For computation of the H_2O cost of C assimilation (3) the $C:N$ ratio used to derive this value from computation (2) was adjusted, in the case of the use of organic acids for OH^- neutralization, for the additional C per unit N in the pH-adjusting organic acids (0.67 mol COH^- per mol N)

N source, location of assimilation, mode of acid–base regulation	(1) Additional H_2O loss with N_2 or NO_3^- relative to NH_4^+ (mol H_2O/ mol N_2)	(2) Total H_2O loss with various N sources (mol H_2O/mol N)	(3) Total H_2O loss with various N sources (mol H_2O/mol C)
(a) NH_4^+	0	14 000	700
(b) N_2 fixation in root nodules; H^+ excretion	1673	15 673	782
(c) NO_3^- reduction in roots; OH^- disposal by excretion to rooting medium	1619	15 519	776
(d) NO_3^- reduction in roots; OH^- disposal by malic acid synthesis	2335	16 335	766
(e) NO_3^- photo reduction in shoots; OH^- disposal by malic acid synthesis and malate salt accumulation	861	14 861	697
(f) As for (e) but with oxalic acid	> 402	> 14 402	> 698

the H_2O cost of growth on the various N sources on the basis of C assimilated into organic matter of the harvested plant.

The predicted H_2O cost per mol N assimilated is greater with N_2 or NO_3^- (regardless of the assumptions made as to the site of NO_3^- assimilation or the mode of acid–base regulation) than with NH_4^+ as N source. However, the predicted H_2O cost per mol C is marginally smaller than, but unlikely to be experimentally distinguishable from, the value for NH_4^+ as N source when NO_3^- is being assimilated in shoots with OH^- disposal by organic acid synthesis and organic anion accumulation. The lower values result from dividing the N-based prediction by a larger $C:N$ ratio incorporating the additional C in the organic acid anions. The C in inorganic ions is unavailable for those uses in the other treatments, as it would negate OH^- dispersal. It therefore does not contribute overtly to selective fitness (resource acquisition and retention; reproduction). A contribution to fitness could result from anti-biophage properties of calcium oxalate, or from the reduced alkalinization of the rooting medium relative to the OH^- efflux mode of acid–base regulation when NO_3^- is being assimilated as a means of aiding P and Fe acquisition (Raven, 1985).

13.4.2 *Experimentally determined costs*

The predictions in column (3) of Table 13.1 can be compared with the measured values for three plant species in Table 13.2. It should be noted that the differences in values for a given N source (NH_4^+, NO_3^-) among the three species, and between the values for a given N source between the prediction and an individual plant, probably relate to differences in plant-to-air vapour-pressure differences which are, however, constant for a species. Accordingly, the comparison between measurement and prediction should relate to whether the estimate for NO_3^- or N_2 as N source differs from that for NH_4^+ in the same direction as do the measurements.

For the only data on N_2 in Table 13.2 it is clear that the H_2O cost of C assimilation is greater than that for NH_4^+, in accordance with the estimate in Table 13.1. The substantially higher value for the H_2O cost of C assimilation with N_2 as N source is found despite the higher molar C : N ratio of the N-grown plants (22.6) than of the NH_4^+-grown plants (15.2) (Allen *et al.*, 1988) which would decrease the increment of H_2O cost of C assimilation due to N_2 rather than NH_4^+ as N source (see Table 13.1).

For NO_3^- relative to NH_4^+, the model suggests that the H_2O cost of C accumulation should be at least as great for NO_3^- as N source as it is for NH_4^+ as N source, while the data show that the H_2O cost with NO_3^- is less than that for NH_4^+ in all three species. The molar C : N ratio of *Phaseolus vulgaris* is 15.9 on NO_3^- and 15.2 with NH_4^+ as N source, while for *Ricinus communis* the values are 15.2 and 12.4, respectively (Allen and Raven, 1987). The lower N content per

Table 13.2. *H_2O costs of plant growth on various N sources*

Plant	Mol H_2O per mol C appearing in plant dry matter with N supplied as:				Reference
	NH_4^+	NH_4NO_3	NO_3^-	N_2	
Helianthus annuus	575	317	374	na	Kaiser and Lewis (1991)
Phaseolus vulgaris	675	nd	609	896	Allen *et al.* (1988)
Ricinus communis	891	nd	714	na	Allen and Raven (1987)

na = not applicable; nd = not determined.

The *Helianthus annuus* data were computed from H_2O loss per gram dry weight (g.d.w.) gain by assuming C in dry weight was $0.45\ g\,g^{-1}$ (see Raven *et al.*, 1992a,b).

unit C for NO_3^--grown than for NH_4^+-grown *Ricinus communis* would reduce the predicted increment of H_2O lost per unit C assimilated caused by NO_3^- rather than NH_4^+ as N source (Table 13.1) but would not, of course, reverse the direction of the effect. For NO_3^-, we note that NO_3^- assimilation in all three species is mainly in the shoot (Allen and Raven, 1987; Allen *et al.*, 1988; Kaiser and Lewis, 1991), so that the predicted values should be closer to 700 than to 766–776 (Table 13.1); this does not qualitatively alter the contradiction with reality (Table 13.2).

13.4.3 *Contributory factors relating theoretical and practical costs*

The explanation of the greater than predicted H_2O cost of C assimilation (and hence of N assimilation) in NH_4^+-grown plants could lie in either a greater H_2O cost of initial photosynthetic CO_2 assimilation, or of a greater respiratory loss of organic C, or both, than is the case for NO_3^--grown plants. This latter explanation is rendered superfluous, at least in the case of *Ricinus communis* (the only species examined in this way) by natural abundance $^{13}C/^{12}C$ ratios (Raven and Farquhar, 1990; Raven *et al.*, 1984). The observation from $^{13}C/^{12}C$ natural abundance ratios was that discrimination (Δ) was lower in NH_4^+-fed than in NO_3^--fed plants (Raven *et al.*, 1984). A lower Δ is consistent both with a higher ratio of gas-phase transport conductance to CO_2 to biochemical conductance, and with a lower fraction of net CO_2 fixation routed *via* non-Rubisco, anaplerotic, carboxylases such as phosphoenol-pyruvate carboxylase (PEPc), especially if anaplerotic fixation occurs mainly in the shoot (Farquhar *et al.*, 1989; Raven and Farquhar, 1990; see also Chapters 12 and 17); these authors also show that the respiratory loss of C causes a minimal change in the Δ of the residual organic C in the plant relative to the Δ of the original products of CO_2 fixation. Raven and Farquhar (1990) used data on the composition of NH_4^+-grown and NO_3^--grown *Ricinus communis* (Allen and Raven, 1987) to show that, while the larger fraction of total CO_2 fixation attributable to PEPc in NO_3^--grown plants changed their Δ in the observed direction, i.e. a decrease relative to NH_4^+-grown plants, this 'biochemical' effect is an order of magnitude too small to account for the observed Δ differences as a function of N-source. The most plausible explanation of the Δ differences is, then, that the NH_4^+-grown plants have a smaller fractional limitation of photosynthesis by CO_2 transport than do the NO_3^--grown plants. This is, in turn, consistent with the observed higher water cost of growth in NH_4^+-grown *Ricinus* being the result of a higher stomatal conductance in NH_4^+-grown plants, resulting in a higher H_2O loss per unit C fixed for the NH_4^+-grown plants under otherwise similar growth conditions.

An obvious test of this explanation of the difference in H_2O cost of organic C accumulation as a function of N source is in terms of the instantaneous H_2O loss rate during net CO_2 fixation in illuminated leaves. This could be measured as the H_2O vapour loss from and CO_2 uptake by illuminated leaves of plants growing with NH_4^+ and with NO_3^- as N source. These measurements are available for *Ceratonia siliqua* (Martins-Loução, 1985) and for *Triticum aestivum* (Leidi *et al.*, 1991a,b). For *Ceratonia* the H_2O loss per unit CO_2 fixed is greater for NH_4^+-grown than for NO_3^--grown plants, while for *Triticum* the H_2O loss per unit CO_2 fixed is similar in plants growing on NH_4^+ and on NO_3^-. Accordingly, the *Ceratonia* data support the interpretation of the *Ricinus* Δ values and H_2O loss per unit C accumulation in the plant, while the *Triticum* values do not. A further consideration in comparing the values in Table 13.2 with the short-term measurements on *Ceratonia* and *Triticum* is the growth rate of the plants. All of the plants in Table 13.2 have a lower specific growth rate on NH_4^+ than on NO_3^-, while the growth rates of *Triticum* are similar on the two N sources, and *Ceratonia* grows faster on NH_4^+ than on NO_3^- (Raven *et al.*, 1992a,b). Thus, the occurrence of higher, similar or lower specific growth rates on NH_4^+ relative to NO_3^- is not a determinant of the short- or long-term H_2O cost of C assimilation. Furthermore, the work of Morgan (1986) shows that the short-term (illuminated leaf) ratio of H_2O lost to CO_2 gained is higher for slower-growing NO_3^--limited *Triticum aestivum* than for faster-growing, NO_3^--sufficient plants; no data are available in this work for other N sources. Clearly, further work is needed to disentangle the interrelations of N source and specific growth rate in determining water loss per unit C fixed in short-term (C fixation = net leaf photosynthesis in the light) and long-term (C fixation = net C accumulation by the plant over many light–dark cycles); such work is under way.

In view of the complications of interpretation it may be premature to seek mechanisms for the apparently frequent higher ratio of stomatal conductance to biochemical conductance in NH_4^+-fed than in NO_3^--fed plants. However, it may be worth noting that what data are available suggest that stomatal conductance is not increased by a higher dissolved leaf $(NH_4^+ + NO_3^-)$ or gaseous NH_3 in and around the leaf within the sort of range likely for NH_4^+-fed relative to NO_3^--fed plants (Farquhar *et al.*, 1980; Whitehead and Lockyer, 1987).

13.5 Parasitic flowering plants

The recent critical and comprehensive review of the physiology of parasitic flowering plants by Stewart and Press (1990) discusses the interaction between nitrogen supply and water supply to flowering plant parasites from their flowering plant hosts. They point out that most photosynthetically competent parasites have higher transpiration rates than their hosts, but that this water-vapour loss is not uncontrolled: their stomata do respond to leaf water potential. The high transpiration rate by the parasite has been related to assuring a nitrogen supply from the host xylem. The high water cost of growth of the parasite is particularly noticeable when the concentration of combined N in the host xylem is low; at higher combined N concentrations in the xylem the water costs of growth by host and by parasite are more nearly equal. Ehleringer *et al.* (1986) suggest that parasite stomatal conductance is regulated by combined N supply from the host xylem. However, Givnish (1986) points out that the change in the difference in the water cost of growth of host relative to parasite with improved N supply to the host is mainly an increase in the water cost of growth of the host, with host stomatal conductance increased by high N availability.

Raven (1983) has previously pointed out that a high water cost of growth by photosynthetically competent angiosperm parasites is rather odd when it is considered that the combined N in the host xylem is not all destined for shoot assimilation even in the unparasitized plant, since a substantial fraction is recycled down the phloem. This is necessarily the case when NO_3^- is the predominant N species in the xylem, where the roots depend on the shoot for organic N; however, there is also some organic N recycling back up the xylem. Some recycling of organic N down the phloem and back up the xylem occurs when inorganic N assimilation occurs in roots and the shoot is supplied with organic N movement up the xylem. Accordingly, interception of host xylem sap by a parasite removes more N per unit water from the host than does N assimilation by the host shoot per unit water transpired. In *Triticum aestivum*, N deficiency may increase the fraction of N moving up the xylem which is recycled down the phloem (Lambers *et al.*, 1982; Larsson *et al.*, 1991; Simpson *et al.*, 1982), which would exacerbate N loss to a parasite from N-deficient plants. Accordingly, since any N recycling down the *parasite* phloem does not, apparently, move back to the host plant, the parasite gets a larger net amount of combined N per unit water in the xylem than does the host shoot. An analogous animal parasite example is that of tape worms located between the site of digestive enzyme secretion into the gut and the site of organic N absorption from the gut; this means that the parasite has access to more N per day than the N available for *net* N absorption by the gut (this argument is based on pp. 128–130 and 159–162 of Cox, 1982).

13.6 Investment of N in nutrient and water acquisition systems

Much attention has been given to cost–benefit analysis of roots, mycorrhizas and other nutrient and water acquisition systems, almost entirely in terms of carbon. Here we outline some further considerations, largely based on nitrogen.

When an organ, such as a root nodule, is adapted to obtain a single nutrient, nitrogen constraints are consequently simple. We calculated that for a spherical nodule, such as those of *Phaseolus vulgaris*, the amount of N invested in *one* nodule up to the stage when N_2 fixation begins, could be used to make 6.75 mm of root of diameter 0.55 mm (Sprent and Raven, 1985). Clearly when *all* nutrients are in short supply roots are a better investment

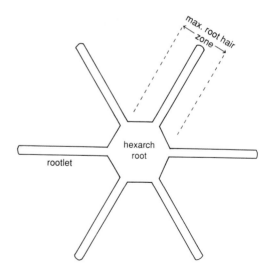

Figure 13.1. *Diagram of a transverse section of an idealized proteoid root, based on* Hakea *species.* (*Lamont, 1972.*)

than nodules. We now consider two types of structure which are more versatile in nutrient acquisition, namely proteoid (cluster) roots and mycorrhizas.

13.6.1 *Proteoid roots*

Although first described in the Proteaceae, proteoid roots occur widely in families represented in nutrient-poor environments (Lamont, 1982). They vary widely in morphology, and the example we take is based on species of *Hakea* as described by Lamont (1972). Rootlets arise opposite protoxylem poles of the subtending root. They are of limited length and densely covered with hairs (Figure 13.1). Longitudinally the clusters range from a few to many millimetres. In the absence of data, we assume that nitrogen concentration in proteoid roots is similar to that of small absorbing roots and therefore base our comparisons on biomass. Proteoid roots of 8 mm in length and with four or six files of rootlets (corresponding to tetrarch or hexarch roots) are calculated to have a surface area of 20–30 mm^2, compared with the root alone at 50 mm^2. This ignores any absorption surface of the subtending root which may remain functional in a proteoid root. Even if this remains functional, the increase in surface area alone afforded by cluster roots would be unimpressive, although magnified by the presence of root hairs. However, if we assume that rootlets are covered with the maximum length hairs observed (0.5 mm) the effective *volume* of soil tapped by a proteoid root is approximately 10 times that of the same length of unmodified root (210 mm^3 vs. 20 mm^3 in the example taken). This greater volume is achieved by an increased biomass investment of around 10%. Thus, assuming that nutrients (including water) can be efficiently obtained throughout the volume enclosed by a proteoid root, these roots represent a very cost effective way of capitalizing on a patchy soil environment, even though proteoid roots are ephemeral (as of course are fine roots).

13.6.2 *Mycorrhizas*

Mycorrhizas enable plants to access very large volumes of soil, albeit at an energy cost to the plant (see, for example, Brundrett, 1991). The C : N ratio of fungal hyphae may generally be slightly lower (~ 8) than that of absorbing roots (~ 12), although critical data are lacking. Thus a unit mass of hypha requires more N than a unit mass of root.

Roots, during their growth, have N supplied by the plant. Is this also true of mycorrhizal hyphae? Almost certainly in those hyphae internal to the root, but in soil we envisage two extreme cases. In the first, hyphal growth is at the expense of plant nitrogen. In the second, most growth is at the expense of soil nitrogen. Fungal hyphae (depending on species and soil type) can utilize almost all known forms of soil N from nitrate through to amino acids and sometimes proteins (Abuzinadah and Read, 1986). In this case, if the plant is to benefit in terms of N as well as other nutrients, soil N must exceed that required for fungal growth. There is the added potential benefit, in some cases, of plant access to N sources (protein) which it could not otherwise use (Abuzinadah and Read, 1986). In terms of access to soil water, mycorrhizas are a good investment. The calculations of Read and Boyd (1986) show that 1 m of external hyphae has a surface area equivalent to 1 cm absorbing root, but since each cm of root may be attached up to 14 m of hyphae, in the case of vesicular-arbuscular (VA) mycorrhizas (Abbot and Robson, 1985) the total available surface area is greatly increased. However, as in the case of proteoid roots, access to a greater volume of soil may be a more important attribute (see Isaac, 1992 for further discussion). Similar arguments apply both to VA and ectomycorrhizas, assuming that, in the latter, N in the hyphal sheath is soil derived. Recent work by Wallander and Nylund (1992) is consistent with this interpretation. Using their data for *Laccaria bicolor* on *Pinus sylvestris*, approximately 10% of the fungal + plant biomass was fungal, and 30% of the fungal biomass was extramatrical. Fungal biomass was greater at 'low' nitrogen levels and smaller at high levels, but the 'low' levels were calculated to allow a 3–4% day^{-1} relative growth rate.

Before leaving this aspect of our discussion, we note that in lower plants rhizoids are an efficient way of maximizing uptake for minimal C + N investment, having some of the geometric attributes of fungal hyphae combined with direct access to both photosynthate and soil nutrients. In terms of water supply, bryophytes deploy rhizoids in many ways, sometimes coupled with either endo- or ectohydry in above-ground organs (Schofield, 1985).

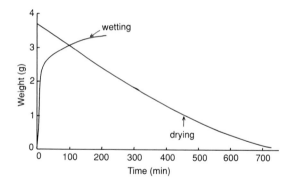

Figure 13.2. *Drying and wetting curves for* Nostoc commune. (*After Scherer* et al., *1984.*)

13.7 Deserts: two examples

When there is a widely scattered rainfall of small individual events, higher plants cannot be supported in the absence of an available ground water source. In such areas cyptogamic crusts are common. Some are effectively monospecific, at least for autotrophs. As an example we take *Nostoc commune* ('earthy ear'!) a desiccation-tolerant cyanobacterial species common in China. In such organisms rehydration is generally more rapid than desiccation, clearly an advantage when water supply is intermittent and unpredictable (Figure 13.2, after Scherer *et al.*, 1984). If we assume a crust of *Nostoc* 1 mm thick when hydrated, then it could (when desiccated) absorb most of a 1 mm rainfall event and carry out significant CO_2 and N_2 fixation before drying again. The major cost of desiccation tolerance may be in the production of protective extracellular polysaccharide, although N may also be invested in water-stress proteins (Scherer and Zhang, 1991). Lichens are also found in desert crusts. Costs here may be in terms of fixed C and N for the fungal partner, rather than extracellular polysaccharide.

When rainfall events are still erratic but individually larger than those considered above, desert succulents come into their own. Their vital attribute is the ability to respond rapidly to water by producing an extensive, ephemeral fine-root system. The special attributes of these plants have been well discussed by Rundel and Nobel (1991) (see also Chapters 10 and 17).

13.8 Epilogue

The limited number of examples discussed here indicate the multitude of interactions between water relations and nutrient acquisition and allocation, and also the great amount of work which remains to be carried out in this area. Such interactions cannot but assume greater significance in the context of global environmental change.

Acknowledgements

Work on N assimilation and metabolism in the authors' laboratory is funded by the AFRC (JAR, JIS).

References

Abbott, L.K. and Robson, A.D. (1985) Formation of external hyphae in soil by four species of vesicular-arbuscular mycorrhizal fungi. *New Phytol.*, **99**, 245–256.

Abuzinadah, R.A. and Read, D.J. (1986) The role of proteins in the nitrogen nutrition of ecto-mycorrhizal plants III. Protein utilization by *Betula*, *Picea*, and *Pinus* in mycorrhizal association with *Hebeloma australinoforme*. *New Phytol.*, **103**, 507–574.

Allen, S. and Raven, J.A. (1987) Intracellular pH regulation in *Ricinus communis* grown with ammonium or nitrate as N source: the role of long distance transport. *J. Exp. Bot.*, **38**, 580–596.

Allen, S., Raven, J.A. and Sprent, J.I. (1988) The role of long-distance transport in intracellular pH regulation in *Phaseolus vulgaris* grown with ammonium or nitrate as nitrogen source, or nodulated. *J. Exp. Bot.*, **39**, 513–528.

Brundrett, M.C. (1991) Mycorrhizas in natural ecosystems. *Adv. Ecol. Res.*, **21**, 171–313.

Cowan, I.R. and Farquhar, G.D. (1977) Stomatal function in relation to leaf metabolism and environment. *Symp. Soc. Exp. Biol.*, **31**, 471–505.

Cox, F.E.G. (ed.) (1982) *Modern Parasitology*. Blackwell, Oxford, pp. xii + 346.

Ehleringer, J.R., Ullmann, I., Lange, O.L., Farquhar, G.D., Cowan, I.R., Schulze, E.D. and Ziegler, H. (1986) Mistletoes: a hypothesis concerning morphological and chemical avoidance of herbivory. *Oecologia*, **70**, 234 – 237.

Evans, J.R. and Seeman, J.R. (1989) The allocation of protein nitrogen in the photosynthetic apparatus: costs, consequences and control. In: *Photosynthesis* (ed. W.R. Briggs). A.R. Liss, New York, pp. 183–205.

Farquhar, G.D., Firth, P.M., Wetselaar, R. and Weir, B. (1980) On the gaseous exchange between leaves and the environment: determination of the ammonia compensation point. *Plant Physiol.*, **66**, 710–714.

Farquhar, G.D., Ehleringer, J. and Hubick, K. (1989) Carbon isotope discrimination and photosynthesis. *Ann. Rev. Plant Physiol. Plant Mol. Biol.*, **40**, 507–537.

Givnish, T.J. (1986) Optimal stomatal conductance, allocation of energy between leaves and roots, and the marginal cost of transpiration. In: *On the Economy of Plant Form and Function* (ed. T.J. Givnish). Cambridge University Press, pp. 171–213.

Hudson, G.S., Evans, J.R., von Caemmerer, S., Arvidsson, Y.B.C. and Andrews, T.J. (1992) Reduction of ribulose-1,5-bisphosphate carboxylase/oxygenase content by antisense RNA reduces photosynthesis in transgenic tobacco plants. *Plant Physiol.*, **98**, 294–302.

Isaac, S. (1992) *Fungal Plant Interactions*. Chapman and Hall, London.

Jarvis, P.G. and McNaughton, K.H. (1986) Stomatal control of transpiration: scaling up from leaf to region. *Adv. Ecol. Res.*, **15**, 1–49.

Kaiser, J.S. and Lewis, O.A.M. (1991) The influence of nutrient nitrogen on the growth and productivity of sunflower (*Helianthus annuus* var. Dwarf Sungold). *S. Afr. J. Bot.*, **57**, 6–9.

Karentz, S., Cleaver, J.E. and Mitchell, D.L. (1991) Cell survival characteristics and molecular responses of Antarctic phytoplankton to ultraviolet-B radiation. *J. Phycol.*, **27**, 326–341.

Kirst, G.O. (1990) Salinity tolerance of eukaryotic marine algae. *Ann. Rev. Plant Physiol. Plant Mol. Biol.*, **41**, 21–53.

Lambers, H., Simpson, R.J., Beilharz, V.C. and Dalling, M.J. (1982) Growth and translocation of C and N in wheat (*Triticum aestivum*) grown with a split root system. *Physiol. Plant.*, **56**, 421–429.

Lamont, B. (1972) The morphology and anatomy of proteoid roots in the genus *Hakea*. *Aust. J. Bot.*, **20**, 155–174.

Lamont, B. (1982) Mechanisms for enhancing nutrient uptake in plants with particular reference to Mediterranean South Africa and Western Australia. *Bot. Rev.*, **48**, 597–689.

Larkum, A.W.D. and Wyn-Jones, R.G. (1979) Carbon dioxide fixation by chloroplasts isolated in glycine betaine. A putative cytoplasmic osmoticum. *Planta*, **145**, 393–394.

Larsson, C.-M., Larsson, M., Purves, J.V. and Clarkson, D.T. (1991) Translocation and cycling through roots of recently absorbed nitrogen and sulphur in wheat (*Triticum aestivum*) during vegetative and generative growth. *Physiol. Plant.*, **82**, 345–352.

Leidi, E.O., Silberbush, M. and Lips, H. (1991a) Wheat growth as affected by nitrogen, type pH and salinity. I. Biomass production and mineral composition. *J. Plant Nutr.*, **14**, 235–246.

Leidi, E.O., Silberbush, M. and Lips, M. (1991b) Wheat growth as affected by nitrogen type, pH and salinity. II. Photosynthesis and transpiration. *J. Plant Nutr.*, **14**, 247–256.

Lorimer, G.H. (1981) The carboxylation and oxygenation of ribulose 1,5 bisphosphate: the primary events in photosynthesis and photorespiration. *Ann. Rev. Plant Physiol.*, **32**, 349–383.

Martins-Loução, M.A. (1985) Estudes fisiologicos e microbiologicos da associação da alfarrobeira (*Ceratonia siliqua* L.) com bactérias do género Rhizobiaceae. PhD Thesis, University of Lisbon.

Morgan, J.A. (1986) The effect of N nutrition on the water relations and gas exchange characteristics of wheat (*Triticum aestivum* L.). *Plant Physiol.*, **80**, 52–58.

Quick, W.P., Schurr, V., Scheibe, R., Schulze, E.D., Rodermel, S.R., Bogorad, L. and Stitt, M. (1991) Decreased ribulose 1,5 bisphosphate carboxylase-oxygenase in transgenic tobacco with 'antisense' rbcS. I. Impact on photosynthesis in ambient growth conditions. *Planta*, **183**, 542–554.

Raven, J.A. (1983) Phytophages of xylem and phloem: A comparison of animal and plant sapfeeders. *Adv. Ecol. Res.*, **13**, 135–234.

Raven, J.A. (1985) Regulation of pH and generation of osmolarity in vascular land plants: costs and benefits in relation to efficiency of use of water, energy and nitrogen. *New Phytol.*, **101**, 25–77.

Raven, J.A. (1991) Responses of aquatic photosynthetic organisms to increased solar UV-B. *J. Photochem. Photobiol. B. Biology*, **9**, 239–244.

Raven, J.A. and Farquhar, G.D. (1990) The influence of N metabolism and organic acid synthesis on the natural abundance of isotopes of carbon in plants. *New Phytol.*, **116**, 505–529.

Raven, J.A., Allen, S. and Griffiths, H. (1984) N source, transpiration rate and stomatal aperture in *Ricinus*. In: *Membrane Transport in Plants* (eds W.J. Cram, K. Janacek, R. Rybora and K. Sigler). Academia, Praha, pp. 161–162.

Raven, J.A., Wollenweber, B. and Handley, L.L. (1992a) A comparison of ammonium and nitrate as nitrogen sources for photolithotrophs. *New Phytol.*, **121**, 5–18.

Raven, J.A., Wollenweber, B. and Handley, L.L. (1992b) Ammonia and ammonium fluxes between photolithotrophs and the environment in relation to the global nitrogen cycle. *New Phytol.*, **121**, 19–32.

Read, D.J. and Boyd, R. (1986) Water relations of mycorrhizal fungi and their host plants. In: *Water, Fungi and Plants* (eds P.G. Ayres and L. Boddy). Cambridge University Press, Cambridge, pp. 181–204

Rundel, P.W. and Nobel, P.S. (1991) Stucture and function in desert root systems. In: *Plant Root Growth: an Ecological Perspective* (ed. D. Atkinson). Special publication number 10 of the British Ecological Society. Blackwell Scientific Publications, Oxford, pp. 349–378.

Scherer, S. and Zhang, Z.-P. (1991) Desiccation independence of terrestrial *Nostoc commune* ecotypes (Cyanobacteria). *Microb. Ecol.*, **22**, 271–283.

Scherer, S., Ernst, A., Chen, T.W. and Böger, P. (1984) Rewetting of drought-resistant blue-green algae: time course of water uptake and reappearances of respiration, photosynthesis and nitrogen fixation. *Oecologia*, **62**, 418–423.

Schofield, W.B. (1985) *Introduction to Bryology*. MacMillan, New York.

Simpson, R.J., Lambers, H. and Dalling, M.J. (1982) Translocation of nitrogen in a vegetative wheat plant (*Triticum aestivum*). *Physiol. Plant.* **56**, 11–17.

Smith, S.V. (1984) Phosphorus versus nitrogen limitation in the marine environment. *Limnol. Oceanogr.*, **29**, 1149–1160.

Sprent, J.I. and Raven, J.A. (1985) Evolution of nitrogen-fixing symbioses. *Proc. R. Soc. Edin. B*, **85**, 215–237.

Stewart, G.R. and Press, M.C. (1990) The physiology and biochemistry of parasitic angios-perms. *Ann. Rev. Plant Physiol. Plant Mol. Biol.*, **41**, 127–151.

Stewart, G.R., Larher, F., Ahmod, I. and Lee, J.A. (1979) Nitrogen metabolism and salt-toler-ance in higher plant halophytes. In: *Ecological Processes in Coastal Environments* (eds R.L. Jeff-eries and A.J. Davy). Blackwell, Oxford, pp. 211–228.

Stitt, M., Quick, W.P., Schurr, V., Schulze, E.-D., Rodermel, S.R. and Bogorad, L. (1991) Decreased ribulose 1,5 bisphosphate carboxylase-oxygenase in transgenic tobacco trans-formed with 'antisense' rbcS. II. Flux-control coefficients for photosynthesis in varying light, CO_2 and air humidity. *Planta*, **183**, 555–566.

Wallander, H. and Nylund, J.-E. (1992) Effects of excess nitrogen and phosphorus starvation on the extramatrical mycelium of ectomycorrhizas of *Pinus sylvestris* L. *New Phytol.*, **120**, 495–503.

Whitehead, D.C. and Lockyer, D.R. (1987) The influence of the concentration of gaseous ammonia on its uptake by the leaves of Italian ryegrass, with and without an adequate supply of nitrogen to the roots. *J. Exp. Bot.*, **38**, 818–827.

Note added in proof

In addition to any effects of N source (Section 13.4) on the water relations of higher land plants, availability of a *given* N-source at concentrations insufficient to support the maximum relative growth rate reduces root hydraulic conduc-tivity. This has been established for NO_3^- as N source, and parallels effects of deficiency of P and S (Chapin, 1988; Karkomer *et al.*, 1991).

Chapin, F.S. III, Walter, C.H.S. and Clarkson, D.T. (1988) Growth response of barley and tomato to nitrogen stress and its control by abscisic acid, water relations and photosynthesis. *Planta*, **173**, 352–366.

Karkomer, J.L., Clarkson, D.T., Saker, L.R., Rooney, J.M. and Purves, J.V. (1991) Sulphate deprivation depresses the transport of nitrogen to the xylem and the hydraulic conductivity of barley (*Hordeum vulgare* L.) roots. *Planta*, **185**, 269–278.

Light-use efficiency and photoinhibition of photosynthesis in plants under environmental stress

Neil R. Baker

14.1 Introduction

When plants are exposed to environmental stress conditions a light-dependent inhibition of photosynthesis can occur, resulting in decreased photosynthetic productivity. Under environmental conditions optimal for growth, the products of the photosynthetic electron transport system (ATP and reductants) will be utilized primarily for carbon assimilation; however, a smaller proportion of these products will be consumed in photorespiration, nitrogen and sulphur metabolism. Under a given light environment, if plants are exposed to a stress that induces a reduction in carbon assimilation, a build-up of the products of photosynthetic electron transport will occur, unless they can be consumed at increased rates by other metabolic processes, and will result in an inhibition of electron transport. Such restrictions on electron transport inevitably lead to rates of excitation of the reaction centres of photosystems 1 and 2 (PS1 and PS2) in excess of those required to facilitate the electron transport rate needed to meet metabolic sink demand for its products. Under such conditions the leaf is confronted with the problem of dealing with an excess of excitation energy in the pigment antennae of the photosystems, which is potentially damaging and can result in the destruction of reaction centre proteins if the excitation energy is not successfully dissipated as heat (see Section 14.4).

It is now well established that thylakoids possess a mechanism for increasing the rate of dissipation of excitation energy by non-radiative, dissipative processes (heat) when the rate of excitation of the reaction centres exceeds the rate of electron transport (Krause and Weis, 1991). This mechanism is associated with PS2 and operates in non-stressed leaves when exposed to light levels above those producing the maximum quantum efficiency of CO_2 assimilation, i.e. at light levels at which an increase in photosynthetically active photon flux density (PPFD) results in a non-linear increase in CO_2 assimilation (non-linear region of the light-dosage response curve of photosynthesis). This decrease in the efficiency of light utilization with increasing light intensity is commonly referred to as a 'down-regulation of photosynthesis', with the implication that the effect is regulatory and not caused by damage to the photosynthetic apparatus. It appears that this regulatory mechanism operates to redirect an increasing proportion of absorbed photons away from the reaction centres, dissipating them directly as heat and, in so doing, protecting the photosynthetic apparatus from photodamage. The mechanism of this light-induced increase in the dissipation of excitation energy as heat has not yet been elucidated, although it appears to be associated with the antennae of PS2. Quenching of excitation energy by zeaxanthin (Demmig-Adams, 1990) and changes in aggregation state of the light-harvesting chlorophyll a/b complex associated with PS2 (Horton et al., 1991) have been suggested to provide the basis of the mechanism; however, such suggestions are as yet based only on correlative evidence.

Recently, rapid, non-destructive, spectroscopic methods for the accurate measurement of the quantum efficiency of light utilization by PS2 in leaves have been developed and have proved fundamental to the elucidation of the role of down-regulation in the physiology of leaf photosynthesis.

14.2 Chlorophyll fluorescence as a probe of the quantum efficiency of PS2 photochemistry in leaves

Chlorophyll fluorescence has been used to probe the fate of excitation energy within the photosynthetic apparatus for many years. Under physiological conditions fluorescence emitted from a leaf emanates primarily from pigment matrices associated with PS2. The usefulness of fluorescence measurements is that this light emission from within the PS2 pigment matrices competes for excitation energy with photochemistry (i.e. photosynthesis) and with non-radiative decay processes. Because of this competition, it is possible to monitor changes in photochemistry and thermal deactivation by measuring fluorescent light emission. Modulated fluorescence measurements (Ögren and Baker, 1985; Schreiber, 1986; Schreiber et al., 1986) permit the separation of fluorescence quenching into photochemical and non-photochemical components. The principle of this method and the distinction between photochemical and non-photochemical fluorescence quenching can be illustrated simply by considering an analysis of quenching during the induction of photosynthesis in a leaf exposed to actinic light (Figure 14.1). When a leaf is held in the dark for an extended period of time (tens of minutes) the quinone acceptors of PS2 (Q_A and Q_B) become maximally oxidized. On exciting the leaf with modulated light of sufficiently low intensity to ensure that the primary PS2

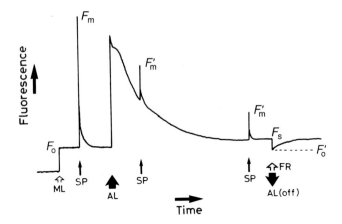

Figure 14.1. *Changes in fluorescence yield of a dark-adapted leaf when exposed to actinic light (AL) and saturating pulses of light (SP). Fluorescence yield was measured with a modulated fluorimeter. F_o is the minimal fluorescence yield produced on exposure of the dark-adapted leaf to the weak modulated measuring light (ML). F_m is the maximal fluorescence yield produced from the dark-adapted leaf by a short, saturating light pulse (SP; PPFD of $10\,000\,\mu mol\,m^{-2}\,s^{-1}$ for 1 s). F_m' is the maximal fluorescence yield produced by the short, saturating light pulse when the leaf is exposed to actinic light (AL). The magnitude of F_m' is dependent upon the period of exposure of the leaf to the actinic light (AL). F_o' is the fluorescence yield obtained in a leaf exposed to actinic light when PS2 reaction centres have bee}n maximally oxidized; this is generally achieved by removing the actinic light and simultaneously exposing the leaf to far-red light (FR; PPFD of c. $100\,\mu mol\,m^{-2}\,s^{-1}$ at wavelengths above 700 nm). F_v is the variable fluorescence yield achieved when non-photochemical quenching processes are minimal in a dark-adapted leaf ($F_v = F_m - F_o$). F_v' is the variable fluorescence yield after exposure of the leaf to actinic light ($F_v' = F_m' - F_o'$). F_s is the fluorescence yield when the leaf is at steady-state photosynthesis. The PPFD of actinic light used in this example was $600\,\mu mol\,m^{-2}\,s^{-1}$. (For further information of the nomenclature of fluorescence quenching analyses see van Kooten and Snel, 1990.)*

quinone acceptor (Q_A) remains oxidized (generally below $1\,\mu mol\,m^{-2}\,s^{-1}$), a minimal level of fluorescence (F_o) is generated. If the leaf is now exposed to a 1 s pulse of very intense light that fully reduces the Q_A pool (typically more than $4000\,\mu mol\,m^{-2}\,s^{-1}$), fluorescence rises to a maximal level (F_m). The rise in fluoresence from F_o to F_m is attributable to the photo-reduction of Q_A in all PS2 complexes; that is, the fluorescence yield increases to a maximum because the competition for excitation by photochemistry is nearly non-existent. This rise in fluorescence associated with Q_A reduction is known as variable fluorescence (F_v; $F_v = F_m - F_o$).

If the dark-adapted leaf is exposed to continuous actinic light then fluorescence will rise rapidly to a peak level (often designated P), which will be lower than F_m if the actinic light is not sufficient to reduce maximally the Q_A pool. Thereafter, fluorescence slowly declines to a steady-state level (F_s). This quenching of fluorescence from P to the steady state is coincident with the induction of CO_2 assimilation in the leaf (Ireland *et al.*, 1984). If the leaf is exposed to the saturating 1 s light pulse during this quenching from P, then a rise in fluorescence is observed to a level F_m', which is considerably less than the F_m level generated from the dark-adapted leaf (Figure 14.1) even though Q_A is essentially fully reduced in both situations. Consequently, the difference between F_m and F_m' is attributable to a non-photochemical quenching of excitation energy in the leaf exposed to actinic light; non-photochemical processes of excitation energy dissipation are involved because this quenching is independent of the redox state of Q_A. This non-photochemical quenching is clearly light-induced in that it does not occur in dark-adapted leaves and the magnitude of the quenching is associated with the rate of electron transport during the induction of photosynthesis. Increases in non-photochemical quenching reduce the quantum yield of PS2 photochemistry (ϕ_{PS2}), and consequently diminish the quantum efficiency of non-cyclic electron transport, which in turn will reduce the quantum efficiency of CO_2 assimilation (ϕ_{CO_2}).

Recently it has been demonstrated that simple fluorescence parameters can be used as a straightforward, rapid measure of ϕ_{PS2}. ϕ_{PS2} depends upon (1) the efficiency with which an absorbed photon can reach a reaction centre to perform photochemistry; and (2) the proportion of reaction centres which are capable of transferring an electron to an acceptor at that point in time, i.e. photochemically 'open' reaction centres (Genty et al., 1989). Since the efficiency of excitation energy capture by 'open' PS2 reaction centres is defined by the ratio of variable to maximal fluorescence. F_v/F_m (Butler and Kitajima, 1975; Kitajima and Butler, 1975), and the photochemical quenching coefficient, qP, is a measure of the proportion of 'open' PS2 reaction centres, Genty et al. (1989) noted that the quantum efficiency of PS2 photochemistry (ϕ_{PS2}) in a leaf is defined by:

$$\phi_{PS2} = F_v'/F_m' \cdot qP \qquad 14.1$$

where qP is determined from $(F_m' - F_s/F_m' - F_o')$; see Figure 14.1 for a definition of the parameters. In practice the relative ϕ_{PS2} for leaves under any given environmental conditions can be made by monitoring the steady-state fluorescence level (F_s) under the conditions of interest, exposing the leaves to a saturating flash and measuring the maximal fluorescence level (F_m') that is attained. The parameter $(F_m' - F_s)/F_m'$ equates to $\{(F_v'/F_m') \cdot qP\}$ for the leaf under the given conditions and thus is a measure of ϕ_{PS2} (Genty et al., 1989).

14.3 Relationship between the quantum efficiencies of PS2 photochemistry and CO_2 assimilation in leaves

In maize leaves, which because they are C_4 have minimal photorespiratory activity, modification of ϕ_{CO_2} in response to changes in light intensity and atmospheric CO_2 concentration or during the induction of photosynthesis in a dark-adapted leaf were observed to be directly proportional to the

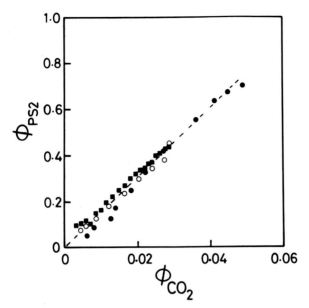

Figure 14.2. *The relationship between the quantum yield of CO_2 assimilation (ϕ_{CO_2}) and the relative quantum yield of PS2 photochemistry (ϕ_{PS2}) for a maize leaf as a function of light intensity over a PPFD range of 40–2700 $\mu mol\, m^{-2}\, s^{-1}$ ($\bullet$), in a range of atmospheric CO_2 concentrations (30–370 p.p.m., $\bigcirc$) at steady state and during the induction of photosynthesis on exposure of a dark-adapted leaf to a PPFD of 650 $\mu mol\, m^{-2}\, s^{-1}$ ($\blacksquare$). (Data from Genty et al., 1989.)*

changes observed in ϕ_{PS2} (Figure 14.2; Genty *et al.*, 1989). Linear relationships between ϕ_{CO_2} and ϕ_{PS2} were also observed in barley and ivy leaves exposed to a range of light intensities under conditions in which photorespiration had been minimized by reducing the atmospheric O_2 concentration to 2% (Figure 14.3; Genty *et al.*, 1989, 1990). Similar relationships have also since been observed between ϕ_{CO_2} (Krall and Edwards, 1990), the quantum yield of O_2 evolution (Keiler and Walker, 1990; Seaton and Walker, 1991; van Wijk and van Hasselt, 1990) and ϕ_{PS2} in a leaves of a range of species. It should be noted that some non-linearity was found at very low light intensities, probably because of differences in the chloroplast populations being monitored for fluorescence and CO_2 assimilation and differences in their excitation rates; light-induced modifications to the organization of the photosynthetic apparatus, such as a state transition, could also account for non-linearity. However, it is clear that under a wide range of physiological conditions, provided that the allocation of ATP and reductants produced by photosynthetic electron transport to processes other than CO_2 assimilation is minimal, a direct proportionality exists between ϕ_{CO_2} and ϕ_{PS2}, which can be defined by:

$$\phi_{CO_2} = \phi_{PS2} \cdot I_{PS2}/I_{abs} \cdot 1/k \qquad 14.2$$

where I_{PS2} and I_{abs} are the photon flux densities absorbed by PS2 antennae and the leaf, and k is the number of electron equivalents required to reduce 1 mol

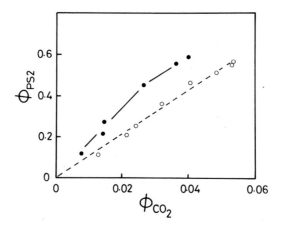

Figure 14.3. *The relationship between the quantum efficiencies of CO_2 assimilation (ϕ_{CO_2}) and PS2 photochemistry (ϕ_{PS2}) for a barley leaf exposed to 2% (○) and 20% (●) O_2 concentrations. Measurements were made at steady-state phytosynthesis over a PPFD range of 50–1600 $\mu mol\ m^{-2} s^{-1}$. (Data from Genty et al., 1990.)*

of CO_2 (Genty *et al.*, 1989). As would be expected for leaves of C_3 plants under conditions favourable for photorespiratory activity where the products of electron transport are apportioned between carboxylation and oxygenation reactions, the strict linear relationship between ϕ_{CO_2} and ϕ_{PS2} breaks down (Figure 14.3; Genty *et al.*, 1990).

Simultaneous measurements of ϕ_{PS2} and the quantum yield of PS1 photochemistry (ϕ_{PS1}) made over a wide range of light intensities on leaves under both photorespiratory and non-photorespiratory conditions have shown a good correlation between these parameters (Figure 14.4; Genty *et al.*, 1990; Harbinson *et al.*, 1989, 1990). This observation, together with the linear relationship found between ϕ_{PS2} and ϕ_{CO_2}, suggests that ϕ_{PS2} is a good measure of the quantum efficiency of non-cyclic electron transport. The conserved linearity between ϕ_{PS1} and ϕ_{PS2} under a wide range of physiological conditions implies that the rate of PS1-mediated cyclic electron transport, if it occurs *in vivo*, is always proportional to the rate of non-cyclic electron transport. This conclusion is consistent with recent photoacoustic spectroscopic measurements of cyclic photophosphorylation in leaves (Herbert *et al.*, 1990). Thus, as mentioned above, it would not appear that modulation of PS1-mediated cyclic electron transport occurs in leaves to meet any changing metabolic demand for ATP and NADPH under the experimental conditions investigated to date.

It is evident from many experiments that quenching of excitation energy in PS2 antennae by non-photochemical processes is a major factor in producing large, light-induced decreases in the efficiency of light utilization for photosynthesis in leaves. Light-induced increases in non-photochemical quenching of excitation energy in a leaf and the observed consequences for ϕ_{PS2} and

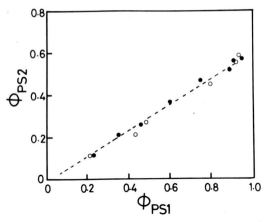

Figure 14.4. *The relationship between the quantum efficiencies of PS1 (ϕ_{PS1}) and PS2 (ϕ_{PS2}) of a barley leaf exposed to 2% (○) and 20% (●) O_2 concentrations over a PPFD range of 50–1600 $\mu mol\, m^{-2}\, s^{-1}$. (Data from Genty et al., 1990.)*

ϕ_{CO_2} are shown in Table 14.1. The parameters F_v'/F_m', qP, ϕ_{PS2} and ϕ_{CO_2} were determined for dark- and light-adapted (30 min at a PPFD of $1500\,\mu mol\,m^{-2}\,s^{-1}$) leaves, then the leaves were exposed to a low PPFD of $34\,\mu mol\,m^{-2}\,s^{-1}$ and allowed to reach steady state. Pretreatment of the leaves in high light induced a 26% decrease in F_v'/F_m', indicative of a large increase in non-photochemical quenching, with concomitant decreases of 12 and 14% in ϕ_{PS2} and ϕ_{CO_2}, respectively. The decrease in ϕ_{PS2} must be attributed to an increase in non-photochemical quenching, since high-light pretreatment results in an increase in qP. Thus in leaves treated with high light a larger proportion of PS2 reaction centres are oxidized than in the dark-adapted leaves when the leaves are exposed to the low actinic light of $34\,\mu mol\,m^{-2}\,s^{-1}$ and measurements made at steady-state photosynthesis. Such an increase in qP would increase ϕ_{PS2} (see Equation 14.1). Clearly the

Table 14.1. *Relationships between the quantum yield of CO_2 assimilation (ϕ_{CO_2}), the quantum yield of PS2 photochemistry (ϕ_{PS2}) and chlorophyll fluorescence quenching parameters in leaves pretreated in the dark and in high light*

Parameter	Dark	High light	% Change
ϕ_{CO_2}	0.051	0.044	−13.8
ϕ_{PS2}	0.492	0.432	−12.2
F_v'/F_m'	0.712	0.526	−26.1
qP	0.690	0.824	+19.4

Leaves of *Silene dioica* were kept in the dark or in a PPFD of $1500\,\mu mol\,m^{-2}\,s^{-1}$ prior to exposure to a PPFD of $34\,\mu mol\,m^{-2}\,s^{-1}$ to measure ϕ_{CO_2}, ϕ_{PS2}, F_v'/F_m' and the photochemical quenching coefficient, qP, at steady-state photosynthesis. The leaves were exposed to 400 p.p.m. CO_2 and 1% O_2 in nitrogen to minimize photorespiration. The decreases in ϕ_{CO_2}, ϕ_{PS2} and F_v'/F_m' after the high-light pretreatment were fully reversible within 1 h of the leaves being placed in the dark. (Data from Genty et al., 1989.)

greater decrease in F'_v/F'_m, rather than the increase in qP, is the dominating factor in determining the change in ϕ_{PS2}, and consequently in ϕ_{CO_2}.

Light-induced, non-photochemical quenching is generally reversible within 1 h of the leaves being placed in the dark, and thus is unlikely to be associated with photoinhibitory damage to PS2 reaction centre proteins (see below) which recover considerably more slowly. It appears that the processes giving rise to light-induced, non-photochemical quenching are a ubiquitous feature of the photosynthetic apparatus of thylakoids. However, it is not clear whether the quantitative relationship between the rates of dissipation of excitation energy by PS2 photochemistry and non-photochemical quenching processes is the same in all plants. It is possible that this relationship is highly conserved within the plant kingdom, thus suggesting that the protective process is an essential mechanism to reduce photodamage to PS2 and optimize the transduction of light energy by the thylakoid membranes, without which the observed photosynthetic efficiencies of photosynthetic organisms could not have been maintained.

From the arguments presented above it can be concluded that ϕ_{PS2}, as measured from chlorophyll fluorescence kinetics, is a good estimate of the quantum efficiency of non-cyclic electron flux through PS2 and PS1 in leaves, although no information is provided by this parameter as to the nature of the terminal electron acceptors involved. If the allocation of the products of electron transport to sinks other than CO_2 assimilation is negligible or constant, then ϕ_{PS2} provides a good relative measure of ϕ_{CO_2}. Although it cannot be stated unequivocally from such studies whether ϕ_{PS2} determines ϕ_{CO_2}, it is unlikely that large changes in ϕ_{CO_2} will not be reflected in parallel changes in ϕ_{PS2}.

14.4 Photodamage to photosystem 2 reaction centres

When leaves receive light in excess of that which can be dissipated by photochemical and non-photochemical processes associated with photosynthesis, damage to PS2 complexes will occur (Critchley 1988; Kyle 1987). Such damage is almost certainly taking place, albeit at a slow rate, in the majority of non-stressed leaves under normal growth conditions. However, no net accumulation of inactivated PS2 complexes will occur in such cases, provided that the rate of repair of the damaged PS2 population is equal to the rate of damage. Only if the capacity for repair is exceeded by the rate of photodamage will a loss in PS2 activity become evident. This situation arises when leaves are exposed to light levels above those normally experienced, or at considerably lower light levels if photosynthetic carbon metabolism is inhibited. In both cases the leaf does not have the capacity to dissipate the excess excitation energy by non-photochemical processes associated with the light-induced down-regulation discussed above. A net loss of photochemically competent PS2 complexes will result in a reduction in the maximal quantum efficiencies and possibly the light-saturated capacity for CO_2 assimilation in leaves (Baker et al., 1988).

Photoinhibitory damage to PS2 complexes appears initially to involve impairment of quinone reduction followed by degradation of the 32 kDa D1 protein of the PS2 reaction centre (Adir et al., 1990; Barber and Andersson, 1992; Styring et al., 1990). In order to repair PS2 complexes containing damaged D1 proteins, it has been speculated that the complexes must migrate from an appressed to a non-appressed thylakoid domain where the protein can be degraded by a membrane-bound protease (Adir et al., 1990; Guenther and Melis, 1990). Newly synthesized D1 polypeptides are then thought to be inserted into the damaged PS2 reaction centre complexes, which then migrate back to appressed regions of the thylakoid membrane and become again fully photochemically competent.

The loss of leaf photosynthetic activity due to photoinhibitory damage to PS2 complexes will not only be dependent upon the amount of damage to D1 proteins, but will also be determined by the ability to degrade the damaged D1 proteins and synthesize and integrate new D1 proteins into the damaged complexes. Consequently, stress-induced modifications to the mobility of PS2 complexes in the thylakoid membrane, to the protease activity required to degrade the damaged D1 protein and to the ability to synthesize and integrate D1 protein into PS2 complexes could be factors in determining the degree and persistence of loss of photosynthetic competence during and after the removal of a photoinhibitory stress. There is evidence that recovery from photoinhibition in leaves requires protein synthesis on chloroplast ribosomes (Greer et al., 1986), consistent with a need to synthesize D1 protein, which is encoded by the chloroplast genome.

It should be emphasized that photodamage to the PS2 reaction centre will only occur under environmental conditions which result in an inability of the leaf to down-regulate and dissipate excess excitation energy in the PS2 antennae as heat. Recovery of a photodamaged PS2 population in a leaf when an environmental stress condition is removed will be dependent on protein synthesis and processing, and consequently the leaf will require considerably longer (generally in the order of hours) to recover from such damage compared to the time required for the relaxation of down-regulation (in the order of minutes).

14.5 Photoinhibition of photosynthesis in the field

Light-induced, non-photochemical quenching has been demonstrated in the leaves of a number of species in the field (Adams et al., 1988; Groom et al., 1990; Ögren, 1988). F_v/F_m in leaves of a winter wheat crop (dark adapted for 15 min prior to measurement) decreased throughout the morning in parallel with the increasing light intensity received by the plants, and then recovered during the afternoon as the light intensity decreased (Figure 14.5). The possibility that the decreases in F_v/F_m at midday observed in winter wheat

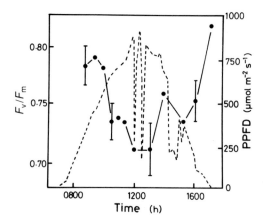

Figure 14.5. *Changes in* F_v/F_m *(●) in leaves in winter wheat in the field during the day. Measurements were made on a crop in north-east Essex, UK, on 23 February 1989. Leaves were dark adapted for 15 min prior to measurement. The changes in PPFD during the day are also shown (– – –). Standard errors of selected mean* F_v/F_m *values are shown. (Data of Groom et al., 1990.)*

leaves might be due to photodamage to PS2 reaction centres was discounted since no significant changes in the ability of thylakoids isolated from the leaves to bind atrazine to the Q_B sites on the D1 proteins of the PS2 reaction centres could be detected throughout the day (Groom, 1990). The close anti-parallel relationship between F_v/F_m and light intensity for leaves in the field during the day has also been observed in willow (Ögren, 1988). Such light-induced decreases in F_v/F_m are clearly associated with decreases in the ϕ_{PS2} at which leaves are operating in the field, and will contribute to a depression in photosynthesis at midday. Light-dependent reductions in photosynthesis of leaves in the field will be imposed via increases in non-photochemical quenching of excitation energy; however, these may be the direct consequence of environmentally induced restrictions to gas exchange through the stomata or, alternatively, due to limitations imposed on photosynthetic carbon metabolism, rather than being a direct effect of a stress on the photosynthetic apparatus of the thylakoids.

14.6 Photoinhibition of photosynthesis and water stress

Reductions in leaf water content result in reduced photosynthetic competence in many plants. Under mild drought stress, decreases in photosynthesis are generally considered to be the result of reduced availability of CO_2 due to stomatal closure (Kaiser, 1987; Schulze, 1986; see also Chapters 8, 9, 12 and 15). It is possible that leaf water deficits can also have effects on chloroplast biochemistry that would contribute to depressions in photosynthetic per-formance. Studies on intact leaves subjected to moderate drought stresses have demonstrated that the potential for thylakoid photochemical activity is not decreased (Ben *et al.*, 1987; Cornic *et al.*, 1989; Genty *et al.*, 1987). Consequently, the effects of low leaf water potentials on the photochemical activities of isolated thylakoids reported previously (Boyer, 1976) are unlikely to be representative of the situation occurring *in vivo* and are probably due either to artefacts caused by the procedures for isolation of thylakoids from stressed leaves or by photoinhibition when electron transport is constrained in the isolated membranes (Cornic and Briantais, 1991). Reductions in intercellular CO_2 concentration and chloroplast biochemical dysfunctions will reduce the rate of CO_2 assimilation in a leaf and would be expected to reduce the quantum efficiency of non-cyclic photosynthetic electron transport. Cornic and Briantais (1991) have demonstrated, using chlorophyll fluorescence measurements similar to those described above, that decreases in ϕ_{PS2} occur during desiccation of a bean leaf, which are accompanied by decreases in F_v'/F_m' and qP (Figure 14.6). From the measurements of ϕ_{PS2}, estimations were made of the rates of non-cyclic photosynthetic electron transport occurring in the leaf at steady-state photosynthesis by multiplying ϕ_{PS2} by the incident PPFD, and it was demonstrated that although decreases in non-cyclic electron transport occur in concert with decreases in the rate of CO_2 assimilation, the reductions in electron transport were proportionally less. It should be noted that in order to estimate rates of non-cyclic electron transport from ϕ_{PS2} it has to be assumed that (1) the absorptivity of the leaf and (2) the proportion of light absorbed by the leaf and distributed to PS2 antennae do not change through the course of the experiment; this may not necessarily be the case during dehydration of a leaf. Under stress conditions the photosynthetic reduction of O_2, via photorespiration and possibly the Mehler reaction, may increase and serve to act as a sink for excess excitation energy in the photosynthetic apparatus, which would normally be dissipated via CO_2 assimilation in the non-stressed leaf. Cornic and Briantais (1991) also estimated the rate of electron flow to O_2 from the measurements of ϕ_{PS2} and CO_2 assimilation, making a number of assumptions, and argued that the allocation of electrons to O_2 increased as the rate of CO_2 assimilated decreased with dehydration of the leaf. Reduction of the O_2 concentration in the atmosphere of a droughted leaf from 21% to 1% resulted in large decreases in ϕ_{PS2} and the estimated rate of non-cyclic electron transport, which were reversible on returning the O_2 concentration to 21%. This demonstrates the importance of photosynthetic O_2 reduction as a sink for excess excitation energy in water-stressed leaves. However, the increase in electron flow to O_2 observed on dehydration cannot dissipate the increased level of excitation energy in the PS2 antennae resulting from the stress-induced reduction in CO_2 assimilation, since F_v'/F_m' declines during this period (Figure 14.6). This decline in F_v'/F_m' is indicative of an increase in the probability of a photon absorbed by PS2 antennae being dissipated as heat, and demonstrates the occurrence of a stress-induced down-regulation of photosynthesis.

In conclusion, it is evident from the above discussions that leaves under optimal growth conditions possess a mechanism by which they can down-regulate photosynthesis to avoid overexcitation of PS2 reaction centres when they

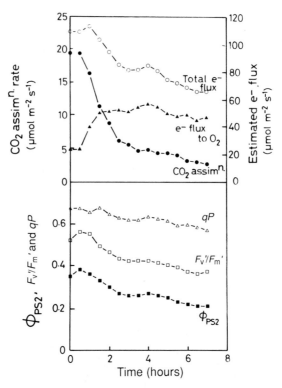

Figure 14.6. *Changes in the rate of CO_2 assimilation (●), the quantum yield of PS2 photochemistry (ϕ_{PS2}, ■), the coefficient for photochemical quenching of fluorescence (qP, △) and F_v'/F_m' (□) during dehydration of a cut bean leaf. Estimations of the rates of non-cyclic electron transport (○) and electron transport to O_2 (▲) were also made as described in the text. (Data of Cornic and Briantais, 1991.)*

are exposed to light levels above those at which maximal quantum efficiencies of photosynthesis can be realized. When leaves of C_3 plants experience decreases in water potential, which result in stomatal closure and a reduction in the rate of CO_2 assimilation (see Chapters 8, 9, 12 and 15), there is an increased flux of electrons to O_2 to dissipate a large proportion of the excitation energy that had previously been utilized to drive CO_2 assimilation. However, increases in the rate of reduction of O_2 will not be sufficient to dissipate the excess excitation energy in PS2 antennae and increased down-regulation of photosynthesis will occur to minimize photodamage to PS2 reaction centres. Under severe water deficits it is possible that electron transport to O_2 and down-regulation will be unable to dissipate the excitation energy in the PS2 antennae and, consequently, photodamage and net loss of the D1 protein of PS2 reaction centres can result.

References

Adams, W.W., Terashima, I., Brugnoli, E. and Demmig, B. (1988) Comparisons of photosynthesis and photoinhibition in the CAM vine *Hoya australis* and several C3 vines on the coast of eastern Australia. *Plant Cell Environ.*, **11**, 173–181.

Adir, N., Schochat, S., Inoue, Y. and Ohad, I. (1990) Mechanism of the light-dependent turnover of the D1 protein. In: *Current Research in Photosynthesis* (ed. M. Baltscheffsky), Vol. II. Kluwer Academic Press, Dordrecht, pp. 490–413.

Baker, N.R., Long, S.P. and Ort, D.R. (1988) Photosynthesis and temperature, with particular reference to effects on quantum yield. In: *Plants and Temperature* (eds S.P. Long and F.I. Woodward). The Company of Biologists Ltd, Cambridge, pp. 347–375.

Barber, J. and Andersson, B. (1992) Too much of a good thing: light can be bad for photosynthesis. *Trend Biochem. Sci.*, **17**, 61–66.

Ben, G.-Y., Osmond, C.B. and Sharkey, T.D. (1987) Comparison of photosynthetic responses of *Xanthium strumarium* and *Helianthus annuus* to chronic and acute water stress in sun and shade. *Plant Physiol.*, **84**, 476–482.

Boyer, J.S. (1976) Water deficit and photosynthesis. In: *Water Deficit and Plant Growth* (ed. T.T. Kozlowski). Academic Press, London, pp. 153–190.

Butler, W.L. and Kitajima, M. (1975) Fluorescence quenching photosystem II of chloroplasts. *Biochim. Biophys. Acta*, **376**, 116–125.

Cornic, G. and Briantais, J.-M. (1991) Partitioning of photosynthetic electron flow between CO_2 and O_2 reduction in a C_3 leaf (*Phaseolus vulgaris* L.) at different CO_2 concentrations and during drought stress. *Planta*, **183**, 178–184.

Cornic, G., Le Gouallec, J.-L., Briantais, J.-M. and Hodges, M. (1989) Effects of dehydration and high light on photosynthesis of two C_3 plants (*Phaseolus vulgaris* L. and *Elatostema repens* (Lour.)). *Planta*, **177**, 84–90.

Critchley, C. (1988) The molecular mechanicsm of photoinhibition – facts and fiction. *Aust. J. Plant Physiol.*, **15**, 27–41.

Demmig-Adams, B. (1990) Carotenoids and photoprotection: a role for the zanthophyll zeaxanthin. *Biochim. Biophys. Acta*, **1020**, 1–24.

Genty, B., Briantais, J.-M. and Viera da Silva, J.B. (1987) Effects of drought on primary processes of cotton leaves. *Plant Physiol.*, **83**, 360–364.

Genty, B., Briantais, J.-M. and Baker, N.R. (1989) The relationship between the quantum yield of photosynthetic electron transport and quenching of chlorophyll fluorescence. *Biochim. Biophys. Acta*, **990**, 87–92.

Genty, B., Harbinson, J. and Baker, N.R. (1990) Relative quantum efficiencies of the two photosystems of leaves in photorespiratory and non-photorespiratory conditions. *Plant Physiol. Biochem.*, **28**, 1–10.

Greer, D.H., Berry, J.A. and Björkman, O. (1986) Photoinhibition of photosynthesis in intact bean leaves: role of light and temperature, and requirement for chloroplast protein synthesis during recovery. *Planta*, **168**, 253–260.

Groom, Q.J. (1990) Photoinhibition of photosynthesis in the natural environment. Ph.D. Thesis, University of Essex, Colchester, UK.

Groom, Q.J., Long, S.P. and Baker, N.R. (1990) Photoinhibition of photosynthesis in a winter wheat crop. In: *Current Research in Photosynthesis* (ed. M. Baltscheffsky), Vol. II. Kluwer Academic Press, Dordrecht, pp. 463–466.

Guenther, J.E. and Melis, A. (1990) The physiological significance of photosystem II hetero-geneity in chloroplasts. *Photosynth. Res.*, **23**, 105–109.

Harbinson, J., Genty, B. and Baker, N.R. (1989) Relationship between the quantum efficiencies of photosystems I and II in pea leaves. *Plant Physiol.*, **90**, 1029–1034.

Harbinson, J., Genty, B. and Baker, N.R. (1990) The relationship between CO_2 assimilation and electron transport in leaves. *Photosynth. Res.*, **25**, 213–224.

Herbert, S.K., Fork, D.C. and Malkin, S. (1990) Photoacoustic measurements *in vivo* of energy storage by cyclic electron flow in algae and higher plants. *Plant Physiol.*, **94**, 926–934.

Horton, P., Ruban, A.V., Rees, D., Pascal, A.A., Noctor, G. and Young, A.J. (1991) Control of the light-harvesting function of chloroplast membranes by aggregation of the LHCII chloro-phyll–protein complex. *FEBS Lett.*, **292**, 1–4.

Ireland, C.R., Long, S.P. and Baker, N.R. (1984) The relationship between carbon dioxide fixa-tion and chorophyll *a* fluorescence during induction of photosynthesis in maize leaves at dif-ferent temperatures and carbon dioxide concentrations. *Planta*, **160**, 550–558.

Kaiser, W.M. (1987) Effect of water deficit on photosynthetic capacity. *Physiol. Plant.*, **71**, 142–149.

Keiler, D.R. and Walker, D.A. (1990) The use of chlorophyll fluorescence to predict CO_2 fixa-tion during photosynthetic oscillations. *Proc. R. Soc. Lond. B.*, **241**, 59–64.

Kitajima, M. and Butler, W.L. (1975) Quenching of chlorophyll fluorescence and primary photochemistry in chloroplasts by dibromothymoquinone. *Biochim. Biophys. Acta*, **376**, 105–115.

Krall and Edwards (1990) Quantum yields of photosystem II electron transport and carbon dioxide fixation in C_4 plants. *Aust. J. Plant. Physiol.*, **17**, 579–588.

Krause, G.H. and Weis, E. (1991) Chlorophyll fluorescence and photosynthesis: the basics. *Ann. Rev. Plant Physiol. Plant Mol. Biol.*, **42**, 313–349.

Kyle, D.J. (1987) The biochemical basis of photoinhibition of photosystem II. In: *Photo-inhibition* (eds D.J. Kyle, C.B. Osmond and C.J. Arntzen). Elsevier, Amsterdam, pp. 197–226.

Ögren, E. (1988) Photoinhibition of photosynthesis in willow leaves under field conditions. *Planta*, **175**, 229–236.

Ögren, E. and Baker, N.R. (1985) Evaluation of a technique for the measurement of chloro-phyll fluorescence from leaves exposed to continuous white light. *Plant Cell Environ.*, **8**, 539–547.

Schreiber, U. (1986) Detection of rapid induction kinetics with a new type of high-frequency modulated chlorophyll fluorometer. *Photosynth. Res.*, **9**, 261–271.

Schreiber, U., Schliwa, U. and Bilger, W. (1986) Continuous recording of photochemical and non-photochemical chlorophyll fluorescence quenching with a new type of modulation fluorometer. *Photosynth. Res.*, **10**, 51–62.

Schulze, E.-D. (1986) Carbon dioxide and water vapor exchange in response to drought in the atmosphere and the soil. *Ann. Rev. Plant Physiol.*, **37**, 247–274.

Seaton, G.G.R. and Walker, D.A. (1991) Chlorophyll fluorescence as a measure of photo-synthetic carbon assimilation. *Proc. R. Soc. Lond. B.*, **242**, 29–35.

Styring, S., Jägerschöld, C., Virgin, I., Ehrenberg, A. and Andersson, B. (1990) On the mechanisms for the photoinhibition of electron transfer and light-induced degradation of the D1 protein in PSII. In: *Current Research in Photosynthesis* (ed. M. Baltscheffsky), Vol. II. Kluwer Academic Press, Dordrecht, pp. 349–356.

van Kooten, O. and Snel, J.F.H. (1990) The use of chlorophyll fluorescence nomenclature in plant stress physiology. *Photosynth. Res.*, **25**, 147–150.

van Wijk, K.J. and van Hasselt, P.R. (1990) The quantum efficiency of photosystem II and its relationship to non-photochemical quenching of chlorophyll fluorescence; the effect of measuring and growth temperature. *Photosynth. Res.*, **25**, 233–240.

15

Plant water deficits in Mediterranean ecosystems*

J.S. Pereira and M.M. Chaves

15.1 Introduction

The Mediterranean climate is characterized by a mildly cold, rainy winter and a hot, dry summer. These contrasting weather conditions during the year have apparently contradictory consequences in terms of environmental stress and disturbance. In addition to soil water deficits, the vegetation is prone to catastrophic fires in summer. Erosion is important in bare soils due to the heavy winter rains. Human occupation with agriculture and deforestation have exacerabated these tendencies. As a consequence many soils have been degraded (as described extensively by, for example, di Castri and Mooney, 1973).

Soil degradation and climate both contribute to the semiarid nature of Mediterranean environments and have shaped the nature of the vegetation. If we disregard riparian vegetation, all plants have to face summer drought. The types of plants that predominate in Mediterranean areas are, to a large extent C_3 drought avoiders, ranging from winter annuals to deep-rooted perennials or summer-deciduous shrubs.

Given the seasonality of rainfall and temperature, we may separate the effects of water deficits before the rooting zone has been thoroughly depleted of water, from the severe dehydration that may occur during the summer. Although the effects of water deficits in the Mediterranean vegetation have been reviewed repeatedly (e.g. di Castri and Mooney, 1973; Tenhunen et al., 1987) most attention has been given to performance during summer drought. In this chapter we will discuss acclimation to water deficits before

---

*We dedicate this work to Prof. O.L. Lange on his sixty-fifth birthday.

the onset of severe water deficits in contrast to drought tolerance when water stress, high irradiances and high temperatures prevail in summer.

15.2 Acclimation to decreasing water availability

In areas with a Mediterranean type of climate the depletion of soil moisture occurs slowly, often intermittently, during the spring months up to the summer. For a given seasonal pattern of water availability, the rate of development of plant water stress varies with depth and extension of the root systems. Whereas shallow-rooted annuals die, after producing seed, by early summer, woody perennials may acclimate to the slowly developing soil water deficits.

A great deal of research has been devoted in past years to the effects of dehydration on leaf tissues. Much less has been done on the acclimation to water deficits before drastic tissue dehydration occurs. Although acclimation to water deficits in perennials is complex, we may hypothesize that to be of adaptive value this must be a response to small, stepwise changes in soil moisture before any important tissue dehydration occurs. This means that the plant should be able to 'sense' minor soil water deficits and possess a feedforward type of response, such as root to shoot signalling. Two processes are important: reduction in growth and increased stomatal closure. Both respond to chemical root signals and both result in a reduction in transpiration (Davies and Zhang, 1991). The situation is clearly illustrated by Figure 15.1. Field-grown grapevine plants had different amounts of water in the soil, either because of soil physical characteristics or due to irrigation. Nevertheless, the level of dehydration at midday in early summer was not significantly different between treatments.

15.2.1 *Effects on growth*

It is well known that primary productivity is linearly related to the amount of solar radiation intercepted by the foliage (Russell *et al.*, 1989). One of the earliest effects during the development of water deficits is a reduction in foliage expansion, and therefore in the amount of solar radiation intercepted or in the quotient of dry matter produced : solar radiation intercepted (ϵ).

The effects of limited water supply in growth of *Eucalyptus globulus* clonal cuttings is shown in Figure 15.2. Half of the plants received a quarter of the water needed to keep the soil near field capacity but they did not suffer any significant dehydration in comparison to the well-watered controls, except

---

Figure 15.1. *Soil water content, with total in profile indicated (a), and diurnal changes in leaf water potential (b) and in stomatal conductance (c) of grapevine plants grown with different water supplies in the field in Santarém, Portugal in the summer of 1991 (Pereira et al., unpublished).*

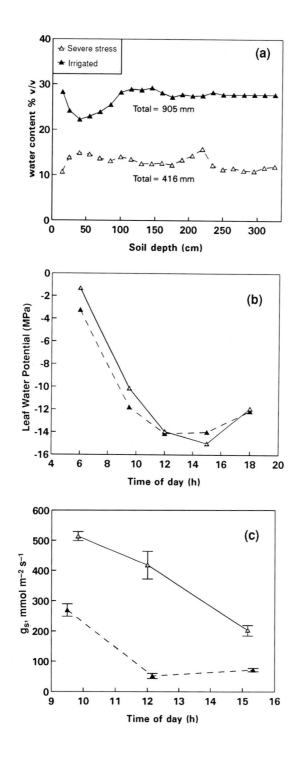

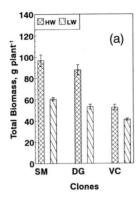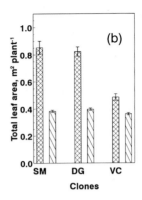

Figure 15.2. Biomass (a) and leaf area (b) production per plant of Eucalyptus globulus clonal cuttings with two levels of water supply: WW, well watered; LW, reduced water supply (see text).

briefly when the watering regime was changed: the pre-dawn leaf water potential, Ψ_d, ranged between -0.17 and -0.27 MPa in all clones irrespective of treatments, and the midday leaf water potential was between -1.2 and -1.6 MPa, again without any significant difference between treatments or clones. A major reduction in leaf area took place in the plants grown with the limited water supply (Figure 15.2) and less radiation absorbed by the foliage was the major cause of reduced growth. Of course, the 'low-water' plants not only received less water but also took up less nutrients, because of the reduction in transpiration rate. However, at least in the case of nitrogen, this did not result in nutrient deficiency because foliar N concentrations in leaf tissues were not significantly different in well-watered and water-deficient plants (Osório and Pereira, unpublished).

In addition to the changes in leaf area, it is common to ascribe reductions in plant growth under the influence of soil water deficits to decreases in the rate of carbon assimilation. Nevertheless, growth may be reduced as a result of modifications in a number of other plant properties, such as an increase in the proportion of assimilates partitioned to roots and non-photosynthetic stems or a decrease in specific leaf area (SLA). In all the eucalyptus clones shown in Figure 15.2 growth was negatively correlated with the root : shoot ratio and SLA. However, reductions in leaf number and leaf size had the greatest importance, as shown in Table 15.1.

A great deal of research has been done on the relationship between cell expansion and turgor (Tomos, 1985) but relatively little is known about the control of leaf area development of many plant species (Davies *et al.*, 1989) and even less about the effects of water deficits on cell division and rates of leaf primordia formation (Gowing *et al.*, 1990; Metcalfe *et al.*, 1989, 1991; Pereira, 1989). It is clear that increased apical dominance (*sensu strictu*) under moderate water deficits has a major influence on the reduction of foliage expansion. The role of accelerated leaf senescence in leaf area reduction with water deficits is quite variable. It may be negligible, as was observed in *E. globulus* seedlings,

Table 15.1. *Effects of water stress in the leaf area of* Eucalyptus globulus *seedlings grown in 10 litre pots subjected to water stress and well watered. The leaf area was measured when stomata were closed for the whole day and 1 week. Intervals of confidence at 95% in parentheses. (From Pereira, 1989.)*

	Well watered	Stressed
Total leaf area per plant at the end of the experiment (cm^2)	1292.2 (296.2)	499.5 (59.1)
Percentage of leaf area in branches	35%	12.4%
Leaf area lost during stress	—	10%
Increase in leaf area of well-watered plants due to larger leaves	62%	—
Increase in leaf area of well-watered plants due to more leaves	96%	—
Relative average leaf area increment (day)	0.091	0.018
Change in predawn water potential during period of study	none	− 1.05 to − 2.22

for example (Table 15.1), but leaf senescence may have a reasonable impact in many species if dehydration becomes severe (Chaves and Pereira, 1992; Pereira and Pallardy, 1989).

The effects of the moderate water deficits on plant growth during the soil dehydration phase in the Mediterranean spring will depend on the type of plant and timing of occurrence of water deficits. In cases in which shoot growth is predetermined in the overwintering bud and phenology limits the period of foliage expansion (e.g. *Quercus* spp.; see Pereira *et al.*, 1987), the number of leaves produced in a growing season may not depend on seasonal rainfall, but leaf size may be reduced during a spring/early summer drought. However, we should point out that there has been very little research on the effects of moderate water deficits on the growth of Mediterranean plants, woody or herbaceous, taking phenology into account.

The stagnation of foliage expansion as a result of moderate water deficits is not the direct result of tissue dehydration, but it goes on even if plant water relations do not change. This was the case of *E. globulus* plants, as shown in Figure 15.2, because a decrease in foliage area was observed, despite the absence of differences in tissue water status between well-watered and water-deficient plants. The same was true in plants grown with partially dehydrated split root systems (Gowing *et al.* 1990) or plants grown in pressurized pots so that tissue turgor was maintained at a high level despite the continued drying of the soil (Passioura, 1988). A clear, although indirect, proof that the plants were responding to hormonal signals from the roots was provided by Gowing *et al.* (1990). When droughted roots were excised from half-dried apple plants there was a significant recovery of leaf growth rate in comparison to plants with the root system partially dehydrated. This recovery occurred in parallel with the recovery of half-dried plants where the droughted roots were rewatered.

15.2.2 *Increased stomatal closure*

The seasonality of leaf gas exchange of Mediterranean vegetation has been described (e.g. Tenhunen *et al.*, 1987). In evergreens, transpiration and carbon assimilation go on during most of the year with minima in winter and summer, as shown in Figure 15.3. Carbon fixed during periods of reduced shoot growth (autumn and winter) is used for root growth and carbohydrate storage. It is likely that spring shoot growth is largely dependent upon the utilization of stored carbohydrates (and nutrients).

During the acclimation phase of dehydration in late spring and early summer, photosynthetic rates decrease, with stomatal closure, as a result of complex and sometimes subtle interactions of soil water deficits, increased vapour pressure deficits of the air and temperature (see Chapter 8, this volume). As the season progresses and soil moisture decreases, partial closure of the stomata occurs for increasingly longer periods during the day (Pereira *et al.*, 1986; Tenhunen *et al.*, 1987). This so-called midday depression in gas exchange simultaneously reduces water losses and daily carbon assimilation, and is one of the features of stomatal behaviour that tends to optimize carbon assimilation in relation to water supply (Cowan, 1981).

The decline in stomatal aperture during the day occurs in mildly water-stressed plants even under a constant environment. When gas exchange parameters of field-grown grapevine plants were measured throughout the day under constant, near optimal conditions of temperature, irradiance and leaf to air water vapour pressure deficit, a continuous decline in g_s and A was detected in the afternoon in stressed plants, but not in well-watered plants. Similarly, an inhibition of quantum yield of non-cyclic electron

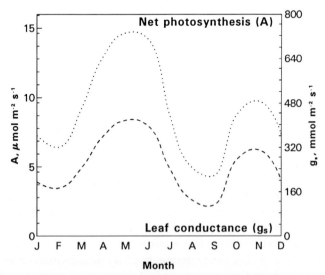

Figure 15.3. *Seasonal variation in maximum daily photosynthetic rate and in maximum leaf conductance in* Quercus suber *plants in Portugal. (Redrawn from Tenhunen et al., 1987.)*

transport (but not of photosynthetic capacity) was detected in the afternoon in the mildly stressed plants of the field. In other species, e.g. *Quercus suber*, a decline in the carboxylation efficiency was also observed during the summer midday depression of gas exchange (Tenhunen *et al.*, 1984). This seems to indicate some kind of internal down-regulatory processes in stressed plants which were absent in well-watered plants and affected both stomata and meso-phyll photosynthetic cells. The effects of soil water deficits on the stomata, as in the case of foliage growth, are very likely mediated by a 'positive root sig-nal', possibly abscisic acid (ABA) (Davies and Zhang, 1991), and the midday depression in stomatal conductance appears to be the result of modulation of the response to ABA concentration in the xylem, or the amount of ABA delivered to the leaves per unit of time (see Chapter 9, this volume). It was also found that stomata of a typical Mediterranean shrub, *Vitis vinifera*, close further for the same concentration of ABA in the xylem in the afternoon than in the morning (unpublished, but see also Chapter 9, this volume).

15.3 Effects during the dehydrated phase

15.3.1 *Short-term effects*

In summer, most plants undergo a lasting water deficiency and leaf tissues become dehydrated. Tolerance to dehydration varies strongly with different species and genotypes, but the degree of damage endured by leaves during this period is critical to support further growth. Photosynthesis at the chloroplast level is generally quite resistant to water deficits, and decreases in carbon assimilation, even with severe tissue dehydration, may be mostly the result of stomatal closure. In fact, no inhibition of net photosynthesis was observed up to a relative water content (RWC) of around 70% when stomatal limitation was overcome in measurements done under saturating CO_2 (Cornic *et al.*, 1989; Kaiser, 1987; Quick *et al.*, 1992). However, some exceptions were reported, as in the case of grapevine, where 'non-stomatal' effects on photo-synthesis have been found at RWCs well above 70% (Quick *et al.*, 1992). Furthermore, reductions in photosynthetic capacity with moderate water deficits may also occur when plants are subjected to the combined effects of high light and high temperature, as will be discussed below.

The tolerance of the photosynthetic machinery to dehydration may also dif-fer with leaf age. Younger leaves were found to be more resistant than older leaves in *Lupinus albus* as shown in Figure 15.4. This increased tolerance to dehydration in younger leaves may be particularly relevant in plants where a severe reduction in the size of leaf canopy occurs as a result of shedding of older leaves, because it allows a fast recovery after rehydration (Chaves, 1991). This is the case for native legumes of the Mediterranean region, such as lupins, where losses of more than 50% of plant leaf area have been observed after a 15 day period without water (Ramalho and Chaves, 1992).

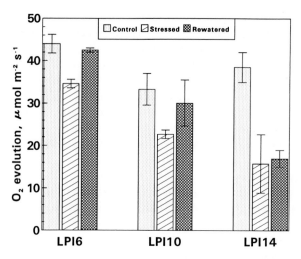

Figure 15.4. *Photosynthetic capacity (rate of photosynthetic O_2 evolution at 25°C and saturating CO_2 and PAR) as a function of leaf plastochron index in well-watered, water-stressed and rewatered plants of* Lupinus albus. *Mean values of four measurements and respective standard errors of the means.*

15.3.2 *Combined effects of high light, high temperature and water stress*

During the 'dry period' in the Mediterranean when stomata are closed for most of the day due to water deficits, the interaction of high light and high leaf temperatures predispose the plants to photoinhibition. Under conditions of high photon flux density and lack of CO_2, the amount of light absorbed by the leaf rapidly leads to the exceeding of the capacity of cells to use the products of the photochemical reactions through photosynthesis and photo-respiration (Powles, 1984). As a first line of defence, dissipative processes occur at the pigment level, namely the increase of thermal deactivation reflected by q_E (Krause and Behrend, 1986). It has been suggested that the conversion of violaxanthin to zeaxanthin upon prolonged illumination is also involved in a process of non-radiative energy dissipation and protects the plants against excess excitation energy (Demmig *et al.*, 1988). In that paper evidence is presented that the photochemical efficiency of *Nerium oleander* leaves is reduced as leaf water content decreases, in a process pre-sumably mediated by the carotenoid zeaxanthin, which acts as a fluorescence quencher. The amount of zeaxanthin increases in the leaves at the expense of violaxanthin when water stress develops. Although the increase in the rate of radiationless dissipation is associated with a decrease in the quantum yield, this disadvantage is counterbalanced by preventing damage to reaction centres.

When plants are subjected to the combined effects of water deficits and high light for long periods, protective effects by non-radiative dissipation are not enough and plants become photoinhibited. Photoinhibition under field conditions has been reported, for example, in the sclerophyllous shrub *Arbutus unedo* and in grapevines, especially during the so-called midday depression. Decreases in the ratio of the variable to maximal fluorescence (F_v/F_m) and in photosynthetic capacity were observed at severe and moderate plant water deficits (Demmig-Adams *et al.*, 1989, Quick *et al.*, 1992).

When high temperatures are superimposed on water deficits and high light, plants become even more susceptible to photoinhibition, because leaves with closed stomata are unable to dissipate absorbed solar radiation as latent heat and consequently the leaf temperature increases further. Although this occurs commonly in Mediterranean-type climates, there are not many studies on the interactive effects of these three types of stresses (Ludlow, 1987). It is known that a mild heat stress may decrease the absorptive cross-section of PS2, which may have a protective effect against photodamage (Sundby and Andersson, 1985) and induce a 'down-regulation' of the ribulose 1,5-bisphosphate (RuBP) carboxylase-oxygenase (Rubisco), whose activation state may decrease substantially with the increase in temperature (e.g. around 40% in cotton leaves when the temperature rises from 25°C to 38°C, as reported in Weis and Berry, 1988).

Recent data from our group (Chaves *et al.*, 1992) showed that the effects of increasing leaf water deficits on photosynthetic capacity of *Lupinus albus* are dependent on leaf temperature. At optimal and suboptimal temperature for photosynthesis (25°C and 15°C, respectively) a statistically significant decrease in photosynthetic capacity occurred when the RWC reached 60%, whereas at supraoptimal temperatures (35°C) the decrease was observed at 80% RWC. Furthermore, when well-hydrated leaves were subjected to 35°C in the dark no changes in photosynthetic capacity were observed and only a small reduction (of about 5%) was measured in the quantum yield of non-cyclic electron transport, estimated by the fluorescence parameter $(F_m' - F_s')/F_m$, according to Genty *et al.* (1989). However when leaves, well-hydrated or previously dehydrated to 60–70% RWC, were subjected to 10 min at 35°C and 1400 μmol photons m^{-2} s^{-1} substantial declines were observed in the quantum yield (66 and 77% of the values of controls in the well-hydrated and dehydrated leaves, respectively) and in photosynthetic capacity (31 and 56%, respectively).

These results indicate that leaves can withstand supraoptimal temperatures] without significant effects on the photosynthetic apparatus. However, when high temperatures are combined with high irradiance and tissue dehydration, leaves may undergo photoinhibition. To avoid this harmful combination of factors some plants, when water stressed, exhibit leaf rolling or other mechanisms of leaf reorientation that protect the leaves from excess irradiation. For example, in the case of *Lupinus albus* subjected to severe water deficits, leaflets close, reducing drastically the area intercepting radiation and exposing the trichome-covered adaxial surface.

15.3.3 *Water deficits and the adjustment of photosynthetic carbon metabolism*

In contrast to the photosynthetic machinery, which is quite resistant to leaf water deficits, the partitioning of newly fixed photosynthate favouring sucrose at the expense of starch is among the early effects of tissue dehydration in some species (see Quick *et al.*, 1989, 1992; Figure 15.5). This increased partitioning of recent assimilates to sucrose has been linked to alterations in sucrose phosphate synthase (SPS) activity and to a general increase in the amounts of metabolites due to a decrease in cytoplasmic volume under drought (Quick *et al.*, 1989). Although some discrepancies have been observed as a consequence of measurements being made with plants either under ambient or saturating CO_2 (compare Quick *et al.*, 1989, with Vassey and Sharkey, 1989, Vassey *et al.*, 1991), it seems that the preferential partitioning of recent assimilates to sucrose in water-stressed leaves is accompanied by an increased activation of SPS, as shown by Zrenner and Stitt (1991) in spinach. It was also shown that starch degradation begins at a higher endogenous sucrose level in water-stressed than in well-watered leaves (Zrenner and Stitt, 1991). Additionally, little or no turnover during the diurnal cycle has been observed in the sucrose pool in water-stressed leaves of various species (Quick *et al.*, 1992), which seems to indicate some kind of inhibition of sucrose export, in spite of the considerably high pool of sucrose in these leaves. This is presumably the result of a reduction in sink demand due to growth cessation.

It is possible to conclude that under water stress the increases in the ratio of newly synthesized sucrose to starch and in the degradation of starch, together with the reduction in sucrose export out of the leaves, explain the constancy, or the increase, in the amount of soluble sugars and the depletion of starch observed in leaves of water-stressed plants as compared with well-watered ones. These events, which have been observed under short-term and long-term water deficits (Zrenner and Stitt, 1991), certainly play a role in osmotic adjustment (Morgan, 1984). However, in some plants, such as grapevines, soluble sugars only accounted for *c.* 45% of the osmotically active solutes (Rodrigues *et al.*, 1992).

When subjected to long-term water deficits, most plants exhibit metabolic changes that suggest the occurrence of regulatory processes at the cellular level, even if they exhibit less tissue dehydration than plants that have been dehydrated quickly. Recent data obtained with grapevines show that plants growing in the field and subjected to a mild water stress ($\Psi_d = -0.4\,\mathrm{MPa}$) suffered changes in the pools of leaf carbohydrates similar to those observed in the more severely water-stressed plants in pots ($\Psi_d = -1.3\,\mathrm{MPa}$), when compared to well-watered controls. They lost most of the starch but maintained soluble sugars throughout the day at a constant level. However, photosynthetic capacity decreased by about 20% in field plants experiencing mild water stress as compared to well-watered plants, whereas a decrease of more

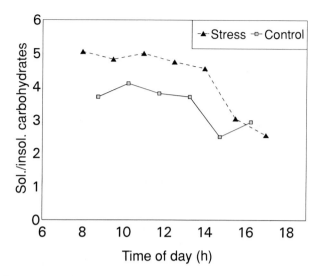

Figure 15.5. *Daily time course of the ratio of radioactivity incorporated into the soluble and insoluble fractions of leaf tissue from well-watered and water-stressed* Lupinus albus *plants. Leaves were exposed for 20 min to* $^{14}CO_2$ *in a leaf disc electrode in an atmosphere of 15%* CO_2*. (From Quick et al., 1992.)*

than *c*. 50% was observed in the more rapidly dehydrated potted plants. The decrease in photosynthetic capacity of water-stressed plants in the field was coincident with a decrease in total activity of RuBP carboxylase-oxygenase (Rubisco). This was not observed in potted plants, thus suggesting that Rubisco activity may change during the process of slow acclimation to water deficits but not necessarily as a result of tissue dehydration.

It seems, therefore, that when plants are water stressed in the field for long periods, and even if they only suffer mild tissue dehydration due to growth adjustment to the water availability, they seem to exert internal regulations at different levels, including the stomata, Calvin cycle enzymes and the size of carbohydrate pools in the cells. Recent studies relating sink demand with the amount of Rubisco protein (Diethelm and Shibles, 1989), sink and developmental changes with the activation state of Rubisco (Hurewitz and Janes, 1987) or the stress-induced changes observed in the amount of mRNA that encodes the small subunit of Rubisco (Ho and Mishkind, 1991), emphasize the importance of the fine regulations that plants undergo in response to environmental and developmental changes. More research is needed in this area for plants growing under the Mediterranean type of climate.

15.4 Conclusions

In spring and early summer in the Mediterranean, native plants as well as crops will suffer moderate water deficits resulting from soil moisture depletion, often slowed down by intermittent rain. The moment and the rate of plant tissue dehydration will depend upon weather conditions and root abundance and depth. Nevertheless, during this phase of acclimation a 'tuning-up' of transpiration with declining water availability in the rooting zone occurs. This results from the combination of decreases in stomatal conductance (short term) and in leaf area (longer term). In both cases root-to-shoot signalling seems to play a dominant role in the physiological responses.

It should be noticed, however, that Mediterranean vegetation is characterized by great diversity, and generalizations are by no means possible. For example, the 'tuning-up' of transpiration and water availability in the rooting zone is most important as an adaptive trait in species of the overstorey (trees, shrubs) because changes in canopy conductance are more likely to influence the transpiration rate of these plants than of understorey plants (Jarvis and McNaughton, 1986). The former are normally deep-rooted evergreens, which normally withstand drought accompanied by high temperatures and high irradiances. At the other extreme are herbaceous plants, often shallow rooted, whose phenology sets a dormant phase (seed, vegetative perenating organs) as soon as (or even before) the uppermost layers of the soil have been depleted of water.

When we analyse the effects of water deficits during the phase in which real tissue dehydration occurs, we have to take into account that the damage endured by leaves during this period is critical to support further growth in perennials. Recent data indicate that photosynthesis at the mesophyll level is generally quite resistant to dehydration unless plants are subjected simultaneously to high temperature and high irradiance. Furthermore, evidence has been accumulating to indicate that under stress conditions regulatory (or 'down-regulatory') processes do occur at the photochemical and biochemical levels, adjusting plant functioning to the environmental constraints.

Acknowledgements

The authors gratefully acknowledge the collaboration of Júlio and Leonor Osório and of Maria-Joäo Correia in data collection.

References

Chaves, M.M. (1991) Effects of water deficits on carbon assimilation. *J. Exp. Bot.*, **42**, 1–16.

Chaves, M.M. and Pereira, J.S. (1992) Water stress, CO_2 and climate change. *J. Exp. Bot.*, **43**, 1131–1139.

Chaves, M.M., Osório, M.L., Osório, J. and Pereira, J.S. (1992) The photosynthetic response of *Lupinus albus* to high temperature is dependent on irradiance and leaf water status. *Photosynthetica*, in press.

Cornic, G., Le Gouallec., J.L., Briantais, J.M. and Hodges, M. (1989) Effect of dehydration and high light on photosynthesis of two C_3 plants (*Phaseolus vulgaris* L. and *Elatostema repens* (Lour) Hall f.). *Planta*, **177**, 84–90.

Cowan, I.R. (1981) Regulation of water use in relation to carbon gain in higher plants. In: *Physiological Plant Ecology. II. Water Relations and Carbon Assimilation*. Encyclopedia of Plant Physiology, Vol. 12B (eds O.L. Lange, P.S. Nobel, C.B. Osmond and H. Ziegler). Springer, Berlin, pp. 589–614.

Davies, W.J. and Zhang, J. (1991) Root signals and the regulation of growth and development of plants in drying soil. *Ann. Rev. Plant Physiol. and Mol. Biol.*, **42**, 55–76.

Davies, W.J., Rhizopoulou, S., Sanderson, R., Taylor, G., Metcalfe, J.C. and Zhang, J. (1989) Water relations and growth of roots and leaves of woody plants. In: *Biomass Production by Fast-Growing Trees* (eds J.S. Pereira and J.J. Landsberg). Kluwer, Dordrecht, pp. 13–36.

Demmig, B., Winter, K., Hruger, A. and Czygan, F.C. (1988) Zeaxanthin and the heat dissipation of excess light energy in *Nerium oleander* exposed to a combination of high light and water stress. *Plant Physiol.*, **87**, 17–24.

Demmig-Adams, B., Adams III, W.W., Winter, K., Meyer, A., Schreiber, U., Pereira, J.S., Kruger, A., Czygan, F.C. and Lange, O.L. (1989) Photochemical efficiency of photosystem II, photon yield of CO_2 evolution, photosynthetic capacity and carotenoid composition during the midday depression of net CO_2 uptake in *Arbutus unedo* growing in Portugal. *Planta*, **177**, 377–387.

di Castri, F. and Mooney, H.A. (eds) (1973) *Mediterranean-type Ecosystems. Origin and Structure*. Springer, Berlin.

Diethelm, R. and Shibles, R. (1989) Relationship of enhanced sink demand with photosynthesis and amount and activity of ribulose 1,5-bisphosphate carboxylase in soybean leaves. *J. Plant Physiol.*, **134**, 70–74.

Genty, B., Briantais, J.M. and Baker, N.R. (1989) The relationship between the quantum yield of photosynthetic electron transport and quenching of chlorophyll fluorescence. *Biochim. Biophys. Acta*, **990**, 87–92.

Gowing, D.J.G., Davies, W.J. and Jones, H.G. (1990) A positive root-sourced signal as an indicator of soil drying in apple, *Malus × domestica*, Borkh. *J. Exp. Bot.*, **41**, 1535–1540.

Ho, T.-Y. and Mishkind, M.L. (1991) The influence of water deficits on mRNA levels in tomato. *Plant Cell Environ.*, **14**, 67–75.

Hurewitz, J. and Janes, H.W. (1987) The relationship between the activity and the activation state of RuBP carboxylase and carbon exchange rate as affected by sink and developmental changes. *Photosynth. Res.*, **12**, 105–117.

Jarvis, P.G. and McNaughton, K.G. (1986) Stomatal control of transpiration scaling up from leaf to region. In: *Advances in Ecological Research*, Vol. 15 (eds A. MacFadyen and E.D. Ford). Academic Press, London, pp. 1–50.

Kaiser, W.M. (1987) Effects of water deficit on photosynthetic capacity. *Physiol. Plant.*, **71**, 142–149.

Krause, G.H. and Behrend, U. (1986) Δ pH-dependent chlorophyll fluorescence quenching indicating a mechanism of protection against photoinhibition of chloroplasts. *FEBS Letts.*, **200**, 298–302.

Ludlow, M.M. (1987) Light stress at high temperature. In: *Photoinhibition* (eds D.J. Kyle, C.B. Osmond and C.J. Arntzen). Elsevier, Amsterdam, pp. 89–109.

Metcalfe, J.C., Davies, W.J. and Pereira, J.S. (1989) Leaf growth of *Eucalyptus globulus* seedlings under water deficit. *Tree Physiol.*, **6**, 221–227.

Metcalfe, J.C., Davies, W.J. and Pereira, J.S. (1991) The control of leaf growth in *Eucalyptus globulus* seedlings. I. Effects of leaf age. *Tree Physiol.*, **9**, 491–500.

Morgan, J.M. (1984) Osmoregulation and water stress in higher plants. *Ann. Rev. Plant Physiol.*, **35**, 299–319.

Ögren, E. and Sjöström, M. (1990) Estimation of the effect of photoinhibition on the carbon gain in leaves of a willow canopy. *Planta*, **181**, 560–567.

Passioura, J.B. (1988) Root signals control leaf expansion in wheat seedlings growing in drying soil. *Aust. J. Plant Physiol.*, **15**, 687–693.

Pereira, J.S. (1989) Whole-plant regulation and productivity in forest trees. In: *Importance of Root-to-Shoot Communication in the Response to Environmental Stress*, B.P.G.R.G. Monograph 21 (eds W.J. Davies and B. Jeffcoat). British Society for Plant Growth Regulation, Bristol, pp. 237–250.

Pereira, J.S. and Pallardy, S. (1989) Water stress limitations to tree productivity. In: *Biomass Production by Fast-Growing Trees* (eds J.S. Pereira and J.J. Landsberg). Kluwer, Dordrecht, The Netherlands, pp. 37–56.

Pereira, J.S., Tenhunen, J.D., Lange, O.L., Beyschlag, W., Meyer, A. and David M.M. (1986) Seasonal and diurnal patterns in leaf gas exchange of *Eucalyptus globulus* trees growing in Portugal. *Can. J. For. Res.*, **16**, 177–184.

Pereira, J.S., Beyschlag, G., Lange, O.L., Beyschlag, W. and Tenhunen, J.D. (1987) Comparative phenology of four Mediterranean shrub species growing in Portugal. In: *Plant Response to Stress. Functional Analysis in Mediterranean Ecosystems* (eds J.D. Tenhunen, F.M. Catarino, O.L. Lange and W.C. Oechel). Springer, Berlin, pp. 503-514.

Powles, S.B. (1984). Photoinhibition of photosynthesis induced by visible light. *Ann. Rev. Plant Physiol.*, **65**, 1181–1187.

Quick, W.P., Siegl, G. Neuhaus, E., Feil, R. and Stitt, M. (1989). Short-term water stress leads to a stimulation of sucrose synthesis by activating sucrose-phosphate synthase. *Planta*, **177**, 535–546.

Quick, W.P., Chaves, M.M., Wendler, R., David, M.M., Rodrigues, M.L., Passarinho, J.A., Pereira, J.S., Adcok, M.D., Leegood, R.C. and Stitt, M. (1992) The effect of water stress on photosynthetic carbon metabolism in four species grown under field conditions. *Plant Cell Environ.*, **15**, 25–35.

Ramalho, J.C. and Chaves, M.M., (1992) Drought effects on plant water relations and carbon gain in two lines of *Lupinus albus*. L. *Eur. J. Agron.*, in press.

Rodrigues, M.L., Chaves, M.M., Wendler, R., David, M.M., Quick, W.P., Leegood, R.C., Stitt, M. and Pereira, J.S. (1993) Osmotic adjustment in water stressed grapevine leaves in relation to carbon assimilation. *Aust. J. Plant. Physiol.*, in press.

Russell, G., Jarvis, P.G. and Monteith, J.L. (1989) Absorption of radiation by canopies and stand growth. In: *Plant Canopies. Their Growth Form and Function* (eds G. Russell, B. Marshall and P.G. Jarvis). Cambridge University Press, Cambridge, pp. 21–39.

Sundby, C. and Andersson, B. (1985) Temperature-induced reversible migration along the thylakoid membrane of photosystem II regulates its association with LHC II. *FEBS Letts.*, **191**, 24–28.

Tenhunen, J.D., Lange, O.L., Gebel, J., Beyschlag, W. and Weber, J.A. (1984) Changes in photosynthetic capacity, carboxylation efficiency and CO_2 compensation point associated with midday depression of net CO_2 exchange of leaves of *Quercus suber*. *Planta*, **193**, 193–203.

Tenhunen, J.D., Catarino, F.M., Lange, O.L. and Oechel, W.C. (eds) (1987) *Plant Response to Stress. Functional Analysis in Mediterranean Ecosystems*. Springer, Berlin.

Tomos, A.D. (1985) Physical limitations to leaf cell expansion. In: *Control of Leaf Growth* (eds N.R. Baker, W.J. Davies and C.K. Ong). Cambridge University Press, Cambridge, pp. 1–13.

Vassey, T.L. and Sharkey, T.D. (1989) Mild water stress of *Phaseolus vulgaris* plants leads to reduced starch synthesis and extractable sucrose phosphate synthase activity. *Plant Physiol.*, **89**, 1066–1070.

Vassey, T.L., Quick, W.P., Sharkey, T.D. and Stitt, M. (1991) Water stress, carbon dioxide and light effects on sucrose-phosphate synthase activity in *Phaseolus vulgaris. Physiol. Plant.*, **81**, 37–44.

Weis, E. and Berry, J.A. (1988) Plants and high temperature. In: *Plants and Temperature* (eds S.P. Long and F.I. Woodward). Company of Biologists, Cambridge, pp. 329–346.

Zrenner, R. and Stitt, M. (1991) Comparison of the effect of rapidly and gradually developing water-stress on carbohydrate metabolism in spinach leaves. *Plant Cell Environ.*, **14**, 939–946.

Water deficits, the development of leaf area and crop productivity

J.B. Passioura, A.G. Condon and R.A. Richards

16.1 Introduction

The responses of crop plants to water deficits have been documented in thousands of papers—biochemical, physiological and agronomic. Many of these papers have been inspired by the hope that there exist simple characters, useful to breeders, that will confer 'drought resistance' on a crop in a wide range of water-limited environments. Examples of such putative characters are proline accumulation, osmotic adjustment, dehydration olerance, leaf rolling and root morphology. Several of the papers in, for example, Mussell and Staples (1979) expertly discuss such characters. More recently there has been much interest, in a similar spirit, in stress-induced proteins.

While much of this research has been valuable in fashioning the conceptual framework we use when thinking about water-limited plants, it has not provided much overtly effective guidance to breeders and agronomists who are concerned with the productivity of dryland field crops. Why not? Because universal characters of this sort probably do not exist. They may be beneficial in one water-limiting environment, detrimental in another and irrelevant in a third.

'Drought resistance' is a term that loses clarity the more closely we examine it, much as a newspaper photograph does when examined with a magnifying glass. In the context of water-limited crop production it is best abandoned. The productivity is much better analysed in terms of physiological economics—that is, in terms of what determines the effectiveness with which a crop uses a limiting supply of water in producing its harvestable yield.

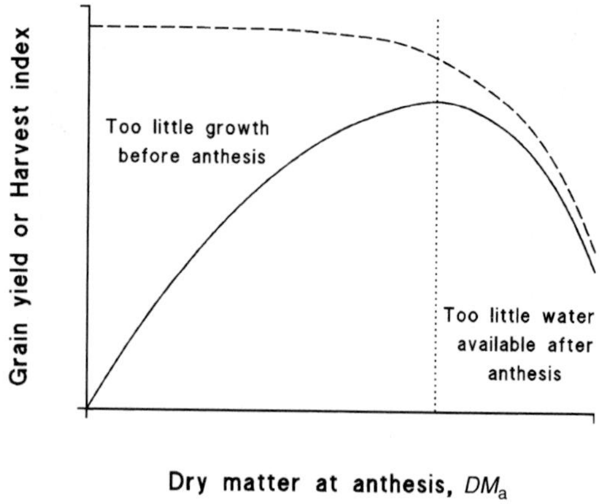
Grain yield or Harvest index
Too little growth before anthesis
Too little water available after anthesis
Dry matter at anthesis, DM_a

Figure 16.1. Schematic relations between grain yield and above-ground dry matter at anthesis, DM_a*, (solid line) and harvest index and* DM_a *(dashed line), in a crop that depends, for setting and filling its grain, on water contained in the soil at anthesis. (After Fischer, 1979.)*

16.2 Modelling the components of yield and water use

We have found it useful to analyse this effectiveness in terms of the following equation:

$$Y = T \times W_T \times HI \qquad 16.1$$

where Y is grain yield; T is the amount of water transpired; W_T is the transpiration efficiency, the ratio of above-ground dry matter to the amount of transpiration; and HI is the harvest index, the ratio of grain yield to above-ground dry matter. To a first approximation the components in this relationship are independent of each other, so an increase in any one of them is likely to increase yield (Condon and Richards, 1992; Fischer, 1979; Ludlow and Muchow, 1990; Passioura, 1977). Conversely, any phenomenon that cannot be readily related to any of these components (e.g. proline accumulation, or the appearance of stress-induced proteins) is unlikely to have much bearing on water-limited yield (see Chapter 13).

Of these components, by far the most important is harvest index, which in turn depends critically on how well the phenology of the crop is suited to the environment. A clear example of the importance of phenology is given by the performance of the first wheat crops grown in Australia 200 years ago. These crops presumably produced a substantial amount of dry matter, but they failed to produce any grain. The cultivars used were adapted to the long growing season in England, and flowered much too late, in the searing summer, when the evaporative demand was huge and when there was probably almost no available water left in the soil. Their harvest index was zero. In the century that followed the flowering date was progressively brought forward by about 3 months, thereby enabling a viable wheat industry.

Figure 16.1, adapted from Fischer (1979), illustrates this point further. Fischer's (1979) paper concerns the performance of wheat in Mediterranean environments, in which the evaporative demand on the crop rises rapidly from about the time of ear emergence onwards, but the principles he discusses apply widely. The most important principle is that a large harvest index can arise only if sufficient water is available in the soil at anthesis for the crop to set and fill a large number of grains. If there is too much growth before anthesis, either because anthesis is too late as in the example in the previous paragraph, or because the plants have been too vigorous in their vegetative growth and have consequently transpired so much water that little remains stored in the soil at anthesis, then harvest index will be low. Studies on a range of lines of wheat in Mediterranean environments show that yield may fall by several per cent for every day that anthesis is delayed (Hamblin, 1993; Perry and D'Antuono, 1989). Delays in anthesis not only reduce the amount of water available for the crop during grain-filling, but also expose it to greater evaporative demands, with a consequent fall in the effectiveness with which the crops can use the water while fixing carbon dioxide.

Figure 16.1 also serves to introduce what we believe to be the next most important determinant of water-limited yield, namely, the development of leaf area through time. Watson (1952) pointed out that a crop's modulation of its leaf area may be rather more influential on yield than its modulation of net assimilation rate. His discussion centred on the capture of light, but his conclusion applies equally well, perhaps more so, to the efficient use of water. There are several influences of leaf area development on yield that are worth noting. Firstly, it largely determines the amount of dry matter produced by anthesis, which in turn strongly influences the potential yield of the crop (that is, the yield if conditions are excellent during grain-filling) (Fischer, 1979). Secondly, it influences the harvest index through its effect on the balance of water use before and after anthesis even when the phenological development is right, that is, when the crops flower at the right time in the given environment. Thirdly, it may influence the amount of water transpired if the soil surface is wet for much of the season—for a high leaf area index will ensure that little water is evaporated directly from a wet soil surface and thereby lost to the plant.

The next section provides some detail on how the trajectory of leaf area index through time may influence the various components of yield embodied in Equation 16.1. We then discuss the physiology of leaf area development in water-limited, or potentially water-limited, crops, and also the prospects for modifying leaf area development, either genetically or through crop management, to improve yield.

16.3 Leaf area development and the components of grain yield

Two examples of how the development of leaf area may influence the yield of a water-limited crop are given below.

16.3.1 *Mediterranean environments*

Here, crops are sown in late autumn at the beginning of the wet season, and

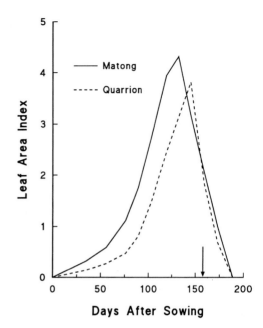

Figure 16.2. *Leaf area index through time of two lines of wheat, Matong and Quarrion, that differ in vegetative vigour.*

the soil surface may remain wet for much of the establishment phase of the crop, when the leaf area index is low. A very high proportion (30–70%) of the total water supply may be lost by direct evaporation from the soil (Condon and Richards, 1992; Cooper *et al.*, 1983; Siddique *et al.*, 1989). The more rapid is the development of leaf area in these circumstances the less will be this evaporative loss and the larger will be the T term in Equation 16.1 (Fischer, 1980; Turner and Nicolas, 1987). W_T will also be larger because less water is traded for carbon dioxide by the leaves under cool conditions. Comparison between two lines of wheat, Matong and Quarrion, that differ in their vegetative vigour illustrates this point (Condon and Richards, 1992). Matong is much the more vigorous. It not only develops leaf area more quickly, but it also has a stomatal conductance that is typically 40–50% larger than that of Quarrion. Figure 16.2 shows the leaf area index through time for the two lines during a field experiment at Wagga Wagga, Australia, in 1989, a season in which the soil surface was wet for several weeks after sowing. Matong had double the green leaf area index of Quarrion throughout the first 100 days after sowing and maintained a large, though proportionately smaller, difference for another month before depletion of soil water and rising evaporative demand brought on rapid leaf senescence. Estimates of transpiration and direct evaporation from the soil were made using the technique of Cooper *et al.* (1983), the accuracy of which was confirmed using more direct techniques (Leuning, Denmead and Dunin, personal communication). These estimates (Figure 16.3, from Condon and Richards, 1992) show that by the time the green leaf area index of Quarrion had caught up with that of Matong, Matong had transpired much more water than had

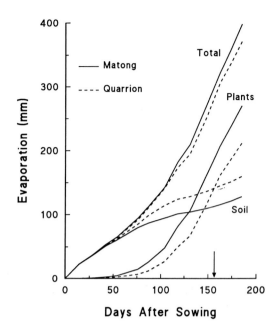

Figure 16.3. *Total water use, and its components transpiration and soil evaporation, through time for the same two lines of wheat depicted in Figure 16.2. (From Condon and Richards, 1992.)*

Quarrion and had allowed much less direct evaporation from the soil. The total water use was essentially the same for the two lines throughout the season. The T term of Equation 16.1 was, by the end of the season, about 60 mm, 25%, larger for Matong than for Quarrion. Interestingly, however, the dry matter produced by Matong was only about 10% larger than that of Quarrion. The low stomatal conductance of Quarrion made it intrinsically more efficient in water use than Matong. That is, its leaves used less water when fixing a given amount of carbon dioxide, with the result that its W_T term was about 15% larger than that of Matong (Condon and Richards, 1992). Consistent with this difference, the carbon isotope discrimination, Δ, of Quarrion was about 2‰ less than that of Matong. This is a point to which we return in the Discussion.

16.3.2 *Semiarid subtropics*

In this environment, crops are often grown in the winter, during which there may be little rain. Such crops rely heavily on water contained in the soil at the time of sowing that accumulated during the summer rains. They yield poorly if they develop leaf area too rapidly when young, for they then transpire so much of their limited water supply before flowering that too little is left during flowering and grain-filling. Such crops may produce substantial dry matter, but their harvest index may be low (Fischer, 1979; Passioura, 1972; Richards and Passioura, 1989)—that is, in terms of Figure 16.1, there is the likelihood that they may lie well to the right of the optimum. Unlike the previous example, losses of water by direct evaporation from the soil are likely to be small.

16.4 Variation in leaf area development

If, then, the rate of leaf area development strongly affects the effectiveness with which crops use a limiting supply of water, what controls this rate and how can we influence it to improve water-limited yields? Firstly, there are intrinsic differences in vegetative vigour that reflect the genetic make-up of the plants and that are amenable to change through breeding. Secondly, there are environmental influences that affect the physiology of the plants and that are amenable to change through management.

16.4.1 *Intrinsic vigour*

The rate of establishment of leaf area after germination strongly influences the whole of the vegetative development of the crop. Even in drought-prone environments it is common for establishing crops to be well supplied with water during their first few weeks of growth. What determines variation in the rate of establishment in such circumstances? It could be differences in the relative rate of leaf expansion, but it looks as though it is variation in the size of the embryo that counts. Ashby (1930, 1937) found this to be so in comparisons between hybrid and inbred lines of maize and tomato. Lopez-Castaneda (personal communication) found the same in a comparison of lines of wheat and barley. Barley typically develops leaf area during establishment at about twice the rate of wheat, but there were no consistent differences in relative leaf expansion rate between the two species. Differences in the size of the leaf primordia in the embryo, that is, in the starting capital for leaf expansion, accounted for most of the variation, with differences in specific leaf weight accounting for the rest. Likewise, variation within each species was associated with differences in the size of leaf primordia and in specific leaf weight. Given the lack of variation in relative leaf expansion rate, it may be possible to select for early vigour by selecting for large first leaves.

Although rapid early development may help a crop to use water that would otherwise be lost by direct evaporation from the soil, there is the danger that, if it is associated with a sustained high relative growth rate, there will be so much vegetative growth before anthesis that the crop will overshoot the optimum shown in Figure 16.1. In principle it should be possible to restrain the mid-season growth by inhibiting the growth of branches or, in the case of cereals, of tillers. There is, for example, a tiller-inhibiting gene in wheat (Richards, 1988) that allows rapid seedling growth and therefore early ground cover by the leaves, without the danger of a blowout in leaf area. The leaf area of plants containing this gene tends to increase linearly rather than exponentially during tillering.

16.4.2 *Influence of leaf water status*

Although it may be common for seedlings to be well supplied with water,

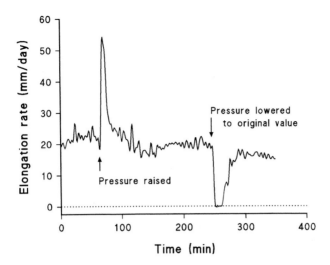

Figure 16.4. *Elongation rate of an expanding barley leaf through time. The water potential of the expanding cells was increased by 100 kPa (by applying a pressure of 100 kPa to the plant's roots, which were contained within a pressure chamber) during the period between the two arrows. (Adapted from Passioura, 1992.)*

eventually, in drought-prone environments, water deficits will start to influence leaf expansion. The mechanisms involved seem less straightforward now than they did a few years ago, and are the subject of vigorous controversy (Kramer, 1988; Munns, 1988; Passioura, 1988b; Sinclair and Ludlow, 1985).

Leaves grow by expanding their young cells. It is widely believed that cell expansion is powered by turgor, because wilted leaves do not grow. The expansion rate is often analysed in terms of Lockhart's (1965) simple model of the plastic deformation of the cell wall, namely:

$$dV/dt = m(P - P_Y) \qquad 16.2$$

where V is cell volume, t is time, m is a coefficient of extensibility, P is the turgor pressure, and P_Y is a minimum value of P below which the wall will not expand irreversibly. The explanatory promise of this equation has not been fulfilled, because it turns out that the supposedly fixed parameters m and P_Y are not fixed at all, and may respond rapidly (within a few minutes) to changes in P so that such changes may have little, if any, sustained effect on the expansion rate (Green *et al.*, 1971; Passioura, 1992; Shackel *et al.*, 1987). Figure 16.4, adapted from Passioura (1992), shows how effectively an expanding barley leaf may control its elongation rate despite sustained changes in its leaf water status.

Despite the inadequacy of Equation 16.2, its intuitive appeal remains strong—surely the plastic expansion of the cell wall will proceed faster if higher turgor induces a higher tension within the wall? But the structure of the expanding wall is complex and necessarily dynamic. New wall material is continuously being deposited as the cell grows, and old material is continuously subject to change as the enzymes within it cleave load-bearing bonds. Fry (1989) has proposed a dynamic structural model of the wall that, at least in principle, explains why the expansion rate need not be controlled by turgor. He proposes that the wall comprises a set of cellulose microfibrils that are coated with a layer of hemicellulose molecules which adhere by hydrogen bonds and which form bridges between adjacent microfibrils. Some of these bridges are load-bearing, while others are relaxed. Expansion of the wall is supposed to occur when the tension in the load-bearing bridges is large enough to break the hydrogen bonds that attach them to the microfibrils. This process leads to strain-hardening, because previously relaxed bridges become load-bearing. Simultaneously, load-bearing bridges are being cleaved enzymically. The strain-hardening would eventually bring expansion to a stop, were it not for enzymic cleavage. Analysis of Fry's model shows that the balance between these two processes would determine the expansion rate, and that apart from transient responses turgor need not affect steady growth at all (Passioura and Fry, 1992).

Despite the intuitive appeal of Lockhart's model of cell wall rheology, the time has come to abandon it. It has no predictive value, and it ignores recent improvements in our understanding of the structure of the wall. It gives rise to the false expectation that local water status necessarily affects growth rate. There is now no good reason to believe that a growing soil water deficit will necessarily inhibit leaf growth by affecting leaf water relations—unless the environment becomes severe enough to cause wilting.

16.4.3 *Influence of soil water status*

Yet it is clear that soil water deficits do inhibit leaf expansion. There is increasing evidence that plants sense conditions in the soil directly, and respond conservatively to deteriorating conditions well before their roots lose the ability to extract water fast enough for the needs of the leaves (Blackman and Davies, 1985; Blum *et al.*, 1991; Davies and Zhang, 1991; Gowing *et al.*, 1991; Passioura, 1988a; Saab and Sharp, 1989; see Chapters 8 and 9). The roots send inhibitory signals to the leaves that may slow leaf expansion and close stomata well before any fall in water status is evident in the shoot. When they do so, they are in effect displaying a feedforward response (see Chapter 12).

When a soil dries many changes take place within it that could affect the behaviour of the roots and that could induce them to send inhibitory signals to the leaves. The soil not only holds water more strongly—it becomes harder, and it transmits water and nutrients less readily. The elongation rate of roots is strongly affected by the hardness of the soil which in turn may change greatly when the soil dries. Soil water content affects the growth rate of wheat plants, through its effect on the hardness of the soil, even when the soil water poten-

Figure 16.5. *Expansion rate of leaves of young wheat plants (as a percentage of that in well-watered controls) growing in drying soil of two bulk densities. The open symbols denote that the soil had a low bulk density; the penetrometer resistance never exceeded 1.0 MPa, a negligible value, at any water content. The closed symbols denote that the soil had a high bulk density; the penetrometer resistance was 2.0 MPa at a water content of 0.25, and rose linearly with decreasing water content until it reached 6 MPa at a water content of 0.15. The starred points denote that the watered and unwatered plants differed significantly in leaf expansion rate. Bulk density did not affect the growth of the well-watered plants. (Redrawn from Passioura and Gardner, 1990, with permission from CSIRO Editorial Publications.)*

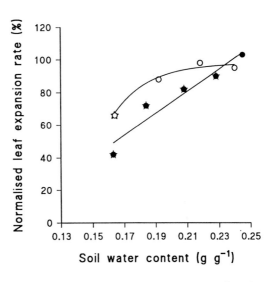

tial is high (Masle and Passioura, 1987). This response may account for the frequent observation that zero-tilled crops lack vigour (Cornish and Lymbery, 1987).

Figure 16.5, adapted from Passioura and Gardner (1990), shows how the development of leaf area of wheat plants growing in drying soil is affected by both the drying and the hardness of the soil, even when the leaf water potential of the plants is prevented from falling. The plants were grown in pots that could be encased in pressure chambers. Applying pressure in these chambers effectively counterbalances the increasing suction in the soil water as the soil dries and prevents the leaf water potential from falling. The pots contained soil that was packed either loosely (bulk density, 1.0 g cm^{-3}) or firmly (bulk density, 1.38 g cm^{-3}). Previous work had shown that the penetrometer resistance of the soil (which is a measure of the difficulty roots have in elongating) was negligible at all water contents at the low bulk density. But at the high bulk density, the penetrometer resistance was on the verge of being significant (in its effect on leaf growth) at the starting water content and was expected to become large as the soil dried. The figure shows that soon after the soil started to dry the growth rate of the plants in the firm soil was affected, thereby implicating soil strength as the inducer of the putative inhibitory signal from the roots to the leaves; but eventually the plants in the loose soil also slowed their growth, thereby implicating low soil water potential as the inducer of the signal. Thus soil drying may affect leaf growth through its effects both on the hardness and on the water potential of the soil (see also Chapters 8 and 9).

The foregoing discussion has emphasized the effects of water deficits on cell expansion, implying perhaps that leaves are smaller because their cells are smaller. This is so when water deficits first bite, but in a sustained drought plants are more likely to modulate cell number (Randall and Sinclair, 1988), thereby producing leaves that are smaller because they have fewer cells, but of normal size. Likewise, tillers and lateral buds may be inhibited, for reasons not yet understood.

16.5 Miscellaneous effects

In the field, there are many other factors that may affect leaf development and hence the effectiveness with which crops use a limiting water supply. Nutrition is an obvious one, and so is temperature. Agronomic variables such as sowing depth are also strongly influential.

16.6 Discussion

We have argued that the trajectory of leaf area index through time greatly influences how effectively a crop uses a limiting water supply. This trajectory depends partly on the intrinsic vegetative vigour of the crop, which is amenable to change through breeding, and partly on environmental influences. We have concentrated on two main environmental examples, Mediterranean, in which crops rely largely on current rainfall and tend to be too slow in their leaf development, and subtropical, in which winter-grown, dry-season, crops rely heavily on water stored in the soil at the time of sowing and which tend to be too fast in their leaf development.

Prospects for improving leaf area development in the Mediterranean environment are particularly promising (Richards, 1991). Earlier sowing increases development rate, because of the generally higher temperatures in early autumn compared with late, and leads to larger leaf area indices during the winter when the soil surface is often wet and prone to lose water by direct evaporation. It also results in a higher W_T because the trade of carbon dioxide for water vapour by the leaves is very economical in the winter. Breeders have recently produced varieties suitable for early sowing that flower at the right time. These varieties give farmers the option of deriving benefits from sowing early whenever the season permits, that is, when planting rains are early.

We noted earlier that Quarrion was much less vigorous in development than was Matong (Figure 16.2) and that it also had a lower Δ and a lower stomatal conductance, which in turn made it intrinsically more water-use efficient. The association between Δ, low stomatal conductance and lack of vegetative vigour can be thought of as a conservative strategy (cf. Chapter 17) and seems to be common amongst wheat varieties; although it is not always found in other species. In environments where crops rely largely on stored soil moisture, or in long duration crops, selection for low Δ or low conductance may not hinder crop improvement. However, if water is limited and soil evaporation is a significant component of evapotranspiration, then breaking the association between low vigour, low conductance and low Δ will be important. It has been argued that this should be possible either through management, by sowing more densely, or through breeding, by increasing early vigour (Condon and Richards, 1992). Then gains in T and W_T (Equation 16.1) resulting from rapid leaf development during establishment can be combined with the large intrinsic water-use efficiency associated with low stomatal conductance (Chapter 12), thereby giving a substantial increase in yield.

These and other ways of managing the rate of development of leaf area offer the prospect of substantially improving the yield of water-limited crops.

References

Ashby, E. (1930) Studies in the inheritance of physiological characters. I. A physiological investigation of the nature of hybrid vigour in maize. *Ann. Bot.* **44**, 457–467.

Ashby, E. (1937) Studies in the inheritance of physiological characters. III. Hybrid vigour in the tomato. *Ann. Bot. NS*, **1**, 11–41.

Blackman, P.G. and Davies,W.J. (1985) Root to shoot communication in maize plants of the effects of soil drying. *J. Exp. Bot.*, **36**, 39–48.

Blum, A., Johnson, J.W., Ramseur, E.L. and Tollner, E.W. (1991) The effect of a drying top soil and a possible non-hydraulic root signal on wheat growth and yield. *J. Exp. Bot.*, **42**, 1225–1231.

Condon, A.G. and Richards, R.A. (1992) Exploiting genetic variation in transpiration efficiency in wheat – an agronomic view. In: *Perspectives on Carbon and Water Relations from Stable Isotopes* (eds J.R. Ehleringer, G.D. Farquhar, A.E. Hall and I. Ting). Academic Press, New York, in press.

Cooper, P.J.M., Keatinge, J.D.H. and Hughes, G. (1983) Crop evapotranspiration – a technique for calculation of its components by field measurements. *Field Crops Res.*, **7**, 299–312.

Cornish, P.S. and Lymbery, J.R. (1987) Reduced early growth of direct drilled wheat in New South Wales: causes and consequences. *Aust. J. Exp. Agric.*, **27**, 869–880.

Davies, W.J. and Zhang, J. (1991) Root signals and the regulation of growth and development of plants in drying soil. *Ann. Rev. Plant Physiol. Plant Mol. Biol.*, **42**, 55–76.

Fischer, R.A. (1979) Growth and water limitation to dryland wheat yield in Australia: a physiological framework. *J. Aust. Inst. Agric. Sci.*, **45**, 83–94.

Fischer, R.A. (1980) Influence of water stress on crop yield in semiarid regions. In: *Adaptation of Plants to Water and High Temperature Stress* (eds N.C. Turner and P.J. Kramer). Wiley, New York, pp. 323–339.

Fry, S.C. (1989) Cellulases, hemicelluloses and auxin-stimulated growth: a possible relationship. *Physiol. Plant.*, **75**, 532–536.

Gowing, D.J.G., Davies, W.J. and Jones, H.G. (1990) A positive root sourced signal as an indicator of soil drying in apple, *Malus × domestica* Borkh. *J. Exp. Bot.*, **41**, 1535–1540.

Green, P.B., Erickson, R.O. and Buggy, J. (1971) Metabolic and physical control of cell elongation rate. *In vivo* studies in *Nitella*. *Plant Physiol.*, **47**, 423–430.

Hamblin, J. (1993) Resource capture by crops. *52nd Nottingham Easter School*, in press.

Kramer, P.J. (1988) Changing concepts regarding plant water relations. *Plant Cell Environ.*, **11**, 565–568.

Lockhart, J.A. (1965) An analysis of irreversible plant cell elongation. *J. Theor. Biol.*, **8**, 264–275.

Ludlow, M.M. and Muchow, R.C. (1990) A critical evaluation of traits for improving crop yields in water-limited environments. *Adv. Agron.*, **43**, 107–153.

Masle, J. and Passioura, J.B. (1987) The effect of soil strength on the growth of young wheat plants. *Aust. J. Plant Physiol.*, **14**, 643–656.

Munns, R. (1988) Why measure osmotic adjustment? *Aust. J. Plant Physiol.*, **15**, 717–726.

Mussell, H. and Staples, R.C. (eds) (1979) *Stress Physiology in Crop Plants*. Wiley, New York.

Passioura, J.B. (1972) Effect of root geometry on the yield of wheat growing on stored water. *Aust. J. Agric.* Res., **23**, 745–752.

Passioura, J.B. (1977) Grain yield, harvest index, and water use of wheat. *J. Aust. Inst. Agric. Sci.*, **43**, 117–121.

Passioura, J.B. (1988a) Root signals control leaf expansion in wheat seedlings growing in drying soil. *Aust. J. Plant Physiol.*, **15**, 687–693.

Passioura, J.B. (1988b) Response to Dr P.J. Kramer's article, 'Changing concepts regarding plant water relations'. *Plant Cell Environ.*, **11**, 569–571.

Passioura, J.B. (1992) The expansion of plant tissues. In: *The Mechanics of Swelling* (ed. T. Karalis). Springer-Verlag, Heidelberg, in press.

Passioura, J.B. and Fry, S.C. (1992) Turgor and cell expansion: beyond the Lockhart equation. *Aust. J. Plant Physiol.*, in press.

Passioura, J.B. and Gardner, P.A. (1990) The control of leaf expansion in wheat seedlings growing in drying soil. *Aust. J. Plant Physiol.* **17**, 149–157.

Perry, M.W. and D'Antuono, M.F. (1989) Yield improvement and associated characteristics of some Australian spring wheat cultivars introduced between 1860 and 1982. *Aust. J. Agric. Res.*, **40**, 457–472.

Randall, H.C. and Sinclair, T.R. (1988) Sensitivity of soybean leaf development to water deficits. *Plant Cell Environ.*, **11**, 835–839.

Richards, R.A. (1988) A tiller inhibitor gene in wheat and its effect on plant growth. *Aust. J. Agric. Res.*, **39**, 749–757.

Richards, R.A. (1991) Crop improvement for temperate Australia, future opportunities. *Field Crops Res.*, **26**, 141–169.

Richards, R.A. and Passioura, J.B. (1989) A breeding program to reduce the diameter of the major xylem vessel in the seminal roots of wheat and its effect on grain yield in rain-fed environments. *Aust. J. Agric. Res.*, **40**, 943–950.

Saab, I.N. and Sharp, R.E. (1989) Non-hydraulic signals from maize roots in drying soil: inhibition of leaf elongation but not stomatal conductance. *Planta*, **179**, 466–474.

Shackel, K.A., Matthews, M.A. and Morrison J.C. (1987) Dynamic relation between expansion and cellular turgor in growing grape (*Vitis vinifera* L.) leaves. *Plant Physiol.*, **84**, 1166–1171.

Siddique, K.H.M., Belford, R.K., Perry, M.W. and Tennant, D. (1989) Growth, development and light interception of old and modern wheat cultivars in a Mediterranean-type environment. *Aust. J. Agric. Res.*, **40**, 473–487.

Sinclair T.R. and Ludlow, M.M. (1985) Who taught plants thermodynamics? The unfulfilled potential of plant water potential. *Aust. J. Plant Physiol.*, **12**, 213–217.

Turner, N.C. and Nicolas, M.E. (1987) Drought resistance of wheat for light textured soils in a mediterranean climate. In: *Drought Tolerance in Winter Cereals* (eds J.P. Srivastava, E. Porceddu, E. Acevedo and S. Varma). Wiley, New York, pp. 203–216

Watson, D.J. (1952) Physiological basis for variation in yield. *Adv. Agron.*, **4**, 101–145.

Gas-exchange implications of isotopic variation in arid-land plants

James R. Ehleringer

17.1 Introduction

At some time during their life cycle, virtually all terrestrial plants are exposed to periods of water deficit that have an impact on leaf gas exchange and possibly also on plant growth. Plants in arid-land ecosystems experience frequent water deficits brought about by extremely low soil water content and high water stress. Deserts, perhaps more than any other ecosystem, are characterized by large shifts in soil moisture availability, ranging from saturating, flood-like conditions after intense rains to extreme soil water deficit during prolonged droughts.

Deserts can be defined as having an average annual precipitation of 250 mm or less, and as regions in which the potential evapotranspiration rate exceeds precipitation input. In the driest deserts, plants are not only exposed to very dry soils, but the duration of this drought is variable, because the predictability of precipitation is low. That is, the coefficient of variation of annual precipitation increases as average annual precipitation decreases. This pattern holds whether the precipitation is considered on an annual (Ehleringer and Mooney, 1983) or on a seasonal (Figure 17.1) basis. The coefficient of variation is a measure of unpredictability, indicating that as annual precipitation decreases along an environmental gradient, the probability of extremes increases, such as long drought lasting more than a year or more, interrupted by unusually wet periods, Plants occurring in arid lands are not only commonly exposed to droughted conditions in which evapotranspiration rates exceed precipitation, but the repeatability of these average environmental conditions from year to year is low, reducing the likelihood that a single adaptive pattern will be optimal. High variability in life-history,

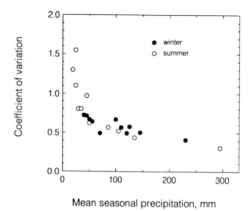

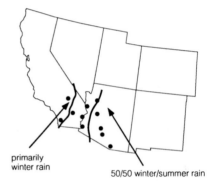

Figure 17.1. *The relationship between mean seasonal precipitation and the coefficient of variation for 10 locations in the arid regions of the south-western United States. Precipitation is divided in two seasons: winter (November–April) and summer (May–October) and is based on US Weather Bureau Records between 1929 and 1979. Shown on the right are the specific locations sampled.*

life-form and physiological parameters characterize many of the world's desert plant communities, and this variation may be an evolutionary response to temporal variation in water availability. It is beyond the scope of this chapter to evaluate plant performance from all deserts throughout the world, and so an effort is made to concentrate on the arid-land plants of western North America, although it is likely that the patterns discussed are applicable to all deserts.

Given the environmental constraints imposed by both low amounts of precipitation and the high interannual variability in that precipitation, how should plants respond? This chapter focuses on one particular aspect of that response, photosynthetic gas-exchange. It will become clear as this chapter develops that variation in particular gas-exchange characteristics are associated with a broader syndrome of metabolic and structural characteristics, and as a result, investigation of gas-exchange patterns can provide information on adaptive variation in life-history patterns in response to fluctuating and diverse arid-land environments. The overall theme is in understanding those key aspects of gas-exchange activity that are important determinates of plant fitness.

17.2 Intercellular CO_2 concentration as a set point for gas-exchange metabolism

17.2.1 *Flux versus set point*

Gas-exchange responses at the leaf level can be viewed from two perspectives: what changes absolute flux rates and what controls flux rates. Changes in maximum photosynthesis (A) and transpiration (E) rates in response to soil moisture availability (or any other measures of plant water status for that matter) have been described in numerous studies (Ehleringer and Mooney, 1983; Lange, *et al.*, 1976; Smith and Nobel, 1986; Smith and Nowak, 1990). While absolute photosynthesis and transpiration rates among species may exhibit substantial variation, these flux rates decrease as water stress increases. Variation in maximum flux capacities is loosely associated with life form, with annuals often having higher rates than perennials (Mooney and Gulmon, 1982; Mooney *et al.*, 1976). However, there are enough counter-examples of annuals having low photosynthetic capacities (Seeman *et al.*, 1980; Werk *et al.*, 1983) and perennials having high photosynthetic capacities (Ehleringer and Björkman, 1978) that generalities of this type cannot be drawn with a high degree of certainty. Changes in flux rates and canopy photosynthetic area almost always show a linear response to plant stress (e.g. water potential), and species vary widely in their capacity to maintain photosynthetic activity under water stress. As a consequence, instantaneous measures of gas exchange activity at a single point in time may provide limited insight into primary productivity and ultimate plant fitness, although the parameters are ultimately linked with each other (Figure 17.2). Continuous monitoring of photosynthetic activity would, of course, provide a stronger correlation with productivity, but is impractical on more than a few individuals. In order to view gas-exchange processes over a broader range of individuals and species, other measures are needed that can be more easily obtained on a wide range of individuals and which encompass an extended period of time.

An alternative approach to examining absolute flux rates and their impact on gas-exchange performance is to examine control points or set points in gas-exchange activity. Set points may be more stable than absolute flux rates, thereby providing more information in response to stress. That is, whereas flux rates will vary greatly in response to resource levels on the short term or to stress levels on the long term, changes in the set point may be substantially less. Photosynthesis requires the simultaneous inward diffusion of carbon dioxide from outside the leaf and its fixation into organic compounds by light and dark reactions within the chloroplast. One set point illustrated in Figure 17.2 is the intercellular CO_2 concentration (c_i), which represents a balance between rates of inward CO_2 diffusion (controlled by stomatal conductance) and CO_2 assimilation (controlled by photosynthetic light/dark reactions). In principle, there is no expected relationship between flux rate

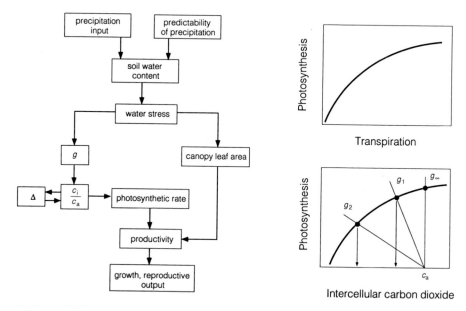

Figure 17.2. *A conceptual model of plant productivity; ultimately growth, reproductive output and plant fitness are influenced by water stress and several of the gas-exchange characters that influence photosynthetic rate. Δ, Carbon isotope discrimination; c_i and c_a are the intercellular and ambient CO_2 concentrations, respectively; E is the transpiration rate; and g is the leaf conductance to water vapour.*

and set point. A primary advantage of set point analysis over flux rate would arise if set points remained relatively fixed among plants under non-stressed conditions, and if, in response to abiotic stresses, there were no changes in the relative rankings of plants. A second advantage would be if there were methods for directly assessing this set point that were more easily measured than continuously monitoring absolute flux rates.

17.2.2 *Carbon isotope discrimination as a measure of intercellular carbon dioxide concentration*

Over extended time periods, the intercellular CO_2 concentration can be estimated through measurement of the carbon isotopic composition of plant material (Farquhar *et al.*, 1989). Carbon isotope discrimination (Δ) in C_3 plants is related to photosynthetic gas exchange; because Δ is in part determined by c_i/c_a, the ratio of CO_2 concentrations in the leaf intercellular spaces to that in the atmosphere (Farquhar *et al.*, 1982; Farquhar and Richards, 1984; Farquhar *et al.*, 1989; see Chapter 12). This ratio, c_i/c_a, differs among plants because of variation in stomatal opening (affecting the supply rate of CO_2), and because of variation in the chloroplast demand for CO_2. Of the models linking C_3 photosynthesis and $^{13}C/^{12}C$ composition, the one developed by Farquhar *et al.* (1982) has been the most extensively tested. In its simplest form, their expression for discrimination in leaves of C_3 plants is:

$$\Delta = a + (b - a)\frac{c_i}{c_a} \qquad 17.1$$

where a is the fractionation occurring due to diffusion in air (4.4‰), and b is the net fractionation caused by carboxylation (mainly discrimination by RuBP carboxylase, approximately 27‰). The result of these constant fractionation processes during photosynthesis is that the leaf carbon isotopic composition represents the assimilation-weighted intercellular CO_2 concentration during the lifetime of that tissue. Farquhar *et al.* (1989) and Ehleringer *et al.* (1992) summarize the data showing that Δ values of leaf material are a reliable estimate of c_i/c_a during the lifetime of that leaf for C_3 species.

The leaf carbon isotopic composition has been used to estimate water-use efficiency (ratio of photosynthesis to transpiration) in C_3 plants (Farquhar *et al.*, 1989). To associate the Δ value only with water-use efficiency is a mistake, since other water relations parameters are also directly related to the c_i value. For instance, both λ, a set point describing the optimal pattern of stomatal behaviour that maximizes carbon gain for a given amount of water loss (Cowan and Farquhar, 1977), and l_s, the extent of stomatal limitation on photosynthesis (Farquhar and Sharkey, 1982; Jones, 1985; see Chapter 12), are directly related to the c_i value of a leaf. Moreover, it is not clear just what water-use efficiency means to plant performance outside of an agronomic context. A more productive approach might be to consider c_i as a measure of the metabolic set point for gas exchange, providing an integrated measure of the multitude of factors that relate both CO_2 uptake and water loss in plants.

17.3 Field observations of carbon isotope discrimination

Is there any evidence to indicate that analysis of c_i values provides new information on gas-exchange metabolism of desert plants; in particular, information suggesting that c_i values are in some way measures of a metabolic set point and associated with specific life-history patterns or a syndrome of morphological and physiological characters?

17.3.1 *Community-level patterns*

In arid-land community-level analyses, carbon isotope discrimination values were related to life expectancy of that species. Ehleringer and Cooper (1988) observed that Δ values were inversely correlated with longevity for Sonoran Desert species. Carbon isotope discrimination values varied by more than 5‰ among different longevity groups (Figure 17.3). Smedley *et al.* (1991) observed a similar pattern between annuals and herbaceous perennials in a Great Basin grassland; similar trends can be extracted from the shrub-tree species data of

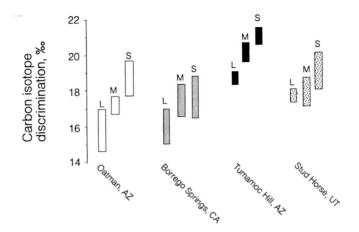

Figure 17.3. *Ranges of carbon isotope discrimination (Δ) values of short-lived (2–5 years), medium-lived (10–40 years) and long-lived (> 50 years) species at four different sites in the deserts of western North America. (Data are from Ehleringer and Cooper, 1988, and Ehleringer, unpublished.)*

DeLucia *et al.* (1988). More recently, Ehleringer (unpublished) extended his earlier observations, demonstrating that similar Δ-value rankings were maintained among different life-history groups across a broad range of desert communities throughout the Mojave and Sonoran Deserts.

Variations in leaf carbon isotope discrimination among species arise because of both genetic and environmental factors (Ehleringer *et al.*, 1993; Farquhar *et al.*, 1989) and it is not clear from Figure 17.3 just how large differences might be in the two components. Although the previous observations represented single snapshots-in-time of the distribution of Δ values within a community, variation in the rankings of species through time appears to be minimal. Ehleringer and Cook (1991) examined 12 species at the same site in the Sonoran Desert 3 years apart; they observed virtually no difference in the rankings among species over that period (Figure 17.4). From other studies with agricultural and rangeland species, carbon isotope discrimination measured over different time intervals within a single season or across seasons has shown interseasonal variation. This variation in the absolute Δ values has

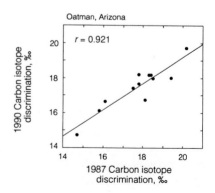

Figure 17.4. *Mean carbon isotope discrimination (Δ) values for different species in 1987 and 1990 at a Sonoran Desert site near Oatman, Arizona. (From Ehleringer and Cook, 1991.)*

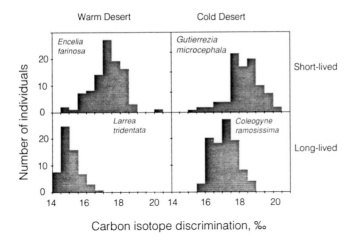

Figure 17.5. *Frequency distribution of carbon isotope discrimination values of subpopulations of species occurring in the Colorado Plateau (cold desert) and Sonoran Desert (warm desert) that differ in their life expectancy. (Data are from Schuster et al., 1992.)*

been attributed to abiotic acclimation, but there were no significant changes in the relative rankings of different plants (Ehleringer, 1990; Ehleringer *et al.*, 1990; Farquhar *et al.*, 1989; Johnson *et al.*, 1990). Thus, if Δ values are used as an indication of a photosynthetic gas-exchange set point, then the set point appears to be relatively stable over time in terms of differences that are maintained between species and/or cultivars within species.

17.3.2 *Intraspecific variation*

A greater understanding of the potential significance of Δ might come from looking at variation among individual plants. Evaluating interplant variation in gas-exchange characteristics has had limited success in the past, mostly because variations in characteristics, such as photosynthesis, are usually so small that it is difficult to detect individual differences and because equipment limitations restrict the number of simultaneous measures that can be obtained. However, stable isotopes offer an approach to overcome those limitations, because of the integrating nature of the measurement. Dawson and Ehleringer (1993), Geber and Dawson (1990) and Schuster *et al.* (1992) all observed significant intrapopulation variation in Δ values. In a study of the extent of population-level variance in Δ values of warm and cold desert ecosystems, Schuster *et al.* (1992) observed that variance was greater in populations of the shorter-lived species than in the longer-lived species (Figure 17.5). If there is a relationship between longevity and Δ value as implied in Figure 17.3, then lower Δ values would be expected in that longer-lived population. Over time, natural selection may favour genotypes with specific Δ values, which would then result in a narrower variance in the longer-lived species. Both trends

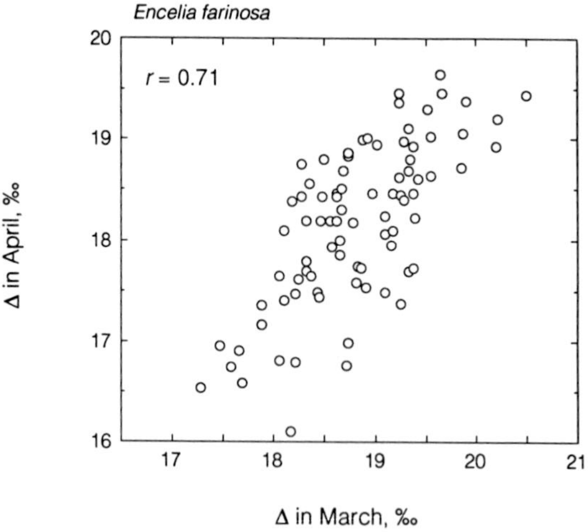

Figure 17.6. *Carbon isotope discrimination values for individuals of* Encelia farinosa *measured under natural conditions in early March and again on newly produced leaves in late April at a Sonoran Desert site near Oatman, Arizona. (Data are from Ehleringer, 1993.)*

are observed in the species comparisons, where variances in Δ values were 0.82 vs. 0.28 and 0.92 vs. 0.47 for the shorter- vs. longer-lived species, respectively, and population mean values were 1.1 and 2.2‰ higher than that of the longer-lived species in the cold- and warm-desert habitats, respectively.

As plants acclimate to variations in soil water availability, there can be changes in Δ values (Ehleringer *et al.*, 1993; Guy *et al.*, 1980; Smedley *et al.*, 1991). When newly produced leaves were sampled at two time intervals in spring on neighbouring plants within a community, Ehleringer (1993) noted that all individuals had somewhat lower Δ values on the second sample date (Figure 17.6). This response is likely to represent an acclimation to drought, as has been noted by others (Farquhar *et al.*, 1989; Guy *et al.*, 1980). With respect to genotype rankings, they remained essentially constant throughout the season. There was a strong, significant correlation between Δ values at the two sampling periods. Again this indicates that at both the intraspecific and interspecific levels (as shown in Figure 17.4), differences in Δ values among different individuals are maintained over time. One possible complication when interpreting these data is that seasonal variations in Δ values could be determined by either abiotic factors (acclimation in response to soil water stress development) or biotic factors (presence of neighbouring plants influencing local soil water conditions). Ehleringer (1993) examined long-term variations in Δ values in a monospecific stand of *Encelia farinosa* before and after neighbour removal (Figure 17.7). The rankings among plants remained constant, reinforcing the notion that Δ is a measure of the relative metabolic set point largely independent of neighbour effects.

The notion of Δ as a measure of set point and an indication of life-history patterns is further supported by studies within different agronomic species. Maturity date among cultivars under uniform garden conditions has been shown to be inversely related to Δ value for several crop species (Ehleringer *et al.*, 1990; Richards and Condon, 1993; White, 1993). Variation in flowering

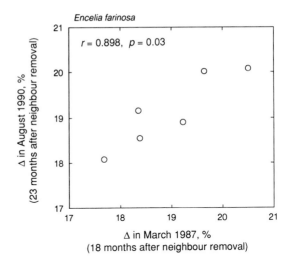

Figure 17.7. *A comparison of carbon isotope discrimination values measured on* Encelia farinosa *individuals in the field 18 months before neighbours had been removed and 23 months following neighbour removal at a Sonoran Desert site near Oatman, Arizona. (Data are from Ehleringer, 1993.)*

date likely reflects carbon gain, allocation and developmental rate differences among genotypes. Since the rankings of Δ values of different genotypes within a crop species remained constant through development within a growing season, the constant Δ rankings among genotypes suggest that the relative set point has remained fixed and would therefore be a useful comparative parameter when examining metabolic differences within a species. That Δ values are tightly correlated with small differences in flowering dates may be coincidental or, more likely, may be indicative of an overall metabolic and morphological syndrome, in which the characters associated with rapid growth and early maturity allow plants to complete their life cycle before the onset of stress.

17.4 Consequences of carbon isotope discrimination differences among neighbours

17.4.1 *Carbon isotope discrimination and growth*

Substantial variation in Δ values of plants within a natural population would arise if

(1) there are substantial differences in Δ values among populations and significant gene flow between populations, even if Δ values are unimportant to overall plant performance and not under any strong selective pressure; or

(2) there is either spatial or temporal variation in environmental characteristics within a small location that selects for divergent Δ values.

In the first possibility, genetic drift would be expected ultimately to fix Δ values within an isolated population, and variance within a population would only be maintained through gene flow with neighbouring populations. In both possibilities, variation in Δ values would be expected, but it is only the latter possibility that offers an explanation for differences in the variance of Δ values between longer- and shorter-lived species. Since Δ values are an indication of a set point for gas exchange that involves water loss, it is conceivable that some aspect of stomatal physiology, water-use efficiency, water-use maximization or drought tolerance is under strong selective pressure, particularly in arid lands.

While the evaporative gradient may place some restrictions on water loss rates, an increased efficiency in the use of soil water is only reasonable if plants can exert some control over the rates of soil water extraction from the soil volume in which its roots are located. If plants are competing for the same limited water, there may be selection against conservative water use, and for rapid resource capture, since that water cannot be stored internally, except in succulents. On this basis, arguments could be constructed to suggest that under competitive situations low Δ-value genotypes may be selected against, and under low competitive environments high Δ-value genotypes would be selected against. From studies of Mojave and Sonoran Desert plants, it is clear that not only is water present in limiting quantities, but that there is strong competition for that water resource (Fowler, 1986). Why then do plants persist that are not conservative in their use of water unless water-use efficiency and growth are unrelated?

Viewed from a different perspective, an alternative argument might be advanced that would lead to the opposite conclusion. For example, if the Δ value is a relatively fixed control point for metabolism, other aspects of growth or carbon allocation may be affected so as to allow the plant to maintain that Δ value. In particular, carbon allocation to root versus shoot growth may be influenced by a plant's Δ value. Virgona *et al.* (1990) provided evidence in support of this idea by showing that in sunflowers there was a strong positive correlation between Δ and the ratio of leaf area to root area (τ). Genotypes with high Δ values also tended to have high τ values.

Consider the possibility that carbon allocation patterns exhibited substantial plasticity, while the Δ value was relatively invariant. If the shoot to root allocation pattern was influenced by a plant's Δ value, then competition for limiting water should influence that pattern. Virgona *et al.* (1990) showed that under well-watered, single-pot conditions τ was positively correlated with Δ as suggested in Figure 17.8. If Δ is a relatively fixed control point, then under competition for water, plants with high Δ values will have to allocate proportionally more carbon to root growth in order to get sufficient water to maintain that Δ value, which is effectively describing the rate of carbon gain to water loss (Figure 17.8). Reallocating carbon to below-ground structures will result in a reduced above-ground growth rate, suggesting that high Δ-value genotypes would be selected against under water-competitive situations. If this model is correct, then high Δ-value plants

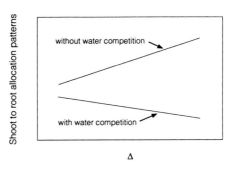

Figure 17.8. *Hypothesized relationship between shoot-to-root areas* (τ) *among genotypes of a species and leaf carbon isotope discrimination* (Δ) *as a function of soil moisture availability. The two lines represent hypothesized responses to situation with and without water competition.*

should exhibit greater plasticity in response to changes in water stress induced either by abiotic or biotic factors.

17.4.2 *Competitive interactions*

To address the possibility that high and low Δ-value genotypes would respond differentially to water availability, Ehleringer (1993) examined relationships between growth, water relations and Δ in monospecific stands of *Encelia farinosa* under natural and neighbour-removed situations. In a series of long-term observations, plant performance was measured on individuals varying in their Δ value. All neighbours were removed around some individuals for a distance of 2 m, which was thought to be the longest possible lateral root extension; other similar-sized plants remained with neighbours as controls. Growth rate was measured as the change in plant size during the experimental period, which was characterized as having average to above average precipitation inputs. While plants with neighbours exhibited essentially no growth during this period, irrespective of their Δ value, plants without neighbours grew substantially and growth was related to the plant's Δ value (Figure 17.9). These data indicate that plants not competing for limiting soil moisture grew at a faster above-ground rate, and that Δ was in some way associated with the capacity of a plant to respond to these less restricting soil moisture conditions. One possibility is that Δ is correlated with leaf conductance, as has been shown for a number of species (Condon *et al.*, 1987; Ehleringer, 1990; Ehleringer *et al.*, 1990). If that were the case, conductance-driven differences among plants could result in increased photosynthesis and growth of high Δ-value plants exposed to reduced water stress. An alternative but not exclusive explanation is that carbon-allocation patterns within plants changed in response to increased soil moisture availability, as hypothesized in Figure 17.8. The available data do not allow us to distinguish among these possibilities, but only to conclude that under reduced competition and greater soil moisture availability high Δ-value plants have a greater growth rate than low Δ-value plants. Ehleringer (1984) and Ehleringer and Clark (1988) showed that carbon gain and growth rate differences in *Encelia farinosa* translated into higher fitness values, indicating a selective

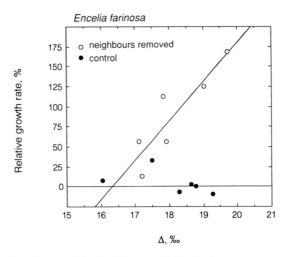

Figure 17.9. *Growth rates of* Encelia farinosa *individuals following neighbour removal and of control plants whose neighbours were not removed as a function of the carbon isotope discrimination value of the plant at a Sonoran Desert site near Oatman, Arizona. (Data are from Ehleringer, 1993.)*

advantage for high Δ-value plants under greater water availability conditions. The question still remains as to why, then, are low Δ-value individuals maintained within a population?

17.4.3 *Impact of long-term drought*

A high Δ value may not be without disadvantages, particularly if the Δ value is strongly correlated with leaf conductance to water loss. Tyree and Sperry (1989) have shown that under low water potentials cavitation events within the xylem increased, leading to a reduced capacity to conduct water (see also Chapter 7). Given equal stem hydraulic conductances between two genotypes, a relatively high stomatal conductance by one should result in an increased transpiration rate and a decreased leaf water potential in that genotype (Figure 17.10). As soil moisture availability decreases, the water potential gradient between leaf and root should increase, and ultimately, under extreme or prolonged drought stress, water potentials may reach the point at which cavitation events occur with high frequency (see Chapter 7). In theory, these cavitation events could restrict water flow sufficiently and result in stem death. This may in part be the explanation for why suffrutescent growth (many stems emerging from a common root base) is so common among arid-land shrubs.

Long-term droughts of 6–18 months occur frequently in arid lands and can result in high shrub mortality. Ehleringer (1993) monitored survival in E. *farinosa* shrubs differing in Δ values through a year-and-a-half drought in the Sonoran Desert. As with the previous competition experiment, plants differing in their Δ value were divided into two groups: control plants in a naturally

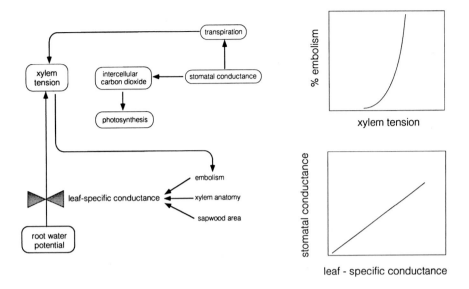

Figure 17.10. *A conceptual model of how higher intercellular CO_2 concentrations driven by increased stomatal conductances may contribute to enhanced xylem embolism rates, particularly under conditions of low root water potentials, as would be expected during periods of extreme soil water deficit. Shown at the right are the expected relationships between xylem tension and extent of embolism and between the leaf-specific conductance and stomatal conductance.*

occurring monospecific stand and individuals within that stand that had their neighbours removed. In response to the drought, there was significant mortality. More than 60% of the individuals in the control population died. On a neighbouring hillside used for long-term demographic studies, almost 50% of the E. *farinosa* population died (Ehleringer, unpublished). Of interest was that, irrespective of Δ value, none of the individuals without neighbours died during the drought. Only those plants competing for water (those with neighbours) died during the drought. Plant size (Student's t test, $p = 0.949$) was not a factor related to mortality or survival through the drought. However, mortality was significantly related to a shrub's Δ value. Plants surviving the drought had significantly lower Δ values than those that died (Student's t test, $p = 0.022$).

These data indicate an advantage to low Δ-value plants and suggest possible trade-offs between conditions favouring high and low Δ-value plants within a natural population. Genotypes with low Δ values may persist through long-term drought or competitive situations better than high Δ-value genotypes. However, under conditions of high resource availability, induced by either microsites without neighbours or possibly wet years, the high Δ-value genotypes significantly outperform the low Δ-value genotypes. Given the temporal variability in precipitation between years and spatial variability in microhabitat quality, variation in Δ values will be maintained in the population.

17.5 Implications of carbon isotopic variation for community structure

The idea of trade-offs associated with Δ value and habitat quality has many ramifications at both the intra- and interspecific levels. At the population level, the previous results imply that variability should be a function of the advantage to be gained by a high growth rate offset by the potential disadvantages associated with drought-induced mortality. Schuster *et al.* (1992) showed that the extent of variability with populations of longer- and shorter-lived desert species was inversely related to the life expectancy of that species. This is exactly the pattern expected in the predictions taken from Figures 17.2 and 17.10. Individuals of longer-lived species will be exposed to more drought events, reducing the likelihood that a high Δ-value individual could persist through time in the population.

At the ecotypic level, Comstock and Ehleringer (1992) have shown that variation in Δ values reflected shifts in habitat quality in *Hymenoclea salsola*, a common shrub in the Mojave and Sonoran Deserts. Under common garden conditions, the isotopic variation was greater than 2‰ and was negatively related to ω, the average leaf-to-air water vapour gradient weighted for periods when soil moisture was available (Figure 17.11). *Hymenoclea salsola* has both photosynthetic twigs and leaves, with twigs always having lower c_i and Δ values. Since leaves and twigs both have small diameters, resulting in

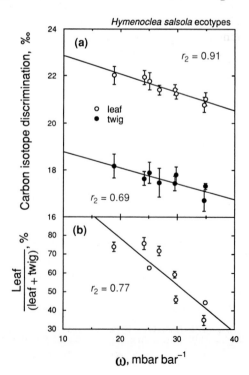

Figure 17.11. (*a*) *Carbon isotope ratio of leaves and twigs from different ecotypes of* Hymenoclea salsola *grown under common garden conditions as a function of ω, the leaf-to-air evaporative gradient weighted for seasonal precipitation input.* (*b*) *The ratio of leaf area as a proportion of total photosynthetic area plotted as a function of ω.*

strong convective exchange and equivalent tissue temperatures, twigs also always have a greater water-use efficiency (Comstock and Ehleringer, 1988). The fraction of leaf to twig photosynthetic areas is also negatively related to ω, resulting in plants from drier habitats (atmospheric drought) having both lower Δ values (higher water-use efficiencies) at the leaf level as well as a greater allocation to the more water-use efficient twig tissues in these environments. Overall, this results in a combined morphological–physiological progression towards canopies of greater water-use efficiency in climates with drier atmospheric conditions. This pattern of decreasing Δ values in plants from drier environments and an increased allocation to photosynthetic twigs is consistent with possible trade-offs between Δ, as a set point for gas exchange, and drought stress. The implication of the Comstock and Ehleringer (1992) study is that the seasonality of soil moisture inputs is important in affecting absolute Δ values; in desert habitats where precipitation occurred during the hotter summer months, plants had lower Δ values than from sites receiving equivalent amounts of precipitation during cooler winter–spring periods of the year. Implicit in this interpretation is that those ecotypes growing in summer-wet habitats have the capacity to utilize summer precipitation. For *H. salsola*, this is the case (Ehleringer and Cook, 1991).

17.6 Responses to seasonal moisture inputs

Recent evidence suggests that not all arid-land plants have the capacity to utilize summer moisture inputs. If summer moisture is a small fraction of total annual precipitation, then plants might not invest in shallow roots and thus would not be expected to utilize much of these infrequent precipitation events. That is, in arid regions with dry summers, carbon may be allocated for deep root growth; roots involved principally in water uptake are not active in the surface layers, but occur only in deeper soil layers where moisture is persistent during summer periods. Surface roots may be involved only in nutrient uptake and represent a relatively small proportion of the belowground structures. As such, water taken up with nutrients would not be detected since it represents such a small fraction of the total water uptake. When infrequent summer precipitation events do occur, the lack of an active upper root layer would prevent significant water uptake from that upper soil layer. As average summer precipitation increases along a precipitation cline, at some point precipitation becomes sufficiently predictable, or achieves some minimal threshold, so that plants develop a dimorphic root system (Figure 17.12) consisting of two zones of active roots, one in the upper soil layers capturing summer precipitation and a second deeper zone for utilizing the more reliable groundwater. The presence of a dimorphic root system depends upon the predictability of moisture sources and the costs associated with producing and maintaining these roots. Although fine root turnover at a high rate may represent a significant cost to plants in their carbon balance (see Chapters 10 and 13), differential root development for water uptake could put plants at

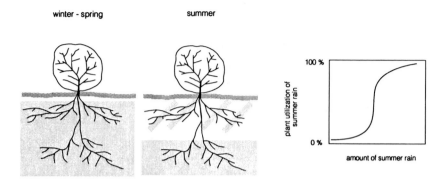

Figure 17.12. *Left: a drawing of the hypothesized distributions of utilized water by perennial shrubs in arid-land regions during winter–spring and summer growing seasons. Specifically, it is anticipated that shrubs growing in regions with infrequent summer rains will have limited capacity to utilize those precipitation events. Right: a conceptual model of the anticipated relationship between the capacity of a desert shrub to utilize summer precipitation as a function of the amount of summer precipitation received during the summer growing season.*

a selective advantage if the reliability of surface moisture is variable. Some roots must remain active near the surface, because of their role in active mineral nutrient uptake rather than water uptake *per se* (functional specialization).

Analysis of the hydrogen isotope composition (δD) of xylem sap in plants provides a succinct measure of water sources. Groundwater and recent precipitation usually differ markedly in δD signal because of the climatic factors controlling evaporation and precipitation at different times of the year depending on geographical location (Ehleringer *et al.*, 1993; Flanagan and Ehleringer, 1991). Flanagan and Ehleringer (1991) evaluated water uptake patterns in a semiarid pinyon–juniper ecosystem site in southern Utah, which received between 30 and 40% of its annual moisture during the summer months. *Chrysothamnus nauseosus* (rabbitbrush) and *Juniperus osteosperma* (Utah juniper) did not utilize summer precipitation during the year of study: the δD of xylem sap remained close to that of the groundwater (effectively the same as the winter recharge precipitation). In contrast, *Pinus edulis* (pinyon pine) and *Artemisia tridentata* (sagebrush) did utilize summer precipitation. In a follow-up study, Flanagan *et al.* (1992) showed that the Utah juniper would take up summer precipitation in some situations, but that the xylem sap of the rabbitbrush retained the same δD as the groundwater source.

Gregg (1991) provided evidence of a cline in the capacity of juniper trees to respond to summer moisture. Along a geographical gradient, where the fraction of summer precipitation varied between 18% and 40%, Gregg (1991) observed that juniper trees on sites with reliable summer precipitation utilized summer precipitation, whereas those on predictably dry summer sites did not respond to summer precipitation (Figure 17.13). It is not surprising that there is ecotypic variation in root structure along a geographical gradient, but the suggestion of a lack of an inducible response to summer rains (Gibson and

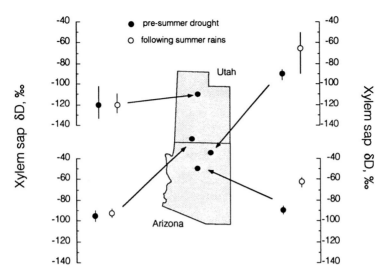

Figure 17.13. *Hydrogen isotope ratios of xylem sap from* Juniperus osteosperma *collected before and after summer rains at various locations along a north-to-south gradient of increasing summer precipitation in the Intermountain West of the United States. (Data are derived from Gregg, 1991.)*

Nobel, 1986) by these arid-land plants is unusual. In a related study from the Mediterranean-climate macchia of Italy, Valentini *et al.* (1992) observed that *Pistacia lentiscus*, *Phyllirea angustifolia* and *Quercus ilex* responded to summer precipitation, whereas *Q. pubescens* and *Q. cerris* utilized soil moisture from greater depths. Species that did not respond to summer precipitation had xylem sap δD values equivalent to that of the groundwater even after a summer rain event.

Another example of differential species response to summer precipitation was given by Ehleringer *et al.* (1991) in an investigation of water sources of desert species in southern Utah, at a site which received 45% of the annual precipitation during the summer. Whereas the annual species and the single crassulacean acid metabolism (CAM) succulent species within the desert community responded and fully utilized summer moisture inputs, that was not the case for perennial species. Herbaceous species utilized summer precipitation more than did woody perennials, and a number of perennials appeared not to utilize summer precipitation at all. Moisture at deeper depths was derived primarily from winter and spring recharge events, and this was reflected in the observed groundwater δD values. Spring measurements of water sources confirmed that all species were using the same water source at that time. Thus, it appears that annual, herbaceous and perennial species within the community compete for the same limiting source of water in one growing season (spring), but that herbaceous species have preferential access to a second water source (surface moisture) during the summer growing season.

17.7 Conclusions

Intercellular carbon dioxide concentration (c_i) can be viewed as a set point for gas-exchange metabolism. The carbon isotope discrimination value (Δ) of a leaf is the assimilation-weighted estimate of c_i. Field examination of Δ values in desert plant species reveals a negative correlation with life expectancy. Intraspecific variation in leaf Δ values appears to represent trade-offs between increased growth capacity of high Δ-value genotypes under water-sufficient conditions and increased survival of low Δ-value genotypes under water-limited conditions.

References

Comstock, J.P. and Ehleringer, J.R. (1988) Contrasting photosynthetic behavior in leaves and twigs of *Hymenoclea salsola*, a green-twigged, warm desert shrub. *Am. J. Bot.,* **75,** 1360–1370.

Comstock, J.P. and Ehleringer, J.R. (1992) Correlating genetic variation in carbon isotopic composition with complex climatic gradients. *Proc. Natl. Acad. Sci., USA,* **89,** 7747–7751.

Condon, A.G., Richards, R.A. and Farquhar, G.D. (1987) Carbon isotope discrimination is positively correlated with grain yield and dry matter production in field-grown wheat. *Crop Sci.,* **27,** 996–1001.

Cowan, I.R. and Farquhar, G.D. (1977) Stomatal function in relation to leaf metabolism and environment. *Symp. Soc. Exp. Biol.,* **31,** 471–505.

Dawson, T.E. and Ehleringer, J.R. (1993) Gender-specific physiology, carbon isotope discrimination, and habitat distribution in boxelder, *Acer negundo. Ecology,* **74,** 798–815.

DeLucia, E.H., Schlesinger, W.H. and Billings, W.D. (1988) Water relations and the maintenance of Sierran conifers on hydrothermally altered rock. *Ecology,* **69,** 303–311.

Ehleringer, J.R. (1984) Intraspecific competitive effects on water relations, growth, and reproduction in *Encelia farinosa. Oecologia,* **63,** 153–158.

Ehleringer, J.R. (1990) Correlations between carbon isotope discrimination and leaf conductance to water vapor in common beams. *Plant Physiol.,* **93,** 1422–1425.

Ehleringer, J.R. (1993) Carbon isotope variation in *Encelia farinosa*: implications for growth, competition, and drought survival. *Oecologia,* in review.

Ehleringer, J.R., and Björkman, O. (1978) A comparison of photosynthetic characteristics of *Encelia* species possessing glabrous and pubescent leaves. *Plant Physiol.,* **62,** 185–190.

Ehleringer, J.R. and Clark, C. (1988) Evolution and adaptation in *Encelia* (Asteraceae). In: *Plant Evolutionary Biology* (eds L. Gottlieb and S. Jain). Chapman and Hall, London, pp. 221–248.

Ehleringer, J.R. and Cook, C.S. (1991) Carbon isotope discrimination and xylem D/H ratios in desert plants. In: *Stable Isotopes in Plant Nutrition, Soil Fertility, and Environmental Studies.* IAEA, Vienna, pp. 489–497.

Ehleringer, J.R. and Cooper, T.A. (1988) Correlations between carbon isotope ratio and microhabitat in desert plants. *Oecologia,* **76,** 562–566.

Ehleringer, J.R. and Mooney, H.A. (1983) Photosynthesis and productivity of desert and Mediterranean climate plants. In: *Encyclopedia of Plant Physiology (New Series),* Vol. 12D. Springer, New York, pp. 205–231.

Ehleringer, J.R., White, J.W., Johnson, D.A. and Brick, M. (1990) Carbon isotope discrimination, photosynthetic gas exchange, and water-use efficiency in common bean and range gasses. *Acta Oecologia,* **11,** 611-625.

Ehleringer, J.R., Phillips, S.L., Schuster, W.F.S. and Sandquist, D.R. (1991) Differential utilization of summer rains by desert plants: implications for competition and climate change. *Oecologia,* **88,** 430–434.

Ehleringer, J.R., Phillips, S.L. and Comstock, J.P. (1992) Seasonal variation in the carbon isotopic composition of desert plants. *Funct. Ecol.,* **6,** 396–404.

Ehleringer, J.R., Hall, A.E. and Farquhar, G.D. (eds) (1993) *Stable Isotopes and Plant Carbon–Water Relations.* Academic Press, San Diego, in press.

Farquhar, G.D. and Richards, R.A. (1984) Isotopic composition of plant carbon correlates with water-use efficiency of wheat genotypes. *Aust. J. Plant Physiol.,* **11,** 539–552.

Farquhar, G.D. and Sharkey, T.D. (1982) Stomatal conductance and photosynthesis. *Ann. Rev. Plant Physiol.,* **33,** 317–345.

Farquhar, G.D., O'Leary, M.H. and Berry, J.A. (1982) On the relationship between carbon isotope discrimination and the intercellular carbon dioxide concentration in leaves. *Aust. J. Plant Physiol.,* **9,** 121–137.

Farquhar, G.D., Ehleringer, J.R. and Hubick, K.T. (1989) Carbon isotope discrimination and photosynthesis. *Ann. Rev. Plant Physiol. Plant Mol. Biol.,* **40,** 503–537.

Flanagan, L.B. and Ehleringer, J.R. (1991) Stable isotope composition of stem and leaf water: applications to the study of plant water-use. *Funct. Ecol.,* **5,** 270–277.

Flanagan, L.B., Ehleringer, J.R. and Marshall, J.D. (1992) Differential uptake of summer precipitation and groundwater among co-occurring trees and shrubs in the southwestern United States. *Plant Cell Environ.,* **15,** 831–836.

Fowler, N. (1986) The role of competition in plant communities in arid and semi-arid regions. *Ann. Rev. Ecol. System.,* **17,** 89–110.

Geber, M.A. and Dawson, T.E. (1990) Genetic variation in and covariation between leaf gas exchange, morphology, and development in *Polygonum arenastrum,* an annual plant. *Oecologia,* **85,** 53–158.

Gibson, A.C. and Nobel, P.S. (1986) *The Cactus Primer.* Harvard University Press, Cambridge, MA.

Gregg, J. (1991) The differential occurrence of the mistletoe, *Phoradendron juniperinum,* on its host, *Juniperus osteosperma* in the Western United States. M.Sc. Thesis, University of Utah, Salt Lake City.

Guy, R.D., Reid, D.M. and Krouse, H.R. (1980) Shifts in carbon isotope ratios of two C_3 halophytes under natural and artificial conditions. *Oecologia,* **44,** 241–247.

Johnson, D.A., Asay, K.H., Tieszen, L.L., Ehleringer, J.R. and Jefferson, P.G. (1990) Carbon isotope discrimination – potential in screening cool-season grasses for water-limited environments. *Crop Science,* **30,** 338–343.

Jones, H.G. (1992) *Plants and Microclimate.* Cambridge University Press, Cambridge, UK.

Lange, O.L., Kappen, L. and Schulze, E.-D. (1976) *Water and Plant Life: Problems and Modern Approaches.* Springer, Berlin.

Mooney, H.A. and Gulmon, S.L. (1982) Constraints on leaf structure and function in reference to herbivory. *BioScience,* **32,** 198–206.

Mooney, H.A., Ehleringer, J.R. and Berry, J.A. (1976) High photosynthetic capacity of a winter annual in Death Valley. *Science,* **194,** 322–324.

Richards, R.A. and Condon, A.G. (1993) Challenges ahead in using carbon isotope discrimination in plant breeding programs. In: *Perspectives of Plant Carbon and Water Relations From Stable Isotopes* (eds J.R. Ehleringer, A.E. Hall and G.D. Farquhar). Academic Press, San Diego, in press.

Schuster, W.S.F., Sandquist, D.R., Phillips, S.L. and Ehleringer, J.R. (1992) Comparisons of carbon isotope discrimination in populations of aridland plant species differing in lifespan. *Oecologia,* **91,** 332–337.

Seemann, J.R., Field, C.B. and Berry, J.A. (1980) Photosynthetic capacity of desert winter annuals measured *in situ. Carnegie Institution of Washington Yearbook*, pp. 146–147.

Smedley, M.P., Dawson, T.E., Comstock, J.P., Donovan, L.A., Sherrill, D.E., Cook, C.S. and Ehleringer, J.R. (1991) Seasonal carbon isotope discrimination in a grassland community. *Oecologia,* **85,** 314–320.

Smith, S.D. and Nobel, P.S. (1986) Deserts. In: *Photosynthesis in Contrasting Environments* (eds N.R. Baker and S.P. Long). Elsevier Science Publishers, Amsterdam, pp. 13–61.

Smith, S.D. and Nowak, R.S. (1990) Ecophysiology of plants in the intermountain lowlands. In: *Plant Biology of the Basin and Range* (eds C.B. Osmond, L.F. Pitelka and G.M. Hidy). Springer, New York, pp. 179–241.

Tyree, M.T. and Sperry, J.S. (1989) Vulnerability of xylem to cavitation and embolism. *Ann. Rev. Plant Physiol. Plant Mol. Biol.* **40,** 19–38.

Valentini, R., Scaracia Mugnozza, G.E. and Ehleringer, J.R. (1992) Hydrogen and carbon isotope ratios of selected species of a Mediterranean macchia ecosystem. *Funct. Ecol.,* **6,** 627–631.

Virgona, J.M., Hubick, K.T., Rawson, H.M., Farquhar, G.D. and Downes, R.W. (1990) Genotypic variation in transpiration efficiency, carbon-isotope discrimination and carbon allocation during early growth in sunflower. *Aust. J. Plant Physiol.,* **17,** 207–214.

Wek, K.S., Ehleringer, J., Forseth, I.N. and Cook, C.S. (1983) Photosynthetic characteristics of Sonoran Desert winter annuals. *Oecologia,* **59,** 101–105.

White, J. (1993) Implications of carbon isotope discrimination studies for breeding common bean under water deficits. In: *Perspectives of Plant Carbon and Water Relations from Stable Isotopes* (eds J.R. Ehleringer, A.E. Hall and G.D. Farquhar). Academic Press, San Diego, in press.

18

Water losses of crowns, canopies and communities

P.G. Jarvis

18.1 Introduction

The partitioning of absorbed energy by vegetation, and hence the fluxes of heat and water vapour and the utilization of soil water, depends on certain properties of vegetation. These properties include area of leaf present, spatial distribution, inclination angle, distribution and optical properties of leaves, as well as the conductance of the stomata, as influenced by photon flux, saturation deficit and CO_2 concentration *at the leaf surface*, and soil water content. Several of these properties are influenced by the commonly accepted aspects of global change, including the increase in atmospheric CO_2 concentration and likely changes in seasonal temperatures.

The evaporation of water from vegetation is of considerable significance from physiological, ecological and hydrological points of view, and has recently become of national importance as a result of the recent prolonged drought in the southern UK (1987–1992) and of possible international importance as a result of global change. The evaporation of intercepted precipitation and the transpiration of a forest represent a large loss of water from a catchment area (Calder, 1990). These losses become a vital concern in times of drought, when water has to be conserved for agricultural, recreational, domestic and industrial purposes. Thus it is important to establish the evaporation losses from different kinds of vegetation in relation to environmental variables and to determine the constraints on evaporation and transpiration resulting from the physiological and structural properties of the vegetation.

Understanding evaporation and transpiration is also relevant to meteorological modelling of weather systems, since the interaction of vegetation with the atmosphere determines both the entrainment of warm and dry air into

Scales of Observation

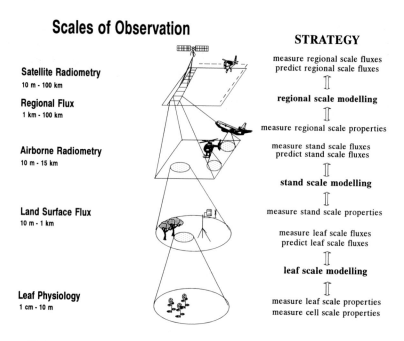

STRATEGY

Satellite Radiometry
10 m - 100 km

measure regional scale fluxes
predict regional scale fluxes

⇕

Regional Flux
1 km - 100 km

regional scale modelling

⇕

measure regional scale properties

Airborne Radiometry
10 m - 15 km

measure stand scale fluxes
predict stand scale fluxes

⇕

stand scale modelling

⇕

Land Surface Flux
10 m - 1 km

measure stand scale properties

measure leaf scale fluxes
predict leaf scale fluxes

⇕

leaf scale modelling

⇕

Leaf Physiology
1 cm - 10 m

measure leaf scale properties
measure cell scale properties

Figure 18.1. Scales of observation and a strategy for associated research.

the mixed layer from above and the input of heat and water vapour into the surface layer from the vegetation. For these reasons in particular, accurate representation of evaporation and transpiration from vegetation is required for global climate models.

Investigations of transpiration are frequently made by plant physiologists in the laboratory on parts of leaves, whole leaves and whole plants in controlled or semi-controlled environmental conditions, and this is reflected in much of the content of this volume. But the investigation of transpiration of individual plants, crops, plantations and forests *in the field* is of crucial importance in determining whether the controls on transpiration established in the laboratory are effective at the larger scale in the natural environment. Because of the rising interest in the impact of global change on all aspects of our society, the emphasis has shifted in recent years to the investigation of processes at the scale of individual plants, communities and stands of vegetation, and landscapes made up of a mosaic of vegetation types as functional units. This involves the use of models to integrate processes measured at one scale up to the next scale, the testing of those models by measurements at the larger scale and consideration of the additional, largely negative, feedbacks that come into play as the scale increases. The basic philosophy underlying this approach is illustrated in Figure 18.1.

In the past, measurements of evaporation and transpiration have, on the whole, been confined to particular scales. Studies of evaporation have been largely at the scale of the stand and have been confined to uniform, often monospecific stands, and associated models have been at the same scale (e.g. Rutter *et al.*, 1975). Investigations of transpiration have frequently been at the single-leaf scale or the potted-plant scale in the laboratory and, less frequently, at the scale of the community or stand in the field, but individual plants have received little attention in the field because of difficulties of measurement. In the past 10 years, aircraft have been used to measure water-vapour fluxes from vegetation at a landscape scale, but the absolute accuracy of the fluxes is in some doubt. So far, few attempts have been made to model landscape-scale fluxes from a knowledge of fluxes from the different areas of vegetation in the landscape. Thus, the coherence of the strategy presented in Figure 18.1 is at present still fragmentary.

This chapter comprises an assessment of the measurement of evaporation and transpiration by individual trees and by stands of trees, drawing on new developments and recent work and utilizing some older work not previously published.

18.2 Evaporation of intercepted water

The direct evaporation of water intercepted by foliage back to the atmosphere is an under-appreciated component of site water balance. In climates in which the rainfall comes in frequent, low-intensity episodes, significant amounts of intercepted water may be evaporated back to the atmosphere without ever reaching the soil. This is particularly the case in windy climates and for forests, which are almost universally well coupled to the atmosphere because of their extreme surface roughness (McNaughton and Jarvis, 1983). In the UK uplands, for example, the average annual interception loss from closed, mainly coniferous forest canopies is about 35% of the annual precipitation, and at some low-rainfall sites this may exceed 40% (Calder, 1990). Comparable annual figures for lowland deciduous woodlands are lower by about 15% (Harding *et al.*, 1992). In climates where the rainfall comes in short, intensive events, and for less rough agricultural crops and native vegetation, the interception loss is much less and may be verging on the insignificant in relation to transpiration, but water intercepted by the foliage of trees and forests is generally returned rapidly to the atmosphere. So far, however, the roles of crown structure, tree spacing and stand structure in influencing the rates of evaporation of intercepted water have not been widely investigated.

On a canopy scale, evaporation of intercepted water can be determined by solving the Penman equation, by measuring the eddy flux directly using the Bowen ratio or eddy covariance methods, or by difference in a mass balance of precipitation and throughfall plus stem flow. For individual or widely spaced trees, however, these methods are difficult to apply.

18.2.1 *Evaporation from the individual tree*

Recently, we devised a method for measuring evaporation and deriving the aerodynamic conductance of individual trees as a function of spacing in an agroforestry system by measuring the weight loss of wetted trees into an atmosphere of appreciable saturation deficit (Teklehaimanot and Jarvis, 1991).

Three spacing treatments were created in an existing stand of Sitka spruce (*Picea sitchensis* (Bong.) Carr.) planted at 2-m spacing to give four 2.4-ha plots with distances of 2, 4, 6 and 8 m between trees (see Teklehaimanot *et al.*, 1991). A complete tree was cut from a nearby stand and hung from a load cell suspended beneath a 16-m-high tripod, and water was sprayed onto the tree from fine spray nozzles. The surface temperature of the tree was measured with thermocouples and with a hand-held radiation thermometer, and nearby atmospheric saturation deficit was measured with an aspirated thermocouple psychrometer. The tripod was moved successively into each spacing, where three successful sets of measurements were completed.

The tree crown was sprayed until thoroughly wet. When spraying stopped there was a period during which water dripped off and this was followed by a period of constant evaporation, during which the tree surface temperature equalled the wet bulb temperature (Figure 18.2). As the crown began to dry, the needle temperature increased until it was again very close to air temperature.

Boundary layer conductance of the trees for the mass transport of water vapour from leaf surface to atmosphere was calculated over the period of steady evaporation from the rate of change in mass of the wet tree and the difference in the partial pressure of water vapour between the tree surface and the free atmosphere at the nearby psychrometer (Equation 18.A1, Appendix I). On a ground-area basis, the stand boundary layer conductance, g_A, of course decreased with decrease in the number of trees, but by no means *pro rata* (Figure 18.3a), whereas boundary layer conductance per tree crown, g_{at}, increased linearly with spacing between trees over the range 2–8 m.

The key result, that g_{at} increased as a function of tree spacing (Figure 18.3b) indicates enhanced turbulent exchange of intercepted water at the wider spacings. This was borne out by studies of turbulent structure made in the same stands and in models of the stands in a windtunnel. Green (1990) showed that below-canopy wind speeds increased with spacing and that there were associated increases in a number of turbulence parameters, such as tangential momentum stress (u/u_*) and turbulence velocity components (σ/u_*), both during the day and at night, in the stands and in modelled stands in the windtunnel (Figure 18.4).

Aside from the practical significance of these results, this experiment has also been very rich in ancillary results that contribute to the theory of interception loss.

The Penman equation for evaporation from wet foliage, E_I, may be written as:

$$E_I = \frac{\varepsilon(Q_n - S)}{(\varepsilon + 1)\lambda} + \frac{g_A D_a}{(\varepsilon + 1)P_a},$$

18.1

where Q_n is the net radiation absorbed, S is soil heat flux, D_a is water-vapour

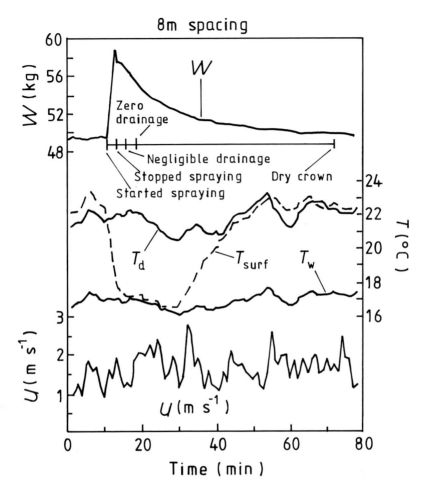

Figure 18.2. *An example from an experiment on one tree in the 8-m-spacing treatment of the change in weight (W) of the wetted tree, dry bulb (air) temperature (T$_d$), wet bulb temperature (T$_w$), mean leaf surface temperature (T$_{surf}$) and windspeed 2 m above the crown (u). (From Teklehaimanot and Jarvis, 1991.)*

saturation deficit of the air, P_a is atmospheric pressure, λ is the molar latent heat of vaporization of water and ε is a coefficient for change in the sensible and latent heat contents of air with respect to temperature. (With constants expressed per mol, rather than per kg, and with conductances in units of mol m^{-2} s^{-1}, this and succeeding equations give the flux density in mol m^{-2} s^{-1}. Data taken from publications, however, are usually expressed in their original units.)

Although this equation can be solved readily for continuous canopies, it is difficult to apply to single trees because of the difficulty in measuring the net radiation absorbed. One of the most useful observations from this series of experiments is that the temperature of the canopy fell to the wet bulb

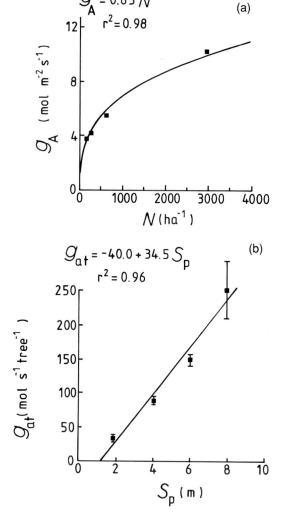

Figure 18.3. *The relationship between (a) stand boundary layer conductance (g_A) per hectare (N) (from Teklehaimanot and Jarvis, 1991) and (b) the linear relationship between boundary layer conductance per tree crown (g_{at}) and spacing distance between trees. (From Teklehaimanot and Jarvis, 1991.)*

temperature during the period of steady evaporation. It is readily shown, through the application of psychrometer theory, that this coincidence leads to a simpler equation for the evaporation of intercepted water that does not include the term containing net radiation (see Appendix I for a derivation), i.e.:

$$E_I = \frac{g_A \mathbf{D}_a}{(\varepsilon + 1) P_a}.$$

18.2

Thus the tree with a wet canopy can be regarded as *the big psychrometer tree*.

We were able to show unequivocally that the rate of evaporation of water from the wet crown was adequately described by this attenuated form of the

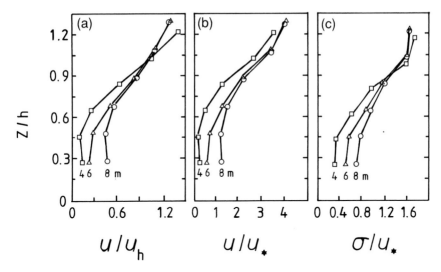

Figure 18.4 *The spatially averaged vertical distributions of (a) mean relative windspeed (u/u$_h$), (b) tangential momentum stress (u/u$_*$), and (c) turbulence velocity (σ/u$_*$) in relation to height (z), relative to tree height (h) at three tree spacings during daytime periods. u is mean horizontal windspeed with standard deviation σ and u$_*$ is the friction velocity. (From Green, 1990.)*

Penman equation—a not unexpected, but useful demonstration. Secondly, we showed that as the crown dried out the rate of evaporation was successfully estimated from the Penman rate for the wet canopy multiplied by W_C/W_S, where W_C is the mass of water on the canopy at the time and W_S is the mass of water stored on the wet canopy after drip has ceased, i.e. the so-called canopy storage capacity. Thus we also verified two key assumptions in the Rutter model of canopy evaporation (Rutter, 1975), as well as defining the dependence of boundary layer conductance on spacing.

18.2.2 *Evaporation from the woodland canopy*

The mass balance method has been widely used for measuring interception loss from closed canopy woodlands of a range of species, both conifers (e.g. Gash *et al.*, 1980) and broadleaves (Harding *et al.*, 1992), in both temperate and tropical forests (Lloyd *et al.*, 1988), and we applied this method to the same stands (Teklehaimanot *et al.*, 1991). Throughfall was measured by point sampling and by Calder sheet gauges (Calder, 1990), and stem flow by collars on a sample of trees stratified by basal area. The use of some tipping bucket gauges allowed the wetting-up and drying periods of individual storms to be delimited. The interception loss over these periods was equated with the evaporation of intercepted water and, by inversion of the attenuated Penman equation (Equation 18.2), aerodynamic conductances were obtained as a function of spacing, in general agreement with those obtained from the hanging-tree method (Table 18.1).

Table 18.1. *Mean weekly results of the mass balance over 10 months. Conductances were obtained by inverting Equation 18.2 (from Teklehaimanot et al., 1991)*

Spacing (m)	2	4	6	8
No. of trees (ha^{-1})	3000	625	277	156
Throughfall (mm)	14.1	19.4	22.1	23.5
Stem flow (mm)	4.34	0.75	0.26	0.13
Interception loss (mm)	7.57	5.87	3.61	2.32
Interception loss (% gross precipitation)	32.9	24.5	14.9	9.2
g_{at} (mol s^{-1} tree^{-1})	25	89	151	180
g_A (mol m^{-2} s^{-1})	7.4	5.6	4.2	2.8

The mass balance method gives an integrated measure of interception loss over periods of days. In principle, the evaporation of intercepted water can be measured over short time-scales of an hour or less, using the micrometeorological methods of Bowen ratio and eddy covariance. However, in practice these methods are difficult to use in wet conditions for several reasons, including: water on instruments such as the net radiometer and sonic anemometer, water droplets or condensation in tubes, sensors with water-sensitive 'windows', very small saturation deficits (<0.1 kPa) and very small gradients of specific humidity. None the less, these methods have been used successfully

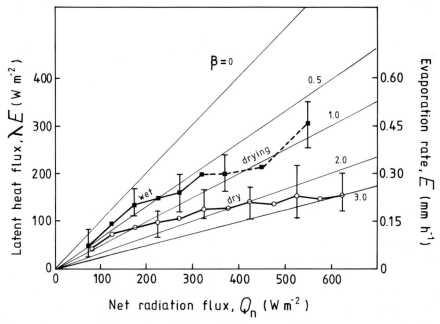

Figure 18.5. *The rate of evaporation of intercepted water from wet and drying foliage (dotted line) (■) compared with the transpiration rate of dry foliage (○) in similar conditions of net radiation over 50 summer days. Each point is the median of 15 to 30 values and the bars indicate interquartile ranges. The lines are of constant Bowen ratio (β) with the values indicated.*

both during and immediately after rainfall has ceased, but when the canopy, and indeed the entire landscape, is still saturated with water (e.g. Stewart, 1977). Figure 18.5 shows that the rate of evaporation from a wet, closed canopy of Sitka spruce forest, similar to the unthinned, 2-m-spacing stand referred to earlier, takes place at a rate somewhat higher than the rate of transpiration from the same canopy with the same net radiation and with a saturation deficit over 10 times larger. At the higher net radiation, the wet canopy is drying out, as indicated by the move towards higher values of Bowen ratio, β. Full details of the site, the instrumentation and the method are given by James (1977) and Jarvis (1993).

18.3 Transpiration

Transpiration can essentially be regarded as either liquid flow within the stem or vapour flux from the canopy. Recently there has been a revival of interest in concepts associated with both of these processes as a result of new and improved technology. Several tracer methods for liquid flow in individual stems have been developed and are now widely used, and there have been several innovations in methods for measuring vapour flux from both individual trees and canopies. These improvements in methodology are proving very useful in both supporting and stimulating new ideas about the transpiration process.

The simplest possible equation for transpiration (Monteith, 1963) expresses transpiration, E, as a function of the saturation deficit *at the surface*, D_{surf}, and implicitly defines what we mean by the conductance of the surface, g_{surf}, i.e.

$$E = g_{surf} D_{surf} / P_a. \qquad 18.3$$

This equation presupposes independent knowledge of the role of the boundary layer in determining the saturation deficit at the surface, but is particularly valuable in emphasizing the essential simplicity of the process. Much of the apparent complexity associated with formulations of transpiration in the vapour phase, as evident in the Penman–Monteith equation, results from treatment of processes in the boundary layers that either implicitly or explicitly lead to the derivation of the saturation deficit at the surface. A derivation of saturation deficit at the surface from that in the ambient air, as a function of the temperature difference between surface and air and the relative magnitudes of stomatal and boundary layer conductances, is given in Appendix II. In studying physiological controls on the fluxes of water vapour (and other scalars), there is much to be gained from relating fluxes to the concentration, or mole fraction, of the scalar at the surface, rather than at some ill-defined point in the surroundings (Aphalo and Jarvis, 1991, 1993; Bunce, 1985).

The Penman–Monteith equation for transpiration by leaf or canopy (Monteith, 1965) can be written as the weighted sum of the equilibrium (E_{eq}) and imposed (E_{imp}) transpiration rates (Jarvis and McNaughton, 1986):

$$E = \Omega E_{eq} + (1 - \Omega)E_{imp}$$

$$= \Omega \frac{\varepsilon}{(\varepsilon + 1)} \frac{Q_n - S}{\lambda} + (1 - \Omega)g_{surf}D_a/P_a, \qquad 18.4$$

where Ω, the decoupling coefficient, is defined as

$$\Omega = \frac{\varepsilon + 1}{\varepsilon + 1 + g_A/g_{surf}} \qquad 18.5$$

and varies between limits of 1.0 for a surface completely decoupled from the ambient atmosphere and 0 for a surface that is tightly coupled to the ambient atmosphere, so that $D_{surf} = D_a$. An essential feature of Penman's derivation was the inclusion of the *net radiation absorbed*, rather than the surface temperature, because the net radiation absorbed by a continuous crop canopy is easier to measure or to estimate than its temperature. (As pointed out by Martin (1989), for conceptual completeness these equations should be written in terms of isothermal net radiation and contain a thermal radiation conductance term: this has been done by McNaughton and Jarvis (1991), but with some loss of clarity.) Expressing transpiration by an equation of this form provides a means of evaluating the relative importance of net radiation and saturation deficit as driving variables, and allows an estimate to be made of the likely consequence of error in the net radiation term or of ignoring it completely. For aerodynamically rough surfaces (woodlands, agroforestry systems) in most climates, the value of Ω lies within the range of about 0.1–0.3, so we may presume that the term containing net radiation can be ignored with impunity for such surfaces when measurement of net radiation is difficult. However, this presumption clearly requires substantiation whenever it is used.

18.3.1 *Transpiration from the individual tree*

A plethora of tracers for liquid water flow in the stems of trees are in current use or have been in recent use. These include heat flux, heat balance, ^{32}P, tritium and deuterium (e.g. Calder, 1992; Waring et al., 1980). Tracer methods for measuring liquid flow in the stems of trees have been widely criticized on the grounds that most such methods yield linear velocities rather than volume flows (see also Chapters 6 and 7). However, volume flows are directly obtainable without empirical calibration if the tracer flux is integrated across the radius of the stem and account is taken of the volume fractions of air, water and wood at the time of the measurement. The heat-pulse velocity method (Edwards, 1980) works well in diffuse-porous stems and in ring-porous conifers over about 2.5 cm in diameter and, if proper care is taken with respect to probe depth, in ring-porous broadleaves as well. Volume flow measured in this way on a range of tree species growing on lysimeters has been found to agree well with weight loss (Figure 18.6) and also, in a range

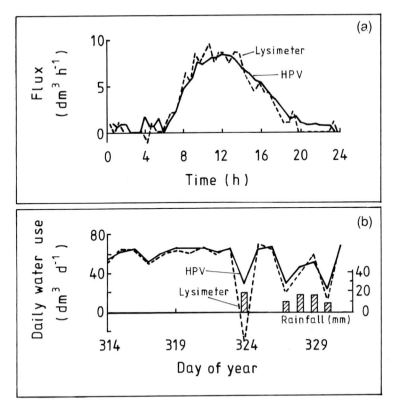

Figure 18.6. *(a) The diurnal curve of transpiration of a* Pinus radiata *tree in a lysimeter measured as weight loss and as the water flux in the stem, measured by the heat-pulse method. (b) A comparison between daily totals of water loss by the tree measured by the lysimeter and by the heat-pulse velocity method (HPV). When it rains, the lysimeter measures the gain in weight, whereas the heat pulse velocity method continues to measure transpiration. (Unpublished data of W. R. N. Edwards.)*

of circumstances, with vapour flux calculated from the Penman–Monteith equation using measured values of stomatal conductance, or determined by Bowen ratio and eddy covariance methods. This method is currently proving extremely useful in evaluating water use by trees in windbreaks and other agroforestry systems in the Sahel (Figure 18.7, Table 18.2).

An alternative method of measuring transpiration from individual trees is to estimate the vapour flux by solution of Equation 18.4. Although measurements of absorbed net radiation and stomatal conductance are certainly not routinely practical in most circumstances, recent attempts to measure all the components of Equation 18.4 in isolated trees are proving very informative in evaluating the relative importance of the driving variables.

The measurement of net absorbed radiation by irregular vegetation is difficult. A horizontally uniform crop or plantation presents a special case for which the measurement is relatively simple: a net radiometer is exposed a few metres above the crop in the horizontal plane and measures the net

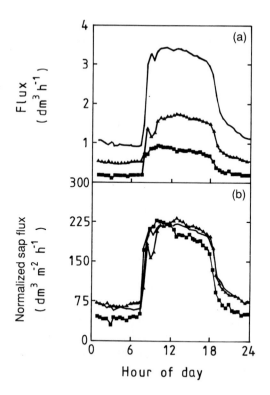

Figure 18.7. *(a) Diurnal sap fluxes for three trees of neem (Azardirachta indica) in a windbreak at the beginning of the dry season in the Sahel, measured by the heat-pulse method. (b) Sap fluxes normalized with respect to stem basal area at 130 cm. (From Brenner, 1991.)*

radiation absorbed by a definable area of vegetation below. Landsberg *et al.* (1975) devised a method for measuring the net radiation absorbed by a hedgerow of trees. They placed four opposed pairs of linear net radiometers parallel with the hedgerow in a cylindrical arrangement around the circumference and, by taking the algebraic sum of the measurements of the individual net radiometers, were able to calculate the net absorbed amount of all wave radiation. This approach was subsequently exploited by Thorpe (1978), who calculated the energy balance and transpiration of hedgerow apple trees, and is being used today to estimate water use by windbreaks of different species in the Sahel (D. M. Smith and P. G. Jarvis). A logical extension of this approach to an isolated tree would be to place individual, hemispherical, net radiometers on the surface of an imaginary sphere enclosing the tree. However, this would require a large number of net radiometers and the structures required to hold them up would undoubtedly interfere with the radiation absorbed.

An alternative solution (Green, 1993; McNaughton *et al.*, 1992) is to mount net radiometers on a lightweight rotatable hoop (the 'whirligig') that encompasses a tree and to measure the net radiation as the hoop is rotated by 360° around the tree (Figure 18.8). As the hoop rotates it describes an enclosing sphere about the tree, with each radiometer following a horizontal, circular path at a fixed latitude on the sphere. The total radiation absorbed by the

Table 18.2. *Transpiration of unit length of a 5- and 6-year-old windbreak of neem (Azardirachta indica) over 6 months of the year in the Sahel, during the wet season and passing into the dry season, based on a sample of trees of different sizes measured with the heat-pulse method (from Brenner, 1991)*

Period	Average daily transpiration			
	1988	1989	1988	1989
	$(dm^3\ day^{-1}\ (m\ windbreak)^{-1})$		$(mm\ day^{-1})$	
Wet	17.59	26.59	2.51	3.80
season	12.53	18.94	1.79	2.71
↓	19.66	29.72	2.81	4.25
	14.74	22.29	2.11	3.18
Dry	10.59	16.02	1.51	2.29
season	10.31	15.59	1.47	2.23

tree is found by taking the weighted sum of the mean net radiation recorded at each latitude.

Figure 18.9a shows the transpiration rate measured in this way using the Penman–Monteith equation, partitioned into the net radiation and saturation deficit dependent components of the transpiration rate, together with measurements by the heat pulse method (Figure 18.9b). It is evident that very good agreement was obtained between the two measurements of transpiration rate and also that, for the greater part of the day, the saturation deficit driven

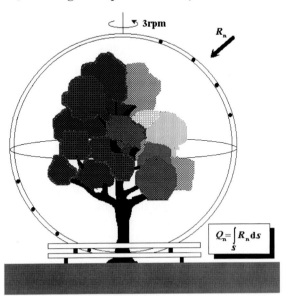

Figure 18.8. *A diagram of the whirligig device mounted around a walnut tree of height 3.5 m. The net radiation absorbed by the tree $Q_n = \int_s R_n\ ds$, where R_n is the net inwards flux density through an element of the surface ds measured by a net radiometer positioned on the surface of the hoop that describes the enclosing sphere. (Unpublished data, S. R. Green.)*

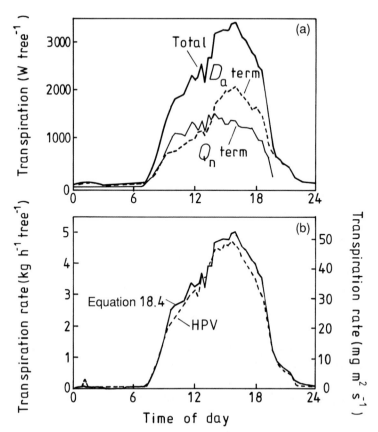

Figure 18.9. (a) The transpiration rate calculated using Equation 18.4 with the net radiation component (ΩE_q) and the saturation deficit component $(1 - \Omega)E_{imp}$ shown separately. The leaf area of the tree was 26.4 m^2. (b) The transpiration rate calculated using Equation 18.4 compared with the stem water flux measured by the heat-pulse velocity method (HPV). (From Green, 1993.)

component considerably exceeded the net radiation driven component. On average over 12 days, the saturation deficit component accounted for 62% of the transpiration and the net radiation component only 38%. Over 3 days with maximum Q_n between 2000 and 3000 W per tree, Ω varied between c. 0.05 and 0.1 at night and c. 0.3 and 0.4 in the daytime (S. R. Green, personal communication). Using the similar but simpler method referred to earlier, Thorpe (1978) found that the net radiation term rarely accounted for more than 30% of the transpiration rate of the hedgerow apple orchard.

By adding quantum sensors to the circular frame, the absorbed quantum flux density can be measured and photosynthesis calculated using an analogous big leaf model for photosynthesis (McNaughton et al., 1992), thus enabling simultaneous calculation of water-use efficiency for the tree crown. The whirligig therefore enables studies of the exchange of water and CO_2 by the crowns of isolated trees to be investigated in a manner not previously possible, and thus opens the door to the test of both commonly accepted and new models.

18.3.2 *Transpiration from the canopy*

Transpiration from canopies is frequently estimated by solving Equation 18.4 for different classes of leaves, stratified with respect to species and dominance of tree, age, ecological type and position in the canopy, particularly on small sites of less than 1 ha in extent. On larger sites that meet fetch and footprint criteria, transpiration has been measured frequently from crops and short natural vegetation, but less frequently from forests, by placing air-intake tubes and sensors above the canopy and utilizing the principles of either the Bowen ratio or the eddy covariance methods (Monteith, 1975).

To complement the study on the evaporation of intercepted water described earlier, details will be given here of the energy balance and transpiration by a 26-year-old stand of pole-stage Sitka spruce, before thinning in a normal summer and after thinning in a somewhat warmer and sunnier summer, by the Bowen ratio method at a site near Aberdeen, Scotland. These results have not previously been presented: full details of the methods and site are given by James (1977) and Jarvis (1993).

The average height of the canopy was 12.0 m (average top height 13.3 m) and there were about 3900 trees ha^{-1}. The closed canopy was continuous and the leaf area index (plan needle area) about 8. After measurements over 3 years, the plots and surrounding areas were thinned, half the trees and c. one-third of the basal area and leaf areas being removed (see Jarvis, 1993, Appendix II).

The mean hourly effluxes of water vapour above the Sitka spruce canopy were calculated from measurements of the concentration gradient of humidity above the forest using the energy-transfer coefficient, K, obtained from Bowen ratio-energy balance theory (Jarvis, 1993, Appendix II), in the present system of molar units, as:

$$E = \rho_a K d(e/P_a)/dz \qquad 18.6$$

and the sensible heat flux density, H, was obtained from the difference in potential temperature $d\theta$ over the same height interval as:

$$H = \rho_a c_P K d\theta/dz. \qquad 18.7$$

where ρ_a is the molar density of air, c_P is the molar heat capacity of air at constant pressure and $d(e/P_a)/dz$ and $d\theta/dz$ are the gradients of water vapour mole fraction and potential temperature, respectively. The gradients were measured over the same 4-m height interval by four replicated gradient-measuring systems, two on each of two 21-m-tall masts, 42 m apart. Each mast carried two 10-m-long horizontal booms normal to the prevailing wind direction, one c. 1 m above tree-top height and the other 4 m above it. The booms on each mast carried two gradient-measuring systems, one on each side of the mast. Each system consisted of eight sampling units, four at each height. Each sampling unit contained a ventilated, shielded platinum resistance thermo-meter and also served as an air-intake port. Each group of four units was interconnected to provide a degree of spatial averaging. The differences in temperature between the two levels of each system were measured directly with a differential Kelvin bridge. The four air intakes at each level of a gradient-measuring system joined a common heated and insulated butyl rubber tube, so that two butyl rubber tubes each carried a continuous flow of air from each system to dew-point meters connected in differential mode. Using solenoid valves, both air samples from the top and bottom levels of each gradient-measuring system were switched together through the dew-point meters so that a differential measurement was obtained directly. The dimen-sionless Bowen ratio, $\beta = H/\lambda E$, was then obtained directly from the ratio of the differential measurements of temperature and vapour pressure over the same height interval.

Monteith (1965) introduced a *climatological* (or *isothermal*) conductance, g_i, to assist in defining the impact of climate on E and β. For the present purpose, this is a convenient grouping of terms and, in the present system of units, can be written as:

$$g_i = \frac{Q_n}{\lambda} \frac{P_a}{D_a}.$$

18.8

Then from Equations 18.4 and 18.8, the canopy conductance can be

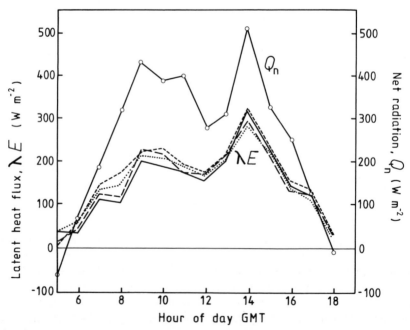

Fig. 18.10. *Agreement amongst the transpiration rates measured by each gradient-measurement system. Q_n is net absorbed radiation, λE is the transpiration flux of latent heat.*

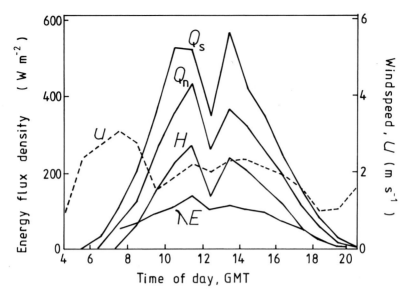

Figure 18.11. *Diurnal energy fluxes for a Sitka spruce stand on a fine sunny day in August, showing solar radiation* (Q_s), *net radiation* (Q_n), *sensible heat flux* (H) *and latent heat flux* (λE). *The sum of soil heat flux, air storage and photosynthetic energy fluxes reached a maximum of 40 W m^{-2} during the morning.*

expressed (Jarvis *et al.*, 1976; Thom, 1975) as:

$$g_{surf} = [(\varepsilon\beta - 1)/g_A + (\beta + 1)/g_i]^{-1}. \qquad 18.9$$

This equation was used to obtain g_{surf} directly from β.

Although there was always some variation amongst the replicate gradient measurements, agreement between fluxes was much better (Figure 18.10). Agreement amongst the four replicate fluxes was generally good for approximately 50% of the measurements, but was less good after long periods of rainfall, when moisture may have entered the tubes.

The energy budget. The components of the energy budget for a typical fine summer day are shown in Figure 18.11. Near midday, the sensible and latent heat fluxes taken together amount to *c.* 95% of the net radiation flux, only a small amount of energy being used by photosynthesis and heat storage.

The Bowen ratio, β. The median values of β obtained for different ranges of solar radiation and D_a are shown in Table 18.3. The largest values of β occurred when solar radiation was high (> 600 W m^{-2}) and D_a was in the range 0.6–0.8 kPa. Median β for these conditions was *c.* 2, although individual values ranged from 1 to 3. These values of β for spruce forest are high compared with β for crops, and for spruce forest in continental rather than oceanic climates (Jarvis *et al.*, 1976). Much higher transpiration rates and much lower Bowen ratios (~ 0.5) result where vapour pressure deficits reach 2–3 kPa. The

Table 18.3. *Median Bowen ratios (β) for a closed canopy of Sitka spruce in dry summer conditions in 1973, between 07.00 and 18.00 h GMT. The figures in parentheses are the number of mean hourly observations*

Water-vapour pressure deficit (kPa)	β		
	Solar radiation ($\mathrm{W\ m^{-2}}$)		
	< 400	400–600	> 600
0–0.2	0.6 (17)	1.1 (12)	0.6 (5)
0.2–0.4	0.8 (26)	1.2 (27)	2.1 (17)
0.4–0.6	1.5 (22)	1.6 (33)	1.8 (18)
0.6–0.8	0.6 (8)	1.6 (17)	2.2 (29)
0.8–1.0	0.5 (4)	0.9 (8)	1.7 (10)

Bowen ratio depends to a considerable extent upon climatic conditions, as well as on the wetness and physiological state of the foliage: both small g_{surf}, resulting from stomatal closure, and large g_i, resulting from small saturation deficits, lead to large values of β, which are, therefore, typically found in oceanic climates (see Jarvis *et al.*, 1976, for further discussion).

Transpiration rate. Figure 18.12 shows transpiration rates as a function of ambient vapour pressure deficit, **D_a**. Rates of transpiration were commonly 70–200 W m^{-2} (0.1–0.3 mm h^{-1}) on fine dry days, but rarely exceeded 250 W m^{-2} (0.4 mm h^{-1}).

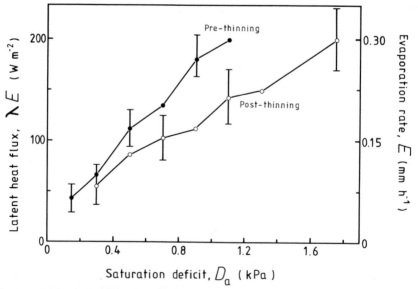

Figure 18.12. *The relation between the rate of evaporation from dry canopies and vapour pressure deficit, **D_a** for 9 days in summer 1973 (o) and 10 days in summer 1975 (O). Each point is the median of 10–20 values; the bars indicate interquartile ranges.*

The difference between transpiration rates in 1973 and 1975 was the result of the thinning carried out in early 1974, when approximately half the trees, together with one-third of the leaf area, were removed from the sample plots near the masts and from the surrounding areas.

Canopy conductance. Median g_{surf} for 32 days with dry canopies showed only a small decrease during the day, with the majority of values in the range 0.4–0.8 mol m^{-2} s^{-1}. However, in the afternoon the variation was large because on some days there was a tendency for g_{surf} to decrease, whereas on others there was a tendency for g_{surf} to increase. Some examples of diurnal curves of g_{surf} on particularly fine, dry summer days are shown in Figure 18.13. It is evident that in general g_{surf} is large when $\boldsymbol{D}_a$ is small, and small when $\boldsymbol{D}_a$ is large. Figure 18.14 shows that a considerable part of the diurnal, daily and seasonal variation in g_{surf} can be accounted for by the influence of saturation deficit. On these particularly dry days, evaporation from bark surfaces and soil was probably negligible so that changes in g_{surf} imply changes in stomatal conductance. The decline in g_{surf} with increasing $\boldsymbol{D}_a$, of about 5% per 0.1 kPa, was similar in each year and is consistent with changes in stomatal conductance measured with a porometer on individual shoots of spruce in the canopy (Watts *et al.*, 1976), and on canopy and seedling shoots in an assimilation chamber (Neilson and Jarvis, 1975; Sandford and Jarvis, 1986), and with experiments on potted spruce plants in a growth chamber equilibrated at constant temperature and different saturation deficits for 24 h (Watts and Neilson, 1978).

Recent experiments have demonstrated the importance of considering the response in terms of the leaf surface saturation deficit $\boldsymbol{D}_{surf}$ rather than $\boldsymbol{D}_a$ (Aphalo and Jarvis, 1991; Bunce, 1985), but in a spruce canopy close coupling of leaves to the atmosphere ensures that $\boldsymbol{D}_{surf}$ is very close to $\boldsymbol{D}_a$. Very similar results were obtained by Tan and Black (1976) with Douglas fir (*Pseudotsuga menziesii* (Mirb.) Franco) in British Columbia. They also found that pronounced changes in canopy conductance during the day could be related to saturation deficit and resulted from a decrease in stomatal conductance, as measured with a porometer.

In Figure 18.15 a comparison is made between the canopy conductance determined from the Bowen ratio method and canopy conductance determined from the integration of measurements made on a sample of shoots with the null-balance diffusion porometer. Although there is considerable scatter, there is no systematic deviation from the 1:1 relationship and the regression is significant at $p = 0.01$. The scatter in the relationship is understandable: both methods are subject to considerable errors. On the one hand, the sources of error are the assumptions and approximations that are involved in the derivation of g_{surf} from the Bowen ratio, and the instrumental errors in measuring the gradients. On the other hand, there are sampling errors in the estimation of both stomatal conductance and leaf area index of the needle classes (see Leverenz *et al.*, 1982), and errors in the calibration and use of

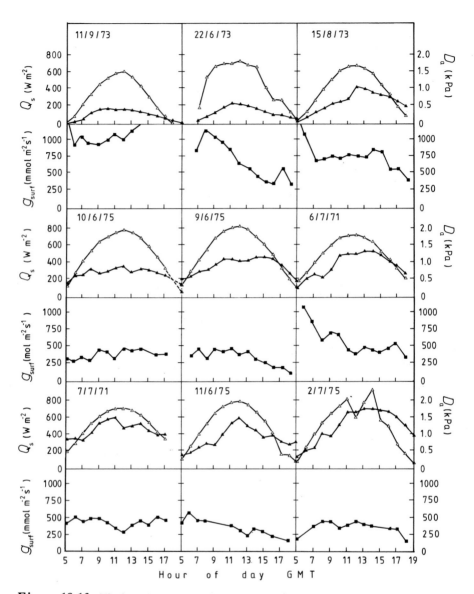

Figure 18.13. *The diurnal course of solar irradiance (Q_s, Δ), saturation deficit within the canopy (D_a, $\blacktriangle$) and canopy conductance (g_{surf}, $\blacksquare$) for a number of summer days when the Sitka spruce canopy was dry.*

the porometer. It is, therefore, reasonable to conclude that changes in canopy conductance are the result of changes in stomatal aperture. A similar set of observations on canopy and stomatal conductance of Douglas fir led to similar conclusions by Tan and Black (1976).

In a Scots pine (*Pinus sylvestris* L.) canopy at Thetford Forest there were much larger decreases in g_{surf} on fine sunny days. A typical decrease was

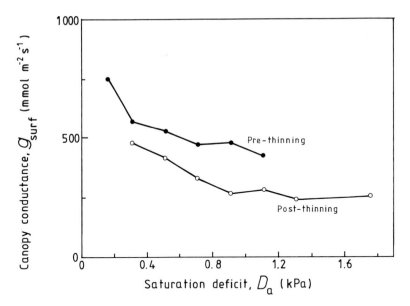

Figure 18.14. *The relationship between conductance (g_surf) and saturation deficit (**D_a**) for dry Sitka spruce canopies in June 1973 before thinning (●) and in July 1975 (○) after thinning in 1974. Each point is the median of 10–20 values.*

from 0.4 mol m^{-2} s^{-1} in the morning to 0.1 mol m^{-2} s^{-1} in the late afternoon (Gash and Stewart, 1975). This was probably the result of the drier climate and sandy soil at Thetford, which frequently leads to much larger values of D_a in the afternoon and to lower leaf water potentials than at Fetteresso.

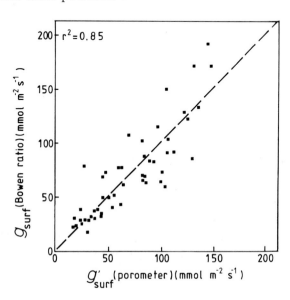

Figure 18.15. *A comparison between canopy conductance of a Sitka spruce stand determined from the Bowen ratio method (g_surf) and from the integration of porometer measurements (g'_surf). The equation for the regression is g_surf = 0.13 + 0.93 g'_surf; r^2 = 0.85.*

Controls on transpiration. On average, the equilibrium evaporation term containing Q_n in Equation 18.4 contributed less than 10% of the total transpiration (Table 18.4), with over 90% attributable to the imposed term containing D_a and moderated by g_{surf}. This is largely because the large boundary layer conductance of the narrow needles of Sitka spruce (< 1 mm) (Landsberg and Thom, 1971) and the roughness and large boundary layer conductance of the canopy (Landsberg and Jarvis, 1973) result in close coupling between temperature and saturation deficit at the needle surface and the ambient temperature and saturation deficit, respectively.

With typical values of the ratio g_A/g_{surf} in the range 10–60 and values of ε of 1.0–2.0, Ω (Equation 18.5) generally lies in the range 0.05–0.2 (Table 18.4). Similar values were reported for dry canopies of Scots pine by Whitehead *et al.* (1984), based on estimation of transpiration from integration of measurements of stomatal conductance of shoots made with null-balance porometers. Thus, transpiration from conifers goes on at a rate which is close to the rate imposed by the prevailing saturation deficit and moderated by the canopy conductance. In general, the most influential driving variable is D_a, which varies from near 0 to no more than 2 kPa in the oceanic climate of Scotland. In practice, large values of D_a are usually correlated with high solar irradiance and net radiation because of the high air temperatures that generally occur at high irradiances. However, when a sudden change in irradiance occurs, the proportional change in transpiration is small compared to the proportional change in sensible heat flux, showing clearly the lack of dependence of transpiration on net radiation (Figure 18.11).

18.4 The upscaling problem

A clear distinction should be drawn between the estimation of fluxes at a larger scale, from parameters or fluxes obtained at a smaller scale, by averaging of parameters or by *summation* of the fluxes, and the *upscaling process* which also takes into account the interdependence of vegetation and the atmosphere. Estimation of canopy fluxes from proper determination of average parameters or from summation of component fluxes is a straightforward arithmetical procedure that has no need to take into account changes in the local atmosphere caused by the fluxes themselves or the impacts of the fluxes from one area of vegetation on the fluxes from an adjacent area of vegetation. In *analyses* of the controls on fluxes from a plant or an area of vegetation at any one time, the driving variables may be treated as independent of the vegetation, although at successively larger scales the driving variables become progressively more dependent on the fluxes themselves through what are largely negative feedbacks between atmosphere and vegetation. Although these feedbacks may be of little consequence for analysis of the processes going on at a moment in time, they must be taken into account when making *predictions* about fluxes at some time in the future. The essential feature of the upscaling process, as

Table 18.4. Midday conditions and fluxes for the unthinned stand on 14 dry days in summer between 11.00 and 12.00 h GMT

	Q_n W m⁻²	u m s⁻¹	T_a °C	D_a kPa	g_i mol m⁻² s⁻¹	λE W m⁻²	g_{surf} mmol m⁻² s⁻¹	β (H/λE)	α (λE/(Q_n − S))	$\Omega E_{eq}/E$ %	$\dfrac{(1-\Omega)E_{imp}}{\Omega E_{eq}}$	Ω
7 June	408	2.03	19.7	0.62	1.46	52	144	6.8	0.13	10.1	8.9	0.05
13 June	501	3.52	10.8	0.63	1.85	85	281	4.9	0.21	6.2	16.1	0.04
14 June	335	1.84	10.6	0.40	1.93	146	783	1.3	0.43	8.1	11.3	0.19
15 June	253	2.15	16.9	0.51	1.12	137	585	0.84	0.54	14.7	5.8	0.16
22 June	526	3.20	20.1	0.64	1.86	181	556	1.9	0.34	9.9	9.1	0.12
24 June	482	3.90	15.0	0.52	2.08	161	631	2.0	0.33	7.9	11.6	0.09
28 June	573	3.90	16.8	0.85	1.54	229	575	1.5	0.40	6.4	14.7	0.09
10 July	427	2.27	18.2	0.76	1.25	158	413	1.7	0.37	7.6	12.2	0.12
24 July	398	1.56	15.1	0.64	1.44	117	353	2.4	0.29	8.5	10.7	0.13
26 July	541	7.06	13.2	0.49	2.47	102	430	4.3	0.19	5.9	15.9	0.03
27 July	581	6.19	13.7	0.51	2.62	124	501	3.7	0.21	6.8	13.7	0.05
14 August	413	2.71	17.2	0.28	3.46	98	600	3.2	0.24	15.9	5.3	0.13
15 August	453	2.59	20.9	0.76	1.32	226	610	1.0	0.50	8.9	10.2	0.16
18 August	507	1.38	15.9	0.77	1.48	175	421	1.9	0.34	10.2	8.8	0.17
Average	457	3.16	16.0	0.60	1.84	142	492	2.2	0.32	9.1	11.0	0.11

distinct from averaging and summation, is the proper inclusion of the feed-backs that enter as scale increases (McNaughton and Jarvis, 1991).

As shown here, the evaporation and transpiration of water from isolated individual plants, particularly tall plants, depends crucially on the ambient saturation deficit and its moderation by the boundary layer conductance of their leaves. Isolated individuals have little impact on the saturation deficit and temperature of the ambient atmosphere, but different boundary layer thicknesses, arising from differences in exposure and in leaf size and clumping, lead to variation in saturation deficit at the leaf surface and may thus influence the flux of water vapour, the stomatal conductance and, consequently, the dependence of transpiration on stomatal conductance.

For plants aggregated into vegetation, the situation is very different. The degree of dependence of transpiration on canopy conductance varies with the *structure* of the vegetation and is not the same for short, closely spaced vegetation, such as heathland and arable crops, as for tall, aerodynamically rough woodlands (Jarvis and McNaughton, 1986). When there are many leaves in a low, dense canopy, the canopy is poorly coupled aerodynamically to the atmosphere in the mixed layer above, so changes in the sensible heat and water vapour fluxes of the vegetation alter the microclimate around the leaves. The saturation deficit in the vicinity of the leaves comes to depend on the heat and water vapour fluxes and may depart widely from that in the air above (e.g. Grantz and Meinzer, 1990), and transpiration goes on at largely the equilibrium rate. This negative feedback reduces both transpiration rate and the sensitivity of transpiration to changes in stomatal conductance: the sensitivity of the leaf-surface saturation deficit to change in the ambient saturation deficit in the air above is reduced, and the saturation deficit close to the vegetation cannot now be regarded as an independent variable (Figure 18.16). Consequently, when scaling up from individual plants to plants in stands or communities, this feedback must be taken into account.

If, in addition, the stomata are sensitive to the saturation deficit at the leaf surface, this feedback will further reduce the sensitivity of transpiration to a change in saturation deficit, while, at the same time, increasing the sensitivity of transpiration to changes in conductance arising from the assimilation system (McNaughton and Jarvis, 1991).

By contrast, tall, aerodynamically rough vegetation, such as the woodlands and plantations considered earlier, is generally so well coupled to the atmosphere above that the surface temperature and saturation deficit are similar to the temperature and saturation deficit of the air in the mixed layer above and relatively insensitive to changes in the local fluxes of heat and water vapour. Transpiration goes on at largely the imposed rate; transpiration rate is open-ended with respect to atmospheric saturation deficit and is highly sensitive to changes in canopy conductance. Thus the structural properties of canopy and stand, as well as physiological properties of the leaves, may have a marked influence on the response of transpiration to changes in both canopy conductance and the environmental driving variables.

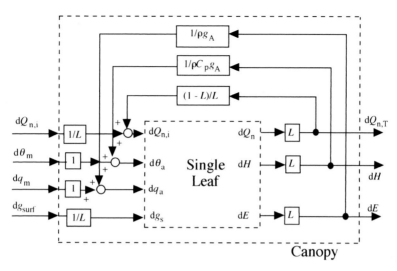

Figure 18.16. *Representation of a single-layer canopy as an assemblage of identical leaves, with total leaf area index L. The canopy aerodynamic conductance, g_A, is the conductance measured from within the canopy airspace (i.e. outside the leaf boundary layer) up to the top of the surface layer. The feedback pathway for net radiation exists because much of the radiation emitted by one leaf is absorbed by others. The diagram shows that the smaller is g_A, the larger is the negative feedback, and so the less sensitive is evaporation to changes in stomatal conductance. This would remain true for more complex representations of a canopy. $d\theta_m$ and dq_m are changes in the potential temperature and specific humidity in the mixed layer. $Q_{n,i}$ is isothermal net radiation and $Q_{n,T}$ is net radiation as a function of surface temperature. (From McNaughton and Jarvis, 1991, with permission from Elsevier Science Publishers BV.)*

In scaling up from the stand to the landscape, additional negative feedbacks come into play. Over larger areas of uniform vegetation, heat and water vapour from the surface warm and humidify the entire convective boundary layer and so change its saturation deficit. Additionally, the convective boundary layer grows vertically during the day by eroding the base of the capping inversion and entraining warmer and drier air from above, driven by the sensible heat flux. As a result, there are two additional negative feedback pathways at the large scale that tend to stabilize transpiration further against changes in canopy conductance (Figure 18.17). Generally, the entrainment pathway dominates (McNaughton and Jarvis, 1991). When the sensible heat flux is large, the convective boundary layer grows rapidly, with the result that the saturation deficit in the mixed layer becomes more like that above the capping inversion, set by large-scale processes and largely unaffected by small changes in fluxes at the surface. Conversely, when the sensible heat flux is small, growth of the convective boundary layer is slow and transpiration is less responsive to small changes in surface conductance.

To complicate the position further, the size of the canopy conductance itself alters the gain in the feedback pathways and so influences the response of transpiration to a change in stomatal conductance. A large canopy conductance stabilizes transpiration by decoupling the saturation deficit at the leaf surface

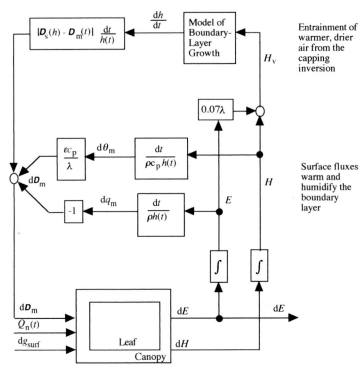

Figure 18.17. *Partial representation of the controls on transpiration operating at regional scale. Some linkages are not shown in the diagram: the height of the boundary layer, which is important in setting the capacity of the layer and so the rate of change of saturation deficit, should be linked to the integrated rate of boundary-layer growth, and the saturation deficit,* $\mathbf{D}_m(t)$*, which appears in the box representing the effect of entrainment, should be shown as the integral of all of the changes* $d\mathbf{D}_m$*. Inputs from the larger-scale weather are also not shown. The general circulation of the atmosphere determines cloudiness, temperature, humidity and stability above the convective planetary boundary layer and so influences* Q_s*,* $\mathbf{D}_s(h)$ *and* dh/dt*. (From McNaughton and Jarvis, 1991, with permission from Elsevier Science Publishers BV.)*

from that above and also by reducing the sensible heat flux and thus the rate of growth of the convective boundary layer.

Control of transpiration at landscape scale is, therefore, analogous to control at smaller scales. When canopy conductance is large, conditions within the convective boundary layer become decoupled from the synoptic scale so energy supply, not stomatal control, is the most important regulator of transpiration. When canopy conductance is small, the conditions within the convective boundary layer are strongly coupled to conditions above by rapid entrainment. These extremes of decoupling and coupling parallel the situation at stand scale but are accentuated at larger scales because there are more negative feedback pathways and because of the additional effect of the sensible heat flux on entrainment.

If the landscape vegetation is not uniform but is a mosaic of types, contrasting in roughness, albedo and canopy conductance, differences in atmospheric transport processes are likely to occur over different parts of the landscape and thereby lead to the development of more complex, three-dimensional meso-scale circulations.

A major current research goal is to develop models at the scale of the individual, the stand and the landscape, as portrayed in Figure 18.1, that take into account both the characteristic *physiology* and *structure* of the vegetation that determine the degree of coupling to the atmosphere, and to represent adequately the feedbacks that occur between the plants and the atmosphere, with increasing complexity as the scale increases. Since leaf area and its spatial distribution are of primary importance to both radiation absorption and canopy conductance (i.e. to both the equilibrium and imposed terms in Equation 18.4), the phenology of bud burst and leaf development is of considerable significance. Changes in phenology are likely as a result of global change in both atmospheric CO_2 concentration and temperature. Consequently, impact studies on phenology in particular are likely to be crucial to the successful modelling of the impact of global change on water use by vegetation at all scales.

Numerous studies over the past 100 years have shown that stomatal conductance decreases with increasing ambient CO_2 concentration, and, more recently, acclimation of stomatal conductance to elevation of the ambient CO_2 concentration has been demonstrated (see, for example, Eamus and Jarvis, 1989): there is also some evidence of seasonal changes in sensitivity of stomatal conductance to ambient CO_2 concentration. In responding to CO_2, stomata respond to the intercellular space CO_2 concentration, but the mechanism by which plants sense CO_2 is still obscure (Mott, 1988, 1990).

Recent studies have shown that leaf initiation, bud burst, rate of leaf development, final leaf size, and leaf senescence are affected by increasing ambient CO_2 concentration. Although studies of the mechanistic bases for these responses are under way and generalities have yet to emerge, empirical observations suggest that ambient CO_2 concentration may well influence energy partitioning and water vapour flux through changing the phenology and dynamics of leaf populations, as well as by affecting stomatal conductance. The effects of temperature on bud burst, both through vernalization in winter and heat sum in spring, are better known than the effects of CO_2; they lead to the expectation of changes in the phenology and dynamics of leaf populations that are likely to affect rates of water use and energy partitioning (Cannell *et al.*, 1989; Murray *et al.*, 1989). At the stand and canopy scale, however, little is known at present about effects of either CO_2 concentration or temperature on physiology or structure. None the less, with an appropriate model, 'what if' type experiments can readily be employed to investigate likely consequences of possible changes in stand structure (e.g. McMurtrie *et al.*, 1992), and the current upsurge of studies at stand scale will surely lead to better descriptions of stand processes that can be used to test the models.

References

Aphalo, P.J. and Jarvis, P.G. (1991) Do stomata respond to relative humidity? *Plant Cell Environ.*, **14**, 127–132.

Aphalo, P.J. and Jarvis, P.G. (1993) The boundary layer and the apparent responses of stomatal conductance to wind speed and to the mole fractions of CO_2 and water vapour in the air. *Plant Cell Environ.*, **44**, 791–800.

Brenner, A.J. (1991) Tree–crop interactions within a Sahelian windbreak system. Ph.D. Thesis, University of Edinburgh.

Bunce, J.A. (1985) Effect of boundary layer conductance on the response of stomata to humidity. *Plant Cell Environ.*, **8**, 55–57.

Calder, I.R. (1990) *Evaporation in the Uplands.* John Wiley and Sons, Chichester.

Calder., I.R. (1992) Deuterium tracing for the estimation of transpiration from trees, Part 2. Estimation of transpiration rates and transpiration parameters using a time-averaged deuterium tracing method. *J. Hydrol.*, **130**, 27–35.

Cannell, M.G.R., Grace, J. and Booth, A. (1989) Possible impacts of climatic warming on trees and forests in the United Kingdom: a review. *Forestry*, **62**, 337–364

Eamus, D. and Jarvis, P. G. (1989) The direct effects of CO_2 increases in the global atmospheric CO_2 concentration on temperate trees and forests (natural and commercial). *Adv. Ecol. Res.*, **19**, 1–55.

Edwards, W. R. N. (1980) Flow of water in trees. Ph.D. Thesis, University of Edinburgh.

Gash, J.H.C. and Stewart, J.B. (1975) The average resistance of a pine forest derived from Bowen ratio measurements. *Bound. Lay. Meteorol.*, **8**, 453–464.

Gash, J.H.C., Wright, I.R. and Lloyd, C.R. (1980) Comparative estimates of interception loss from three coniferous forests in Great Britain. *J. Hydrol.*, **35**, 385–396.

Grantz, D.A. and Meinzer, F.C. (1990) Stomatal response to humidity in a sugarcane field: simultaneous porometric and micrometeorological measurements. *Plant Cell Environ.*, **13**, 27–37.

Green, S.R. (1990) Air flow through and above a forest of widely spaced trees. Ph.D. Thesis, University of Edinburgh.

Green, S.R. (1993) Radiation balance, transpiration and photosynthesis of an isolated tree. *Agric. For. Meteorol.*, **64**, 201–221.

Harding, R.J., Neal, C. and Whitehead, P.G. (1992) Hydrological effects of plantation forestry in north-western Europe. In: *Responses of Forest Ecosystems to Environmental Change* (eds A. Teller, P. Mathy and J.N.R. Jeffers). Elsevier Applied Science, London, pp. 445–455.

James, G.B. (1977) Exchanges of mass and energy in Sitka spruce. Ph.D. Thesis, University of Aberdeen.

Jarvis, P.G. (1993) Resource capture by coniferous forests – A case study. In: *Resource Capture, Proceedings of Nottingham Easter School 1992* (eds K. Scott, M.H. Unsworth and J.L. Monteith), in press.

Jarvis, P.G. and McNaughton, K.G. (1986) Stomatal control of transpiration: scaling up from leaf to region. *Adv. Ecol. Res.*, **15**, 1–49.

Jarvis, P.G., James, B.G. and Landsberg, J.J. (1976) Coniferous forest. In: *Vegetation and the Atmosphere*, Vol. 1 (ed. J.L. Monteith). Academic Press, London, pp. 171–240.

Landsberg, J.J. and Jarvis, P.G. (1973) A numerical investigation of the momentum balance of a spruce forest. *J. Appl. Ecol.*, **10**, 645–655.

Landsberg, J.J. and Thom, A.S. (1971) Aerodynamic properties of a plant complex structure. *J.R. Meteorol. Soc.*, **97**, 565–570.

Landsberg, J.J., Beadle, C.L., Biscoe, P.V., Butler, D.R., Davidson, B., Incoll, L.D., James, G.B., Jarvis, P.G., Martin, P.J., Neilson, R.E., Powell, D.B.B., Slack, E.M., Thorpe,

M.R., Turner, N.C., Warrit, B. and Watts, W.R. (1975) Diurnal energy, water and CO_2 exchanges in an apple (*Malus pumila*) orchard. *J. Appl. Ecol.*, **12**, 659–684.

Leverenz, J., Deans, J.D., Ford, E.D., Jarvis, P.G., Milne, R. and Whitehead, D. (1982) Systematic spatial variation of stomatal conductance in a Sitka spruce plantation. *J. Appl. Ecol.*, **19**, 835–851.

Lloyd, C.R., Gash, J.H.C., Shuttleworth, W.J. and Marques, A. de O. (1988) The measurement and modelling of rainfall interception by Amazonian rain forest . *Agric. For. Meteorol.*, **43**, 277–294.

McMurtie, R.E., Comins, H.N., Kirschbaum, M.U.F. and Wang, Y.-P. (1992) Modifying existing forest growth models to take account of effects of elevated CO_2. *Aust. J. Bot.*, **40**, 657–677.

McNaughton, K.G. and Jarvis, P.G. (1983) Predicting effects of vegetation changes on transpiration and evaporation. *Water Def. Plant Growth*, **7**, 1–42.

McNaughton, K.G. and Jarvis, P.G. (1991) Effects of spatial scale on stomatal control of transpiration. *Agric. For. Meteorol.*, **54**, 279–302.

McNaughton, K.G., Green, S.R., Black, T.A., Tynan, B.R. and Edwards, W.R.N. (1992) Direct measurement of net radiation and PAR absorbed by a single tree. *Agric. For. Meteorol.*, **62**, 87–107.

Martin, Ph. (1989) The significance of radiative coupling between vegetation and the atmosphere. *Agric. For. Meteorol.* **49**, 45–53.

Monteith, J.L. (1963) Gas exchange in plant communities. In: *Environment Control of Plant Growth* (ed. L.T. Evans). Academic Press, New York, pp. 95–112.

Monteith, J.L. (1965) Evaporation and environment. *Symp. Soc. Exp. Biol.*, **19**, 205–234.

Monteith, J.L. (1975) *Principles of Environmental Physics*. Edward Arnold, London.

Mott, K.A. (1988) Do stomata respond to CO_2 concentrations other than intercellular? *Plant Physiol.*, **86**, 200–203.

Mott, K.A. (1990) Sensing of atmospheric CO_2 by plants. *Plant Cell Environ.*, **13**, 731–737.

Murray, M.B., Cannell, M.G.R. and Smith, R.A. (1989) Date of bud burst of 15 tree species in Britain following climatic warming. *J. Appl. Ecol.*, **26**, 693–700.

Neilson, R.E. and Jarvis, P.G. (1975) Photosynthesis in Sitka spruce (*Picea sitchensis* (Bong.) Carr.). VI. Response of stomata to temperature. *J. Appl. Ecol.*, **12**, 879–891.

Rutter, A.J. (1975) The hydrological cycle in vegetation. In: *Vegetation and the Atmosphere*, Vol. 1 (ed. J.L. Monteith). Academic Press, London, pp. 111–114.

Rutter, A.J., Morton, A.J. and Robins, P.C. (1975) A predictive model of rainfall interception in forests. II. Generalization of the model and comparison with observations in some coniferous and hardwood stands. *J. Appl. Ecol.*, **12**, 367–380.

Sandford, A.P. and Jarvis, P.G. (1986) The response of stomata of several coniferous species to leaf-to-air vapour pressure difference. *Tree Physiol.*, **2**, 89–103.

Stewart, J.B. (1977) Evaporation from the wet canopy of a pine forest. *Water Resources Res.*, **13**, 915–921.

Tan, C.S. and Black, T.A. (1976) Factors affecting the canopy resistance of a Douglas fir forest. *Bound. Lay. Meteorol.*, **10**, 475–488.

Teklehaimanot, Z. and Jarvis, P.G. (1991) Direct measurement of evaporation of intercepted water from forest canopies. *J. Appl. Ecol.*, **28**, 603–618.

Teklehaimanot, Z., Jarvis, P.G. and Ledger, D.C. (1991) Rainfall interception and boundary layer conductance in relation to tree spacing. *J. Hydrol.*, **123**, 261–278.

Thom, A.S. (1975) Momentum, mass and heat exchange of plant communities. In: *Vegetation and the Atmosphere*, Vol. 1 (ed. J.L. Monteith). Academic Press, London, pp. 57–109.

Thorpe, M.R. (1978) Net radiation and transpiration of apple trees in rows. *Agric. Meteorol.*, **19**, 41–57.

Waring, R.H., Whitehead, D. and Jarvis, P.G. (1980) Comparison of an isotopic method and the Penman–Monteith equation for estimating transpiration from Scots pine. *Can. J. For. Res.*, **10**, 555–558.

Watts, W.R. and Neilson, R.E. (1978) Photosynthesis in Sitka spruce (*Picea sitchensis* (Bong.) Carr.). VIII. Measurements of stomatal conductance and $^{14}CO_2$ uptake in controlled environments. *J. Appl. Ecol.*, **15**, 245–255.

Watts, W.R., Neilson, R.E. and Jarvis, P.G. (1976) Photosynthesis in Sitka spruce. VII. Measurements of stomatal conductance and $^{14}CO_2$ uptake in a forest canopy. *J. Appl. Ecol.*, **13**, 623–638.

Whitehead, D., Jarvis, P.G. and Waring, R.H. (1984) Stomatal conductance, transpiration and resistance to water uptake in a *Pinus sylvestris* spacing experiment. *Can. J. For. Res.*, **14**, 692–700.

Appendix I The big psychrometer tree

Evaporation of intercepted water, E_I, from a wet tree with the foliage at wet bulb temperature T_w, can be expressed as:

$$E_I = g_A[e*(T_w) - e_a]/P_a \qquad \text{18.A1}$$

where e_a and $e*$ are the water-vapour pressure and the saturation water-vapour pressure of air, respectively.

The basic equation for a wet bulb psychrometer is:

$$e_a = e*(T_w) - \gamma \Delta T \qquad \text{18.A2}$$

where γ is the psychrometer constant $(c_P P_a/\lambda$, where c_P is the molar heat capacity at constant pressure) and ΔT is the difference between wet bulb and dry bulb temperature, T_d.

Substituting e_a from Equation 18.A2 into Equation 18.A1 gives:

$$E_I = g_A \gamma \Delta T/P_a. \qquad \text{18.A3}$$

The water-vapour saturation deficit of the ambient air, D_a, is

$$D_a = e*(T_d) - e_a. \qquad \text{18.A4}$$

Substituting again for e_a from Equation 18.A2, and following Penman's procedure of a linear approximation over a short interval of the $e*$ vs. T function, gives:

$$D_a = (s + \gamma)\Delta T = (\varepsilon + 1)\gamma \Delta T \qquad \text{18.A5}$$

where $s = de*(T)/dT$ and $\varepsilon = s/\gamma = s\lambda/(c_P P_a)$.
Replacing $\gamma \Delta T$ in Equation 18.A3 from Equation 18.A5 gives the evaporation of intercepted water as

$$E_I = \frac{g_A}{\varepsilon + 1} \frac{D_a}{P_a} \qquad \text{18.A6}$$

for the condition when the foliage is at the wet bulb temperature.

Appendix II Saturation deficit at the surface

The simplest possible representation of transpiration rate (Monteith, 1963) is

$$E = g_{surf} D_{surf}/P_a, \qquad 18.A7$$

where D_{surf} is the saturation deficit at the surface.

A common alternative representation, taking the boundary layer into account, is

$$E = \frac{(e^*(T_{surf}) - e_a)/P_a}{1/g_{surf} + 1/g_A} = \frac{g_{surf}(e^*(T_{surf}) - e_a)}{(1 + g_{surf}/g_A)P_a}. \qquad 18.A8$$

Following Penman and adding and subtracting $e^*(T_a)$ to the right-hand term in the numerator and making a linear approximation over a short interval of the e^* vs. T function gives:

$$E = \frac{g_{surf}(D_a + s(T_{surf} - T_a))}{(1 + g_{surf}/g_A)P_A}. \qquad 18.A9$$

Equating Equation 18.A7 and Equation 18.A9:

$$D_{surf} = \frac{D_a + s(T_{surf|} - T_a)}{1 + g_{surf}/g_A}. \qquad 18.A10$$

Similarly, by equating Equation 18.A7 with Equation 18.4, an expression for D_{surf} in relation to D_a in terms of $(Q_n - S)$ and Ω is obtained.

The representation of vegetation in large-scale models of the atmosphere

A.J. Dolman

19.1 Introduction

General circulation models (GCMs) describe the three-dimensional structure of the atmosphere and are being used increasingly for studies of climate and land-use change. Experiments with these models have shown a great sensitivity in global circulation patterns to changes in the representation of the land surface (e.g. Rowntree, 1986, 1991). Table 19.1 lists a number of more recent modelling experiments in which GCMs were used to assess the impact of tropical deforestation and desertification on climate. Tropical deforestation experiments generally consist of a comparison of a control simulation run for a forest with a simulation run in which the land surface is replaced by pasture or savannah. The climates of the two runs are then analysed for a change in rainfall, surface temperature, radiation, etc. Most deforestation experiments so far indicate that a weakening of the hydrological cycle may result from large-scale deforestation of the Amazon. Table 19.1 shows this phenomenon as a decrease in rainfall over the Amazon Basin.

In most of the desertification experiments the description of the land surface is very simple. The land surface is represented as a bucket with a finite water-holding capacity. Depending on how much water is available in the bucket, the potential evaporation rate is reduced to an actual rate. In the experiments listed in Table 19.1, the formulation of this relationship is changed in the desertification run. A decrease in the availability of soil moisture in these experiments leads to a decrease in Sahelian rainfall.

It is now becoming clear that the use of bucket models is not adequate in GCM land-use and climate-change studies. A number of new land-surface

Table 19.1. *Changes in rainfall following a change in land surface characteristics for a number of desertification and deforestation experiments*

Experiment	Anomaly		Change in rainfall in Amazon Basin (mm day^{-1})
	Area	Change in land use	
Deforestation			
Lean and Warrilow (1989)	Amazon	Forest–pasture	−1.34
Dickinson and Henderson-Sellers (1988)	Amazon	Forest–pasture	0
Nobre *et al.* (1991)	Amazon	Forest–pasture	−1.76
		Change in B (wet–dry)[a]	Change in rainfall at 15°N in Africa
Desertification			
Suarez and Arakawa (see Mintz, 1984)	Global	1-0[b]	−4
Carson and Sangster (see Rowntree, 1991)	Global	1-0[c]	−3
Cunnington and Rowntree (1986)	Africa (north of 10°N)	1-0[c]	−3

[a] $B = E/E_p$ (actual divided by potential evaporation).
[b] B is a time-dependent function of soil moisture.
[c] B is a time-dependent function of soil moisture; the values quoted are the initial values.

parameterizations have been developed, which recognize several vegetation types and attempt to model the interaction of the vegetated land surface with the atmosphere in a more realistic manner (see, for example, Dickinson, 1984; Noilhan and Planton, 1989; Sellers *et al.*, 1986; Verseghy, 1991; see also Chapter 18). The deforestation experiments listed in Table 19.1 use these more sophisticated models.

Despite the complexity of a land-surface model, all the models describe the land-surface atmosphere interaction in a one-dimensional sense and assume homogeneous surface cover. The Earth's surface, however, is never homogeneous and consequently these models have in some way to take into account the land-surface heterogeneity. Two- and three-dimensional numerical models are increasingly used to study the aggregation process on scales smaller than those of a typical GCM gridsquare (10 000 km^2) (Blyth *et al.*, 1993; Mason, 1988; Noilhan *et al.*, 1991; Wood and Mason, 1992).

This chapter aims to present a brief overview of current soil vegetation atmosphere transfer (SVAT) schemes and speculates on how fully interactive biomes may, in the future, be incorporated in GCMs. It also reviews briefly some recent aggregation studies and suggests some simple practical rules for the averaging of heterogeneous land-surface cover.

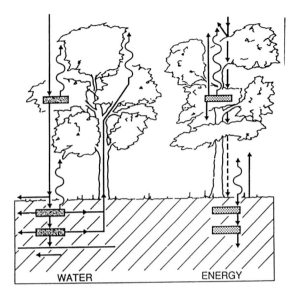

Figure 19.1. *Schematic diagram of a soil–vegetation–atmosphere transfer scheme. (Redrawn after Shuttleworth, 1991.)*

19.2 SVAT schemes

Figure 19.1 shows a diagram of the structure of a typical SVAT scheme. The soil and vegetation are represented by storage reservoirs for energy and water. The schemes represent the transfer of heat, water vapour, liquid moisture and momentum in a one- dimensional manner. Precipitation reaching the canopy fills up a canopy store, which is subsequently emptied by evaporation and drainage (see Chapter 18). A fixed proportion of the precipitation and the drainage of the canopy form the throughfall, reaching the soil moisture store. From there it may evaporate or penetrates further down into the soil profile. Typically, a GCM soil model has one to four layers. The fluxes of latent and sensible heat are usually parameterized as

$$F = (c_{surf} - c_{air})/r \qquad 19.1$$

where F is a flux, c_{surf} a scalar quantity or windspeed at the surface and c_{air} at reference level, r is a resistance, in the case of momentum r_{am}, in the case of water vapour $r_{av} + r_s$, the aerodynamic resistance to water vapour plus the canopy surface resistance, and r_{ah} in the case of sensible heat transfer. The differences between the SVAT schemes are mainly the number of layers used to describe the canopy and soil and the details of the parameterization of the resistances (for a recent review see Avissar and Verstraete, 1990). The BATS and Simple Biosphere (SiB) models (Dickinson, 1984; Sellers *et al.*, 1986) are more complex, whereas that of Noilhan and Planton (1989) and the current version of the UK Meteorological Office GCM (Warrilow *et al.*, 1986) form an inter-

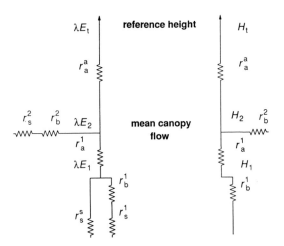

Figure 19.2. *Resistance diagram for the MITRE model. Subscripts a, b and s refer to aerodynamic, boundary layer and surface resistances, respectively. The superscripts t, s, 1, 2 refer to total, soil and vegetation canopy 1 and 2, respectively. H and λE are sensible and latent heat fluxes.*

mediate level of complexity between these two models and the simple bucket model.

19.2.1 *The MITRE model*

As an example of a modern SVAT scheme, the recently proposed MITRE model (Dolman, 1993) is described. This model is intended to replace the land-surface scheme in the UK Meteorological Office GCM in the next few years. Realizing that 70% of the world's canopies have a leaf area index of less than unity and can be considered sparse, a new SVAT scheme was developed which explicitly models the interaction between bare soil and vegetation. It is designed to model three different canopy architectures within the same model framework. They represent a homogeneous agricultural crop (Type I), a two-component system like a tropical savannah, with a well-defined two-dimensional structure (Type II), and a closed canopy such as a forest (Type III).

The MITRE model is a generalization of the Shuttleworth–Wallace equation (Shuttleworth and Wallace, 1985), which describes the interaction of a soil and vegetation layer through the introduction of a within-canopy temperature and vapour pressure deficit. Figure 19.2 shows a resistance diagram of the model. The model allows the full incident radiation to reach each of the two components separately, whereas in the original Shuttleworth–Wallace equation the radiation received by the soil was obtained by attenuating the radiation at the top of the canopy through Beer's law. In the case of the closed-canopy limit the model is equivalent to the Penman–Monteith equation (Monteith, 1965; Penman, 1948; see Chapter 18). A software switch allows the model structure to collapse to the Shuttleworth–Wallace equation for agricultural crops. Evaporation and sensible heat flux are given as the weighted sum of the two components, but are calculated to

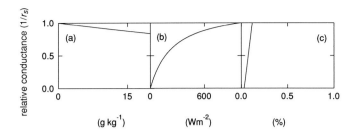

Figure 19.3. *Stress functions for humidity deficit (a), solar radiation (b), and soil moisture deficit (c) in the surface conductance formulation of the MITRE model.*

allow interaction between the two components. The emphasis in the MITRE model is on adequate modelling of the two-component surface or canopy resistances. A minimum resistance (maximum conductance) is increased by a number of environmental stress functions. The stress functions defining the environmental interaction of the stomata with humidity deficit, solar radiation and soil moisture are shown in Figure 19.3. The model is calibrated for three biome types at present and further calibrations are under way.

As an example, some of the preliminary calibration results for a tropical savannah are described. In the case of savannah (Type II) a distinct herbaceous understory is interspersed by woody shrubs. The data used to calibrate the fallow savannah were obtained during the Sahelian energy balance experiment (Gash *et al.*, 1991; Wallace *et al.*, 1990). Observations of latent and sensible heat and momentum flux were made at the start of the dry season in 1988 at a site near Niamey, Niger ($13°15'$N, $2°17'$E). The calibration was a trial-and-error optimization procedure in which, for a series of four consecutive days when the vegetation dried down, those values of the parameters were chosen which gave the best fit with the observations. Sellers *et al.* (1989) applied a more objective parameter-optimization procedure using non-linear optimization packages, but unfortunately the present data did not allow such a procedure to be applied. For the tropical forest biome (Type III) this procedure was successfully applied for the MITRE model configuration (Dolman *et al.*, 1991).

Figures 19.4 and 19.5 show a comparison of modelled and observed evaporation for a period just after the last rain and 6 weeks later, respectively. In general the agreement between observed and modelled evaporation is good. Regression analysis for these two periods gave a slope of 1.00 ± 0.02 and 1.04 ± 0.02 with correlation coefficients of 0.94 and 0.90, respectively. The difference in transpiration between the two periods is striking, with 4 mm day^{-1} dropping to 1–2 mm day^{-1} 6 weeks later. The main cause for this reduction in evaporation is likely to be an increasing soil moisture deficit during this period, coupled with natural senescence of the herbaceous understory. This is accurately modelled by the soil moisture response curve (Figure 19.3). However, it was found that the current model is not able to

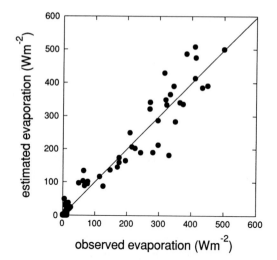

Figure 19.4. *Observed versus modelled evaporation for tropical savannah for 4 days immediately following the last rains.*

model the daily trend in surface resistance accurately when there is a high atmospheric demand coupled with a low soil moisture availability. In these cases the root system cannot supply enough water to the transpiring leaves, which respond by closing their stomata (see Chapters 8, 9 and 12). A daily limit to transpiration was introduced to simulate this effect but, clearly, further work, such as the explicit modelling of root water uptake (Chapter 10) and the introduction of capacitances in the system (Chapter 11), is needed in this area.

19.2.2 *Seasonality*

The increase in surface resistance obtained for the savannah as a result of an increase in soil moisture deficit coincides with a general decline in green leaf

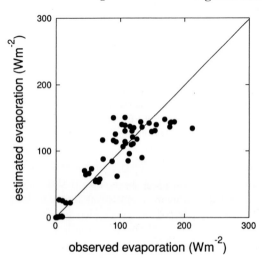

Figure 19.5. *Observed versus modelled evaporation for tropical savannah for 4 days, 6 weeks after the last rains.*

area of the herbaceous understory and bushes. This raises the important issue of seasonality in GCM–SVATs. To obtain a realistic diurnal and seasonal behaviour of vegetation in GCMs, it is necessary to include the effects of leaf growth and senescence of the vegetation in the models. SiB (Dorman and Sellers, 1989) and BATS (Dickinson, 1984) prescribe this seasonality by assigning different values to the parameters during the season. At present this is the only realistic way to obtain seasonality in the models.

This requires large, multi-year datasets to calibrate the biomes. Such datasets are hard to obtain for remote areas of the world such as rainforests, savannah and tundra. Several international programmes are currently attempting to combine national resources to obtain these data in a number of large-scale international experiments (see Shuttleworth, 1991, for a recent review). In future, the seasonality will have to be calculated interactively in the models through prescribed physiological responses to temperature, radiation and precipitation. This is all the more important as the effects of an increase in atmospheric CO_2 might not only be found in changes in the amount of leaf area but also in the timing of germination and senescence of vegetation (see Chapter 18).

19.2.3 *Biome change*

On a longer time scale than seasonality, the issue of possible changes in biomes as a result of climate change, needs consideration. The more sophisticated SVAT schemes recognize a limited number of biome types for which parameters are specified in the model. A typical list of biome types for SVAT models is given in Table 19.2. Such a characterization of the world's vegetation into a limited number of types is based on the assumption that large-scale momentum and energy-exchange processes are mainly determined by physiognomic characteristics of vegetation, such as leaf-area, height and leaf-area density distribution. Such an assumption is justifiable at present,

Table 19.2. *Biome classification in SVAT models*

Number	Description
1	Tropical rainforest
2	Coniferous forest
3	Deciduous forest
4	Deciduous needleleaf forest
5	Tall grass
6	Short grass
7	Tropical savannah
8	Broadleaf shrubs/soil
9	Tundra
10	Desert
11	Agricultural crop
12	Swamp

but is, in the longer term, unsatisfactory in climate-change experiments because biome zones may shift in response to a change in climate. Consequently, predicting what biomes may result from a change in climate is an active research area (e.g. Emanuel *et al.*, 1985; Henderson-Sellers; 1990; Henderson-Sellers and Pittman, 1992; Prentice, 1990). Emanuel *et al.* (1985) were among the first to attempt to predict the possible changes in the distribution of biomes using a GCM. They used the Holdridge (1947) classification, which relates the presence of a biome to total annual precipitation, mean annual biotemperature and potential evaporation, to predict the present-day global distribution of biomes. They showed that the use of such a scheme could reasonably well predict the global distribution of biomes, although in some cases a rather large discrepancy was found between the predicted and observed present-day biomes. Similar results were obtained by Prentice (1990) after modifying a number of global classification schemes. The major drawback of this approach is that the biome distribution is calculated off-line from the real climatological data or model results, allowing no feedback between the land surface and the atmosphere. Another drawback is that the Holdridge scheme is essentially a statistical relationship between climatological variables and biome distribution, which may change if the climate changes.

Woodward (1987) has developed a simple model which predicts the physiognomy of biomes on the basis of the use of soil moisture by vegetation. The basic assumption in his model is that vegetation optimizes the leaf area index (the total projected leaf area per unit area of ground) in response to soil moisture. By drawing up a water balance, and formulating the evaporation in terms of leaf area index and stomatal conductance, he was able to obtain a reasonable distribution of biome types for a number of localities around the globe. This is shown in Figure 19.6 for the location of Brisbane in Australia. No recharge, and consequently perpetual drying, is predicted for leaf area indices greater than 5, so the optimum leaf area index of a sustainable vegetation for this site should be between 3 and 5. Such an approach, if refined and adapted, has the advantage that it can be used in transient-climate experiments, in which GCMs are used to assess the atmospheric response to slowly increasing levels of CO_2. Leaf area index alone is, unfortunately, not enough to predict a specific biome type, as several biomes may have similar leaf area indices but completely different physiognomy. Attempts are therefore under way to predict more structural characteristics of biomes, such as canopy height. If, for instance, both canopy height and leaf area index can be predicted, specific parameters which now have to be calibrated for each biome may be calculated interactively in the land-surface model. Some of the more recent schemes, such as SiB and BATS, already have detailed subroutines that calculate albedo from a knowledge of distribution of leaf area within a canopy. Roughness length, the prime parameter determining momentum exchange, has also been shown to be related to canopy leaf area distribution and height (Shaw and Pereira, 1982). A combination of leaf area index and

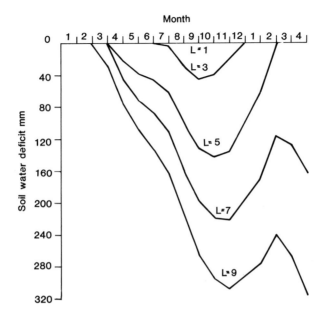

Figure 19.6. *Modelled soil water deficits for a number of leaf area indices (L) for Brisbane, Australia. (Redrawn after Woodward, 1987, with permission from Cambridge University Press.)*

height may be able to discriminate between the major biome types of the world. Figure 19.7 shows an initial attempt at this, showing how eight out of the 12 vegetation classes of Table 19.2 can be distinguished on the basis of leaf area index and height.

19.3 Aggregation of spatially inhomogeneous land cover

All the models described so far consider the interaction of vegetation with the atmosphere in a strict one-dimensional way, i.e. they assume spatial uniformity. Given the resolution of GCMs (typically 300 by 300 km) this is an assumption that is hard to justify. Experiments with mesoscale models have shown that significant mesoscale circulations may develop as a result of differences in land-surface cover, within areas the size of a typical GCM gridsquare (Mahfouf *et al.*, 1987; Segal *et al.*, 1988). Because of the non-linear relation between the land surface characteristics and fluxes, this implies that simple averaging of parameter values does not necessarily yield the correct fluxes (Dolman, 1992). To investigate the aggregation of land-surface parameters and to define *effective* parameters, two- and three-dimensional models of the atmosphere are used at a range of spatial scales. *Effective* parameters are defined as those parameters which give the correct area-averaged flux, resulting from the calculation of, for instance, a two-dimensional boundary layer model or a mesoscale model, which would incorporate and resolve all the relevant scales. André *et al.* (1989) and Shuttleworth (1991) suggested

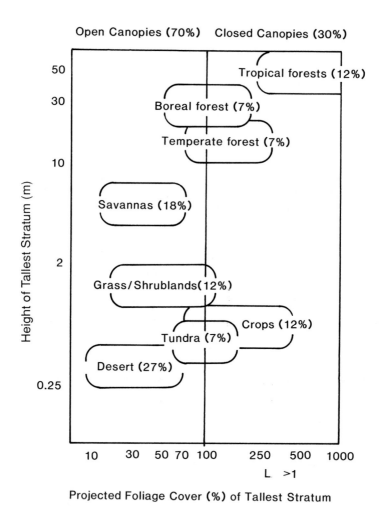

Figure 19.7. *Schematic diagram of the major biomes of the world as a function of height and leaf area index. (Redrawn after Graetz, 1990.)*

that two types of land-surface organization should be distinguished. Type A where the typical length scale of land-surface cover is less than 10 km, and a single effective boundary layer develops over the region; and Type B where the length scale is greater than 10 km and an organized response in the boundary is often observed. This distinction provides a way forward in understanding the aggregation processes at these scales.

19.3.1 *Small-scale aggregation*

In small-scale aggregation, the individual surface patches are not large enough to give an organized response in the atmospheric boundary layer and consequently the boundary layer only 'sees' an average surface. To predict the average surface properties, Wieringa (1986) and Mason (1988) introduced the concept of a blending height. This is the height at which the surface fluxes are roughly in local equilibrium but the properties of interest, such as windspeed, temperature and humidity, are almost spatially homogeneous. Mason (1988) used this concept successfully to predict average roughness lengths for momentum transport. Blyth *et al.* (1993), Claussen (1991) and Wood and Mason (1991) have extended the concept to investigate the aggregation processes of heat and water-vapour fluxes.

Following Blyth *et al.* (1993), it is relevant to analyse the aggregation process of surface fluxes in a resistance framework. The case of the surface energy balance is particularly illuminating for the problems encountered when scaling up from small patches of heterogeneous land to large areas. It is common practice to write the aerodynamic resistance of homogeneous vegetation to heat and water vapour transport (r_{ah}, r_{av}) as a combination of two terms, a momentum term, r_{am}, and an excess term, r_e, which compensates for the fact that transport of heat and water vapour is mainly a molecular diffusion process, whereas momentum transport also includes transport by pressure forces:

$$r_{av} = r_{ah} = r_{am} + r_e. \qquad 19.2$$

This relation is well documented for homogeneous vegetation (e.g. Verma, 1989). The excess resistance in these cases can be parameterized as

$$r_e = \ln\left(z_0/z_{0t}\right)\left(ku_*\right)^{-1} \qquad 19.3$$

where z_0, z_{0t} are the roughness length for momentum and heat, respectively; k is von Kármán's constant (0.41); and u_* is the friction velocity. In homogeneous terrain, it is realistic to assume that the ratio of the two roughness lengths is $1:10$ (Brutsaert, 1982).

In heterogeneous terrain, these simple relationships break down and it becomes necessary to perform a more detailed analysis. Mason (1988) and Wood and Mason (1991) have shown how the introduction of individual rough elements in a domain *increases* the overall roughness length by a disproportionate amount. This is due to the fact that individual roughness elements are very effective momentum absorbers. In terms of aerodynamic resistance, this implies that the inclusion of rough elements *decreases* the overall aerodynamic resistance by a disproportionate amount. This is illustrated in Figure 19.8. Wood and Mason (1992) subsequently included in their analysis a heat flux that was assumed constant over the domain. As a consequence of this assumption the total effective aerodynamic resistance to heat transport remains constant over the domain and the effective excess resistance (r_e) has to increase (Figure 19.8).

In reality the sensible heat flux is a component of the surface energy balance, so evaporation has to be considered as well (see Chapter 18). Assume, for instance, that the low surface resistances are associated with rough elements

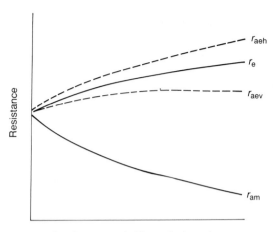

Figure 19.8. *Hypothetical behaviour of the aerodynamic and excess resistances to heat and water vapour in heterogeneous landscapes:* r_{aeh} *is the excess resistance to heat transport and* r_{aev} *the excess resistance to water vapour transport in heterogeneous landscapes;* r_e *is the excess resistance for both heat and water vapour transport in homogeneous landscapes;* r_{am} *is the aerodynamic resistance to momentum transport.*

in the domain as for well-watered trees that can evaporate freely. Evaporation in this case is controlled mainly by the low surface resistance, and the effective excess resistance to evaporation is reduced. However, as a result of high evaporation, low sensible heat fluxes now occur, and the excess resistance for heat transport increases by a different amount. The relation (Equation 19.2) which held for homogeneous conditions, now is clearly invalid in heterogeneous conditions, and aerodynamic resistances to heat flux and water-vapour flux have to be considered separately. This example illustrates how relations between parameters may change when the landscape changes from homogeneous to heterogeneous.

Blyth *et al.* (1993) further showed how the concept of blending height could successfully be used to estimate effective values for surface resistance. They also found that in most cases a simple arithmetic average of the parallel and series resistance gave a very good approximation of the effective resistance, as calculated from a two-dimensional boundary layer model. In cases where some parts of the domain were covered with water, this simple approximation was not valid and use had to be made of the blending height concept. They also highlighted a few cases where the concept of an effective resistance became meaningless, for instance when a negative sensible heat flux balanced exactly a positive flux over the domain (in this case the effective resistance approaches infinity).

19.3.2 *Large-scale aggregation*

Mesoscale models are increasingly being used to define effective parameters on the scale of a GCM gridsquare (Noilhan *et al.*, 1991; Noilhan and Lacarrère, 1992). These atmospheric models have gridscales of the order of 10–50 km and can resolve typical small-scale phenomena such as sea-breezes. They are particularly useful for filling in the gap between the boundary layer studies

Table 19.3. *Simple aggregation rules*

Parameter	Rule	Equation used
α_e	$\Sigma \alpha_i f_i$	$S_{surf} = S_d(1 - \alpha)$
L_e	$\Sigma L_i f_i$	various
r_s^e	$\Sigma f_i / r_s^i$	$\lambda E = \rho_a c_p (q_s - q_r)/r_s$
z_0^e	$\Sigma f_i / (\ln(z_b/z_0^i))^2$	$u = \ln(z/z_0)u_*/k$

f_i, fraction of domain covered by vegetation type i; α is the shortwave albedo; S_{surf}, shortwave radiation at the surface; and S_d, downward shortwave radiation; L, leaf area index; r_s^e, effective resistance specific heat and density of air; ρ_a, density of air; c_p, specific heat of air; q_s, q_r, specific humidity at the surface and at the reference level; λE, evaporation, z_0^e, z_0^i, effective roughness length for momentum and for the ith vegetation type; u, windspeed; u_*, friction velocity; z_b, the blending height; the subscript e stands for effective.

mentioned in the previous section and the GCM gridbox. Noilhan and Lacarrère (1992) investigated the usefulness of a number of simple aggregation techniques for an area in southern France. These simple aggregation rules (see also Shuttleworth, 1991) consist of applying averaging rules in a manner which is consistent with the way in which the parameters are used in the equations in the model. Table 19.3 lists a number of these rules, together with the relevant equations in which they are used.

For albedo and leaf area index, a simple linear average can be taken. For the surface resistance of vegetation, a linear average of the conductances is required (see Equation 19.1) and for momentum roughness length the proposal of a blending height by Mason (1988) can be used. Effective parameters for soil properties were obtained by Noilhan and Lacarrère (1992) by relating the soil parameters to the fraction of sand and clay over an averaging domain.

Two types of experiments were performed, one where effective parameters were obtained by the above-mentioned rules, and one where a dominant vegetation cover was defined. The land surface defined by these parameters was then used in a one-dimensional column of the atmosphere to calculate fluxes. These results were compared with the results of the full three-dimensional mesoscale model. It was found that the use of a dominant type gave the largest errors in evaporation, sensible heat flux and transpiration. This was caused by the relatively large differences in surface cover over the domain: forest as opposed to agricultural crops (forest was the dominant cover in the domain). Furthermore, since the soil hydraulic and thermal properties depend on soil type in a very non-linear way, a high sensitivity to the soil parameters was found. Mesoscale models are known to be very sensitive to the specification of soil moisture (Bougeault, 1991), although this sensitivity is also found in some GCMs. However, the most important conclusion from this work is that, using the full information content of the land surface and applying sensible aggregation rules, it is possible to derive realistic values for effective parameters.

19.4 Conclusions

An overview of some of the more recent developments in soil vegetation atmosphere transfer schemes has been presented. As an example of a SVAT scheme, the MITRE model was presented and discussed in terms of the calibration of the tropical savannah biome. Seasonality was introduced as a feature that is currently not properly represented in SVAT models. Biome change was identified as another area requiring further attention, and the application of some biome prediction schemes was discussed. The incorporation of land-surface spatial heterogeneity in large-scale models of the atmosphere was discussed and suggestions were made to define effective parameters on both small and large scales.

Acknowledgement

The author is pleased to acknowledge the support of a CEGB Senior Research Fellowship funded by the Joint Environmental Program of National Power and PowerGen.

References

André, J.-C., Bougeault, P., Mahfouf, J.-F., Mascart, P., Noilhan, J. and Pinty, J.-P. (1989) Impacts of forests on mesoscale meteorology. *Proc. R. Soc. Lond. B*, **324**, 407–422.

Avissar, R. and Verstraete, M.M. (1990) The representation of continental surface processes in atmospheric models. *Rev. Geophys.*, **28**, 35–52.

Blyth, E.M., Dolman, A.J. and Wood, N. (1993) Effective resistance to sensible and latent heat flux in heterogeneous terrain. *Quart. J. Roy. Meteor. Soc.*, **119**, 423–442.

Bougeault, P. (1991) Parameterization schemes of land-surface processes for mesoscale atmospheric models. In: *Land Surface Evaporation: Measurement and Parameterization* (eds T. Schmugge and J.-C. André). Springer, New York.

Brutsaert, W. (1982) *Evaporation into the Atmosphere*. D. Reidel, Dordrecht, pp. 55–92.

Claussen, M. (1991) Estimation of areally-averaged surface fluxes. *Bound. Lay. Met.*, **54**, 387–410.

Cunnington, W.M. and Rowntree, P.R. (1986) Simulations of the Saharan atmosphere – dependence on soil moisture and albedo. *Quart. J. R. Meteor. Soc.*, **112**, 971–999.

Dickinson, R.E. (1984) *Modelling Evapotranspiration for Three Dimensional Climate Models* (eds J.E. Hansen and T. Takahashi), Geophysical Monographs No. 29. American Geophysical Union, Washington, DC, pp. 58–72.

Dickinson, R.E. and Henderson-Sellers, A. (1988) Modelling tropical deforestation: a study of GCM land-surface parameterizations. *Quart. J. R. Meteor. Soc.*, **114**, 439–462.

Dolman, A.J. (1992) A note on areally-averaged evaporation and the value of the effective surface conductance. *J. Hydrol.*, **138**, 583–589.

Dolman, A.J. (1993) A multiple-source land surface energy balance model for use in general circulation models. *Agric. For. Meteor.*, in press.

Dolman, A.J., Gash, J.H.C., Roberts, J.M. and Shuttleworth, W.J. (1991) Stomatal and surface conductance of tropical rainforest. *Agric. For. Meteor.*, **54**, 303–318.

Dorman, J.L. and Sellers, P.J. (1989) A global climatology of albedo, roughness length and stomatal resistance for atmospheric general circulation models as represented by the Simple Biosphere Model (SiB). *J. Appl. Meteor.*, **28**, 834–855.

Emanuel, W.R., Shugart, H.H. and Stevenson, M.P. (1985) Climatic change and the broad-scale distribution of terrestrial ecosystem complexes. *Clim. Change*, **7**, 29–460.

Gash, J.H.C., Wallace, J.S., Lloyd, C.R., Dolman, A.J., Sivakumar, M.V.K. and Renard, C. (1991) Measurements of evaporation from fallow Sahelian savannah at the start of the dry season. *Quart. J. R. Meteor. Soc.* **117**, 749–760.

Graetz, R.D. (1991) The nature and significance of the feedback of changes in terrestrial vegetation on global atmospheric and climatic change. *Clim. Change*, **18**, 147–173.

Henderson-Sellers, A. (1990) Predicting generalized ecosystem groups with the NCAR CCM: first steps towards an interactive biosphere. *J. Clim.*, **3**, 917–940.

Henderson-Sellers, A. and Pittman, A.J. (1992) Land-surface schemes for future climate models: specification, aggregation and heterogeneity. *J. Geoph. Res.*, **97 D3**, 2687–2696.

Holdridge, L.R. (1947) Determination of world plant formations from simple climatic data. *Science*, **105**, 367–368.

Lean, J. and Warrilow, D.A. (1989) Simulation of the regional impact of Amazon deforestation. *Nature*, **342**, 411–413.

Mahfouf, J.F., Richard, E. and Mascart, P. (1987) The influence of soil and vegetation on the development of mesoscale circulations. *J. Clim. Appl. Meteor.*, **26**, 1483–1495.

Mason, P.J. (1988) The formation of areally-averaged roughness lengths. *Quart. J. R. Meteor. Soc.*, **114**, 399–420.

Mintz, Y. (1984) The sensitivity of numerically simulated climates to land surface boundary conditions. In: *The Global Climate* (ed. J.T. Houghton). Cambridge University Press, Cambridge, pp. 79–105.

Monteith, J.L. (1965) Evaporation and environment. *Symp. Soc. Exp. Biol.*, **IX**, 205–234.

Nobre, C.A., Sellers, P.J. and Shukla, J. (1991) Amazonian deforestation and regional climate change. *J. Clim.*, **4**, 957–988.

Noilhan, J. and Lacarèrre, P. (1992) GCM gridscale evaporation from mesoscale modelling. *J. Clim.*, submitted.

Noilhan, J. and Planton, S. (1989) A simple parameterization of land surface processes for meteorological models. *Month. Weath. Rev.*, **117**, 536–549.

Noilhan, J., Lacarèrre, P. and Bougeault, P. (1991) An experiment with an advanced surface parameterization in a mesobeta-scale model. Part III. Comparison with the HAPEX-MOBILHY dataset. *Month. Weath. Rev.*, **119**, 2393–2413.

Penman, H.L. (1948) Natural evaporation from open water, bare soil and grass. *Proc. R. Soc. Lond. A*, **193**, 120–145.

Prentice, K.C. (1990) Bioclimatic distribution of vegetation for general circulation model studies. *J. Geoph. Res.*, **95 D**, 11811–11830.

Rowntree, P.R. (1986) Review of General Circulation Models as a basis for predicting the effects of vegetation change on climate. In: *Forests, Climate and Hydrology: Regional Impacts* (eds E.R.C. Reynolds and F.B. Thompson). Kefford Press, Singapore, pp. 162–193.

Rowntree, P.R. (1991) Atmospheric parameterization schemes for evaporation over land: basic concepts and climate modelling aspects. In: *Land Surface Evaporation: Measurement and Parameterization* (eds T. Schmugge and J.-C. André). Springer, New York, pp. 5–30.

Segal, M., Avissar, R., McCumber, M.C. and Pielke, R.A. (1988) Evaluation of vegetation effects on the generation and modification of mesoscale circulations. *J. Atmos. Sci.*, **5**, 2268–2292.

Sellers, P.J., Mintz, Y., Sud, Y.C. and Dalcher, A. (1986) A simple biosphere model (SiB) for use within general circulation models. *J. Atmos. Sci.*, **43**(6), 505–531.

Sellers, P.J., Shuttleworth, W.J., Dorman, J.L., Dalcher, A. and Roberts, J.M. (1989) Calibrating the Simple Biosphere Model for Amazonian tropical forest using field and remote sensing data. Part I: Average calibration with field data. *J. Appl. Meteor.*, **28**, 727–759.

Shaw, R.H. and Pereira, A.R. (1982) Aerodynamic roughness of a plant canopy: a numerical experiment. *Agric. Meteor.*, **26**, 51–65.

Shuttleworth, W.J. (1991) The Modellion concept. *Rev. Geoph.*, **29**, 585–606.

Shuttleworth, W.J. and Wallace, J.S. (1985) Evaporation from sparse crops – an energy combination theory. *Quart. J. Roy. Meteor. Soc.*, **111**, 111–855.

Verma, S.B. (1989) Areodynamic resistances to tranfers of heat, mass and momentum. In: *Estimation of Areal Evaporation* (eds A. Black, D.L. Spittlehouse, M.D. Novak and D.T. Price). IAHS Press, Wallingford, pp. 13–20.

Verseghy, D.L. (1991) Class – a Canadian land surface scheme for GCMs. I. Soil model. *Int. J. Clim.*, **11**, 111–133.

Wallace, J.S., Wright, I.R., Stewart, J.B. and Holwill, C.J. (1991) The Sahelian Energy Balance Experiment (SEBEX): ground based measurements and their potential for spatial extrapolation using satellite data. *Adv. Space Res.*, **11**, 131–141.

Warrilow, D.A., Sangster, A.B. and Slingo, A. (1986) *Modelling of Land Surface Processes and Their Influence in European Climate.* Meteorological Office, Bracknell, DCTN 38.

Wieringa, J. (1986) Roughness dependent geographical interpolation of surface windspeed averages. *Quart. J. R. Meteor. Soc.*, **112**, 867–889.

Wood, N. and Mason, P. (1991) The influence of static stability on the roughness lengths for momentum and temperature. *Quart. J. R. Meteor. Soc.*, **117**, 1025–1056.

Woodward, F.I. (1987) *Climate and Plant Distribution.* Cambridge.University Press, Cambridge.

Index